ENVIRONMENTAL IMPACT ASSESSMENT IN AUSTRALIA

AF757432

ENVIRONMENTAL IMPACT ASSESSMENT IN AUSTRALIA

THEORY AND PRACTICE

Sixth Edition

Mandy Elliott

THE FEDERATION PRESS
2014

Published in Sydney by:
The Federation Press
PO Box 45, Annandale, NSW, 2038
71 John St, Leichhardt, NSW, 2040
Ph (02) 9552 2200 Fax (02) 9552 1681
E-mail: info@federationpress.com.au
Website: www.federationpress.com.au

First edition 1996
Second edition 1998
Third edition 2001
Fourth edition 2005
Reprinted 2007
Reprinted 2008
Fifth edition 2009
Sixth edition 2014

National Library of Australia
Cataloguing-in-Publication entry

Environmental impact assessment in Australia: theory and practice
Mandy Elliott

6th ed.
Includes bibliographical references and index.
ISBN 978 1 86287 945 4

Environmental impact analysis - Australia.
Environmental impact statements - Australia.
Environmental monitoring - Australia.

333.7140994

© The Federation Press

This publication is copyright. Other than for the purposes of and subject to the conditions prescribed under the Copyright Act, no part of it may in any form or by any means (electronic, mechanical, microcopying, photocopying, recording or otherwise) be reproduced, stored in a retrieval system or transmitted without prior written permission. Enquiries should be addressed to the publishers.

Typeset by The Federation Press, Leichhardt, NSW.
Printed by Griffin Press, Adelaide, SA.

Author Note

Environmental Impact Assessment in Australia: Theory and Practice was conceived and written by Ian Thomas across its first three editions. For the fourth and fifth editions he was joined by Mandy Elliott, who has now assumed sole authorship of the sixth edition.

Preface to the Sixth Edition

This edition, similar to previous editions, of *Environmental Impact Assessment in Australia: Theory and Practice* has been revised in the context of evolving Environmental Impact Assessment (EIA) practice around Australia and includes a number of case studies throughout the text to complement the theory. The case studies should not be viewed as the only way of doing EIA (or a component of EIA), but rather as examples to demonstrate key points within the text.

The historical context and evolution of environmental impact assessment and the research in this area of EIA provided in previous editions by Ian Thomas largely remains, although updated where required. I have focused this edition on introducing more context to the evolving theory and practice regarding impact assessment, specifically strategic environmental assessment, the role of EIA in developing countries, the current discussion regarding the Commonwealth's *Environment Protection and Biodiversity Conservation Act* and the issues surrounding post-EIA monitoring.

Practical experience working in EIA for State government (Victoria) for approximately 10 years, as well as more recent experience as a consultant working on EIA projects, has provided insight into how EIA is undertaken in practice, how good assessments can lead to better decisions as well as how politics plays an integral role in EIA. I have endeavoured to reflect this experience throughout the sixth edition.

Chapter 6, which summarises the Commonwealth, State and Territories' EIA procedures, has largely been revised, as many of these jurisdictions have been through some type of reform to their EIA procedures, including amendments to their EIA legislation, or to existing EIA Guidelines/administrative procedures. Notwithstanding these changes, the way in which EIA is undertaken still remains. That is, screening, scoping, EIS production, public review and assessment.

What has changed since the fifth edition? In some respects, not a lot – the broad approach of EIA has not been challenged and its highly political nature remains. A most noticeable difference is the evolving relationship between the Commonwealth's procedures (under the *Environment Protection and Biodiversity Conservation Act 1999*) and the equivalent State and Territory legislation. At the time of publication, all States and Territories have signed Assessment Bilateral Agreements with the Commonwealth government to help in reducing duplication, among other matters, of EIA procedures. Negotiations have commenced between the Commonwealth and the States and Territories to agree to Approval Bilateral Agreements, with the intention of having these in place by the end of 2014.

EIA activity is reported worldwide, and there are continuing reports of the formal adoption of EIA procedures through legislation, indicating that internationally EIA is becoming more uniformly established. Further update and discussion of the role of the World Bank and other international funding bodies has been included in this edition.

The sixth edition of *Environmental Impact Assessment in Australia: Theory and Practice* has the same structure as before, but has expanded the discussion of assessment approaches and their operation, and reports the current (late 2013) EIA procedures across Australia. Generally the material of the chapters focuses on broad principles and processes and does not date. Nonetheless, as the experience of EIA grows we gain insight into these practices, and the jurisdictions modify their procedures.

In recognition of the developing interest in Strategic Environment Assessment, the discussion in Chapter 3 has been expanded to recognise the growing emphasis that it is attracting both internationally and in Australia. A case study on Melbourne's Urban Growth Boundary is included.

In considering the future of EIA, in Chapter 10, I have particularly commented on the connections of EIA with environmental management. Recent developments and their future directions in SEA are also reviewed. This edition continues the discussion on the issue of post-EIA monitoring, which has been a focus for practitioners and academics for a number of years, and the little evidence of much post-EIA assessment occurring. Post-EIA monitoring remains a critical issue to be addressed to ensure outcomes of EIA are implemented and issues managed as they were proposed to be.

This edition of *Environmental Impact in Australia: Theory and Practice* responds to both the evolution in EIA thinking and to the maintenance of the basic approach. In terms of the general application of EIA concepts and principles nothing has really changed, but additional experience may have been gained of specific aspects of EIA. The list of references is already large and there is no need to duplicate material. Readers interested in obtaining the latest texts on EIA will find plenty to choose from in libraries and on the internet. Nonetheless, the reference list has been updated to reflect the changes discussed, and to include key internet sites, such as those which provide information about both the Australian and international EIA context.

The emphasis in *Environmental Impact Assessment in Australia: Theory and Practice* is on developing understanding of the EIA processes. This is intended to place the reader in the position of being able to apply these processes to any situation – while being aware of the strengths and limitations of the approaches being used and the assumptions being made.

As already emphasised, the practice and administration of EIA procedures are far from being static. It is important that those involved in EIA appreciate this dynamism and make sure they know the most recent procedures when participating in EIA.

As in previous editions, I acknowledge the cooperation of the officers of relevant government agencies in supplying information about the specifics of EIA processes throughout Australia, particularly where significant reviews have occurred in their jurisdictions. Further, I am indebted to numerous supporters for their constructive comments on a number of matters associated with the book including the extensive research and the dedication to *Environmental Impact Assessment in Australia: Theory and Practice* by Ian Thomas not only in previous editions of the text but as an ongoing mentor to myself.

Mandy Elliott
January 2014

Contents

Abbreviations

ANZECC	Australia and New Zealand Environment Council
BCA	benefit-cost analysis
CBA	cost-benefit analysis
CER	consultative environment review
CIA	cumulative impact assessment
EARP	Environmental Assessment Review Process
EcIA	economic impact assessment
EHIA	environmental health impact assessment
EIA	environmental impact assessment
EIS	environmental impact statement
EPBC	*Environment Protection and Biodiversity Conservation Act 1999* (Cth)
EP(IP)A	*Environment Protection (Impact of Proposals) Act 1974* (Cth)
EPA	Environment Protection Agency
ERMP	environmental review and management program
ESD	ecologically sustainable development
EU	European Union
GATT	General Agreement on Tariffs and Trade
HIA	health impact assessment
HIS	health impact statements
IIA	integrated impact assessment
IGAE	InterGovernmental Agreement on the Environment
LCA	life cycle analysis
NEPA	*National Environmental Policy Act 1970* (US)
NOI	notice of intention
OECD	Organisation for Economic Cooperation and Development
PEA	*Local Government (Planning and Environment) Act 1990* (Qld)
PER	public/preliminary environmental report/review
PPA	post-project analysis
RA	risk analysis
RIA	regulatory impact assessment
SDA	*State Development and Public Works Organisation Act 1971* (Qld)
SEA	strategic environment assessment
SIA	social impact assessment
TA	technology assessment
UNESCO	United Nations Educational, Scientific and Cultural Organization

Introduction

In introducing the sixth edition of *Environmental Impact Assessment in Australia: Theory and Practice*, the author emphasises that little has changed in the format of the text, however, the sixth edition has been revised in the context of evolving environmental assessment theory and procedures from across the world. This edition provides some case studies not to define how Environmental Impact Assessment should be undertaken, but to illustrate key points or elements, of the EIA process. EIA has become an established fact of environmental planning and policy. Intuitively, it is a straightforward concept to help us make decisions about whether we can live with the consequences of proceeding with an action.

There has been some discussion of EIA among those who are regularly exposed to it, and more frequently now through the media we hear about an environmental impact statement (EIS) being required or released. But, in Australia at least, there is little readily available information for those who want to know more than the media provide, to prepare themselves for involvement in an Australian EIA.

Simplistically, EIA can be considered on two levels. First, there is the theory that has evolved over the four decades since EIA officially began. This principally describes the experiences over that time, along with some theory as to what EIA should be aiming for. At the other end of the spectrum is the practice of EIA, as evidenced by EIA legislation and procedures.

As in other fields of endeavour, it is easiest to concentrate on one aspect, such as "getting on with the job", and not worrying or even thinking about any theory, such as the role of EIA or what experience could illustrate. But practitioners and academics ignore each other's interests at their peril. Each has something to learn from the other. The title *Environmental Impact Assessment in Australia: Theory and Practice* makes the point that the package which EIA represents is best viewed as a whole, rather than one part.

In light of the above, the purpose of *Environmental Impact Assessment in Australia: Theory and Practice* is fivefold:

1. To bring together the many and varied aspects of EIA;
2. To establish the context for EIA as undertaken within Australia;
3. To illustrate the practice of EIA so that newcomers to the field will have a handbook to assist them;
4. To provide a resource that will give a starting point for anyone wanting more detail on particular aspects of EIA (hence the heavy emphasis on references);
5. To remind the reader that EIA is a "social construct" in that we have manufactured it – as a result it relies on value judgments, and it has become part of our political process.

The historical setting for EIA is briefly outlined in Chapter 1. This provides the setting for the discussion in Chapter 2 of the current objectives and approaches, and the experience of EIA. Chapter 3 expands the investigation to review a range of other forms of assessment, all of which have much in common with EIA, and with each other, and which can be explored to improve aspects of EIA. It could be said that EIA has provided a catalyst for the vastly increased levels of public involvement we have seen, so Chapter 4 explores the role of community participation in general, and with particular reference to EIA.

That EIA has worldwide acceptance is demonstrated in Chapter 5 where the role of the United States of America is discussed and the extent of procedures in other countries is noted. Updates have been provided where relevant. In Chapter 6, the discussion is extended to Australia, where the legislation and procedures operating at Commonwealth and State and Territory levels are outlined.

The practice of EIA is the focus of subsequent chapters. Content of the EIS (the public documentation of EIA) is covered in Chapter 7 to assist the practitioner to compile the EIS. Chapter 8 deals with the difficult, and "thorny", subject of identifying and evaluating impacts by discussing the various methods, which have been developed in an attempt to establish objective evaluations. To condense all this information and to put it in the sequence for a typical EIA, Chapter 9 summarises the main points in a handbook format.

EIA continues to evolve and its role within the community requires review. Chapter 10 considers these issues and particularly looks at EIA's future directions.

Throughout these chapters there is the underlying reminder that EIA is a process which provides a basis for decision-making. As such it is involved in the politics of society, and to an extent is a political process in itself. Consequently EIA opens up questions of values – the values of decision-makers, authors of the EIS, and society in general. The extent to which values are involved is compounded by the "inexact" nature of assessing environmental impacts, making it even easier for claims of bias to be made.

The value-laden nature of EIA is not something to be embarrassed about, however. Unlike most other technical, economic, social or political processes, EIA provides the opportunity to bring these values into the open.

Chapter 1

THE BACKGROUND OF EIA

1.1 A Historical Perspective

Like every other aspect of society, Environmental Impact Assessment (EIA) has not suddenly appeared. Rather, over many years social evolution has led to the need for the environmental assessment of development on the environment and society, which is in turn evolving to meet the demands of an expanding technological society. This short historical review focuses on the changing environmental awareness of civilisations, which is at the basis of EIA.

The civilisation of Mesopotamia provides a convenient starting point for this review. These people were effective at constructing cities and large-scale agricultural projects to provide food to support the city dwellers. Unfortunately, their irrigation schemes and approach to the land resulted in salting, flooding and destruction of the countryside.

It is significant that the first urban societies were also the first societies to abandon a religious attitude of oneness with nature and to adopt one of separation. As Hughes (1975) points out, the dominant myth and reality in Mesopotamia was the conquest of chaotic nature by divinely ordained human order. Such societies were ultimately unsuccessful in maintaining a balance with their environment.

In contrast, Greek religion recognised humanity's oneness with nature. The ancient Greeks could not conceive of a sacred area without trees; for instance, on the rocky summit of the Acropolis when the Parthenon was built, holes were excavated in the solid rock and trees were planted in rows to flank the temple. Sacred groves of trees were preserved and hunting was forbidden in these groves. While it is stretching the point to see these as national parks in anything like the modern sense, it is clear that the Greeks had a respect for nature.

The Romans, on the other hand, generally believed that the world was there for human use, and they proceeded in a very pragmatic way to find uses for natural resources. They sought knowledge for its own sake only rarely, always seeming to demand a practical application for intellectual endeavour. Closely allied to practicality was a profound desire for order. No other people in ancient times, Hughes (1975) points out, imposed such a rigorous and artificial structure upon their natural environment as did the Romans. Like other groups, though, the Romans considered that the environment shapes the people who live within it. This served as an explanation and vindication of their rule over other people, and nature.

Two characteristics which are important for environmental appreciation are an interest in ecology and a curiosity to understand how nature works. Romans were interested in nature and curious about it, but they were motivated by pragmatism and seldom were they true scientists. Early Romans worshipped nature and were inhibited from making major changes in the environment by religious taboos, but the Romans in the middle and late Republic and the Empire were increasingly utilitarian and willing to exploit their natural resources.

More enduring than this Roman view, the Judaeo-Christian concepts of God and the order of nature were often combined by the early Church Fathers to create a conception of the habitable world. Glacken (1967) notes that this interpretation of life, that nature and the earth were limited had the force and resilience for the vast majority of the peoples of the Western world to endure until the mid-19th century. The Judaeo-Christian idea of humanity's domination over nature is usually attributed to the passage from Genesis: "Be fruitful, and multiply, and replenish the earth, and subdue it: and have dominion over ... every living thing that moveth upon the earth" (1: 28). This domination emphasised a creativity, activity and technical advancement which did not come from humans, but had been ordained by God. Humans were made in God's image, but were not part of nature in the way that plants and animals were; they were more stewards of God.

Stewardship progressively gave way to outright control when during the Middle Ages attention was given to magic, astrology and alchemy as bodies of knowledge worthy of study. As the works of Glacken and Hughes indicate, these gifts promised much in terms of control over nature and over humans. When evidence that undesirable changes in nature were caused by humanity began to accumulate in the 18th and 19th centuries, the philosophical and theological underpinning of the classical, and later of the Christian, idea of stewardship was threatened. If torrents and soil erosion followed rapid forest clearing and if wildlife disappeared because of relentless hunting, humanity seemed to be failing in its appointed task, going a way of its own, and selfishly defying the will of God and Nature's plan. As a result, questions about humanity's treatment of nature became important in the late 19th century, particularly in the United States where Marsh (1967) wrote about the need to protect forests and wildlife. It was at this time that movement was made towards protecting areas of wildlife with the first national park being proclaimed in the United States. Yellowstone National Park was proclaimed in 1872 and with it a new concept was born with a new way for people to preserve and protect the best of what they had for the benefit and enjoyment of future generations (National Park Service (2008)).

There are differing interpretations of the precise views and actions of past generations. Clearly, as today, we cannot assume that these categorisations include everyone. In any community there will always be a variety of views. However, the summary above indicates the dominant views, and is broadly supported by the detailed work of Marshall (1992).

Since the early years of the 20th century, the concerns of the environmental movement in Western countries have become well known. The concept of "ecology", which valued aesthetics and biology, began to penetrate the public mindset during the post-World War II years. Growth of cities, pollution of the food chain and urban water supplies, and the deterioration of air quality started to get the public's attention. Hence, there was a powerful reaction to Rachel Carson's 1962 classic *Silent Spring*, a quietly shocking book about the widespread pesticide poisoning of man and nature. *Silent Spring* caused the public to press the United States government into action to protect the natural environment for its values (United States Environment Protection Agency (1992)).

It is fair to say that a universal feeling that nature is there to be exploited regardless of the effects has now gone. Nonetheless, current debates about the reality of climate change, the protection of biodiversity, the use of nuclear power, water resource management, air pollution from motor vehicles and other environmental issues indicate that environmental issues generally do not receive the top priority. Climate change and water resources are becoming more of priority issue for governments as the effects of drought are becoming more prominent and are affecting everyone – not just farmers. Nonetheless, while books such as Neider's *Man Against Nature* (1954) are published and read, there will be voices raised in support of ignoring environmental considerations in preference using natural resources and giving little regard to the consequences.

In some cases the stewardship approach has been extended to fit the situation of "development with an environmental conscience". For example, in the former Soviet Union it was said that concern for the natural environment was not limited to the protection of resources and maintenance of natural processes. Rather, Kalensnik and Pavlenko saw that their society was helping nature to become still better and richer. This was being achieved by measures related to the transformation of nature, being "deliberate changes for the benefit of people, enrichment of natural resources for their fuller utilisation, for the increased ... productivity of the environment, and for the suppression of harmful natural processes" (1976: 128).

The strong inference from this is that nature is there to be used, an attitude held in most Western countries and reviewed in O'Riordan's (1976, 1985) works. Elsewhere the emphasis on "use" may not be so explicit, but it is frequently embodied in policies and actions. As an example, the Australian 1983 National Conservation Strategy gave considerable emphasis to development, and ways to reduce the effects on conservation issues, while the 1992

National Strategy for Ecologically Sustainable Development (ESD) did little to change this focus.

So far we have discussed typical Western attitudes, principally because these form the background for many Australians, and we have much in common with Western societies (particularly the United Kingdom, the United States and European countries). Other philosophies and attitudes to nature are evident in Eastern cultures (eg, in Buddhist societies) and the approaches of American Indian and Australian Aboriginal societies (eg, see Marshall (1992)). While the attitudes of these societies are more attuned to a conservation viewpoint they have had little impact on the majority of Western traditions, or institutional arrangements (eg, EIA).

1.2 The Need for EIA

The speed, scope and magnitude of effect on the natural environment are determined by the level of technology available to a community. Using human and animal power, and the energy of water, wind and fire with the relatively simple tools and machines that had been invented, the ancient peoples constructed monuments that still impress us. But, as Hughes (1975) comments, their level of interaction with the natural environment was relatively low compared with that of modern industrial society. The changes wrought by ancient civilisations have been massive, but took centuries or millennia to accomplish. Today more significant changes take place in months or years (or seconds in the case of nuclear explosions).

Of all the ancient peoples, the Romans possessed the most highly developed technology. Hughes (1975) comments that their machines for war, construction and industry foreshadowed some that are still used today. The fact that ancient peoples absorbed and survived successive changes in the technology of war and peace cannot be much comfort to us, as the rapidity, size and power of changes today are of an entirely different order of magnitude from anything experienced in ancient times.

Another factor determining the way a community will interact with the natural environment is the degree of organisation and control the community possesses. This is because environmental ends desired for the good of the community may involve sacrifices on the part of its individual members, sacrifices that they would not make without some degree of social encouragement or coercion. In this context, Hughes notes how ancient civilisations were able to exert a considerable degree of social control because the vast majority of ancient people regarded themselves primarily as part of their societies, and only secondarily as individuals. Each person had a place in the social hierarchy that was rigidly defined and rarely changed.

The problems of human communities affecting the natural environment did not begin with the ecological awakening of the 1960s, or with the onset of the Industrial Revolution, or in the Christian Middle Ages. People have had to find a way of living with nature from the earliest times, and many of our answers to that habitual challenge would have received their first discussion within ancient societies. However, the personal freedom inherent in current Western democracies, coupled with technology, enable us to have a considerable effect on nature, which may well affect others in the community or be out of step with what the community wants. To avoid such unwanted effects we have to develop ways to curb the excesses.

1.3 Before EIA

EIA is an approach developed to afford greater consideration to nature and the environment. While in concept and scope EIA is very different from previous approaches to considering the environment, it has evolved from an increasing interest in the environment (and an increasing level of government control). In particular, there are three fields that have provided a background for EIA.

1.3.1 Pollution

Pollution has been with human society wherever large groups of people have assembled together, and inevitably some law has been developed to try to control the problem.

A municipal law under Julius Caesar prohibited all wheeled vehicles in Rome between sunrise and two hours before sunset so as to improve the situation for pedestrians. This law, Hughes (1975) notes, fell into disuse in the 3rd century AD and Roman writers complained about the level of noise pollution generated by traffic in the streets. Ashby (1976) points out that more recently, in 1810, Napoleon issued a decree that divided noxious occupations into categories: those to be removed from habitation; those permitted on the outskirts of towns; and those tolerated close to houses. This was an early form of land use planning aimed at reducing odour and air pollution in some areas of Paris.

Australia's first pollution regulations related to the protection of water quality in the Tank Stream, Sydney's only water supply for the first 40 years. Within the first five years of settlement, regulations were passed to forbid the felling of trees within 50 feet (15 metres) of the stream. Powell comments that this was insufficient to protect water quality, and so Governor King proclaimed that if any person were found "throwing any filth into the Stream of fresh water, cleaning fish, washing, erecting pigsties near it ... on conviction before a Magistrate, their houses will be taken down and forfeit five pounds for each offence to the Orphan Fund" (1975: 35). A year later, King also announced a fine of 5 shillings per tree felled within two "rods" (approximately 10 metres) of the water line of the Hawkesbury River in an attempt to protect the banks from flood erosion.

The modern day equivalents of these actions are the policies and regulations associated with environmental protection legislation. Throughout Australia, and in most other countries, governments have established agencies to enforce environmental protection to manage air, water and land contamination.

1.3.2 Public health

Early pollution legislation probably was as concerned with odour and appearance as it was with health, while recent legislation concentrates on the health aspect with pollution control regulations being an obvious example. But, as Cullingworth (1974) discusses, there was also a strong element of concern about health in early town-planning laws, through the separation of noxious industries, the provision of parkland, and the arrangement of streets to provide light and fresh air to houses. More particularly, Factory Acts in the 19th century and recent occupational health and safety legislation, which regulate working conditions, were designed to mitigate the worst excesses of exploitation of human resources through control of (some) aspects of worker health.

Twentieth century "health" Acts have extended the philosophy of earlier legislation to establish further protection of the human "resource". At the beginning of the 21st century the link between human health and environmental conditions was made explicit in the National Environmental Health Strategy 2000.

1.3.3 Naturalists

In the late 19th century, a number of North Americans expressed concern over the reduction in the area of forests and in the number of native animals in the United States. Marsh's book *Man and Nature* (1967) helped to mobilise support for a greater concern for the natural environment, resurrecting the concept of stewardship. In the Australian context, Dingle (1984) records that "defenders of the bush" and "nature lovers" appeared on the Victorian scene through 1860 to 1880. Leisure time available to urban dwellers and the improved transport of the time helped people to develop an interest in nature.

As a result of this interest, and no doubt the increasing wealth of the country, the "national parks movement" developed, largely to protect areas of bush for the enjoyment of people. Over the last 100 years we have also seen the passing of legislation that recognises the value of plants and animals; that is, national parks Acts, forestry Acts, and Acts to protect particular plants and animals. The first National Park proclaimed in Australia was Royal National Park (NSW) with legislation being passed in 1879.

1.4 Evolving Assessment Practices

An appreciation of our cultural and philosophical history provides an understanding of the origins of EIA. There is no single reason for the development of modern society's increasing concern for the environment, or for the importance of the assessment of impacts. However, the important point is that these concerns exist, and have been brought together in the design of impact assessment procedures.

Much legislation relating to aspects of the environment, such as pollution control or protection of biodiversity (flora and fauna species), has been enacted. This can be read as indicating a general environmental awareness. The problem has been that while the awareness may be real enough, these environmental aspects have not been considered to have been a high priority, and they have been evaluated individually. The result has been that they have not then been compared on an equal footing with (particularly) economic issues. EIA is an attempt to redress this imbalance to ensure social, economic and environmental impacts are assessed concurrently.

The theory and practice of impact assessment, especially EIA, are addressed in detail in the following chapters. However, it should be remembered that these formalised procedures will be effective only if there is the will (political, legislative, and cultural) to apply them.

Ultimately, if there is a willingness to think about the consequences of actions, before they are undertaken, formal procedures will not be necessary. In other words, impact assessment can be done by anyone, at any time for any scale of project. Individuals can always accept the responsibility of reviewing the effects of a proposal; they do not have to rely on regulations and bureaucracies. EIA is not just a tool used for large proposals deemed to have potentially significant impacts on the environment (although some practitioners may argue this now seems to be the case) – rather, it is a process that can, and should, be used in all situations where negative environmental change is possible.

As a result, the focus of the discussion is not so much the intricacies of the techniques that have developed to serve EIA, although these issues will be covered as they provide some useful guides for someone embarking on an assessment task. Rather, the main emphasis will be given to two issues which are the basis for the ways in which EIA, and other assessment processes, will be applied in future. First, the political nature of EIA will be a recurring theme. EIA is political in terms of the way in which governments legislate for EIA, and the ways in which value judgments and political decisions, at the level of the individual, permeate virtually every element of EIA. Second, EIA, and assessment processes in general, is intrinsically simple and requires neither great intelligence nor sophisticated understanding. It follows that anyone should be able to take and apply the ideas. This is to be encouraged so that this skill can be applied to the wide range of decisions we make about policies, projects and proposals which can change the environment. If everyone were empowered to undertake such assessments, we would not need formal EIA procedures, and the environment would inevitably be given a higher status.

To assist this empowerment, the following material has been designed to be accessible to the interested reader. While the language and terms of EIA cannot be avoided, the relatively simple concepts of the assessment processes should be readily apparent. For those who wish to delve into the intricacies of the processes, particularly EIA, the material provides guidance to reference materials and internet sites.

Chapter 2

EIA: CONTEXT AND CONTENT

2.1 Initial Intentions of Impact Assessment

Concern over pollution, public health and the natural environment helped to raise awareness about the environment, and it seems likely that it was the catalyst that led to EIA. More recently the concern about finite resources and the impacts of climate change have provided an additional reason for thinking about the effects of proposals.

For most of its existence, Western civilisation has been concerned with the discovery, conquest and subsequent exploitation of empty spaces (or those inhabited by "inferior" species which could be subdued). Also, in the past, there was never a general feeling of reverence for what exists. The prevailing attitude was that anything that was old was useless, probably unhealthy and ripe for removal if it stood in the way of progress. Only where superstition (or faith) stood in the way did temples and holy places remain unscathed. The idea of a "museum" is recent, and the community's interest in protection of historic monuments and buildings, let alone landscapes or townscapes, is even more recent.

Within these social contexts the significance of EIA as a technique has been outlined by Caldwell:

1. Beyond preparation of technical reports, EIA is a means to a larger end – the protection and the improvement of the environmental quality of life.
2. It is a procedure to discover and evaluate the effects of activities (chiefly human) on the environment – both natural and social.
3. It is not a science, but uses many sciences (and engineering) in an integrated interdisciplinary manner, evaluating relationships [and risks] as they occur in the real world.
4. It should not be treated as an appendage ... to a project, but regarded as an integral part of project planning ...
5. EIA does not "make" decisions, but its findings should be considered in policy – and decision-making and should be reflected in final choices ...

6. The findings of EIA should focus on the important or critical issues, explaining why they are important and estimating probabilities in language that affords a basis for policy decisions. (1989: 9)

However, Eversley (1976) points out that had our forebears known, or cared, about pollution, visual intrusion and historic structures to the exclusion of other interests, such as science, we may not have had the development that has given us the comfortable and relatively secure life we now have. These issues have had a substantial influence on EIA, for while it has been developed as a social process to examine the environmental impacts of proposed development projects, it is not a process which allows decisions to be made on the basis of environmental considerations alone.

Formby (1987) observes that very few proposals considered under the Australian Commonwealth EIA procedures have been prevented, but many proposals have been modified and advocates of EIA would argue in favour of this process for robust examination of proposals before any decision was made about their approval. This comment remains current and is likely to apply to the EIA procedures of the State and Territories as well. As with other situations where there are contrasting goals (eg, those of Table 2.1), the outcome of an EIA is not necessarily going to favour one goal.

Initially, those who feared that it would be a costly step in the development of projects viewed the introduction of EIA with concern. However, a survey by the Bureau of Resource Economics found that the direct compliance cost of EIA for large Australian companies was not a significant problem, especially where the environmental assessment was integrated with feasibility studies; these costs were found to be only a small proportion of the total cost of the project.

Similarly, Wathern (1988a) reports that a study by the United States Environment Protection Agency around 1980 showed that there were significant improvements in waste-water projects through the EIA process. Also, costs associated with the process, and savings accruing from the project modifications, more than covered delays.

2.2 The Objectives of Environmental Impact Assessment

As nations mature, and this implies that they become wealthier, they change their political priorities. Broadly speaking, O'Riordan (1976) sees that the trend is towards improving the general social welfare and increasing the equality of social opportunity. An example is the number of international conferences held during the 1960s (such as UNESCO's Biosphere Conference) that crystallised the concern of academics and government scientists about the lack of consideration given to the environmental effects of projects. EIA is a result of this maturity. As a formal process it could only be conceived and developed in societies with an existing system of government administration, and where there is enough wealth to support it; with any administrative

process there has to be someone there to oversee it, and information on which to base the administrative decisions has to has to be provided to that person. This all means a commitment of resources (monetary and people).

EIA then is generally considered as the formal (administrative) process that was developed within the 1970s (first in the USA) as a result of the concerns raised about environmental pollution and the environmental effects of projects during the 1950s and 1960s. In particular, Barlett (1989) notes that impact assessment was seen as a major innovation in policy-making and administration. While the power of EIA was not emphasised in its early years, its potential was recognised.

This potential is indicated in Hollick's (1986) discussion of the goals of EIA, and the related evaluation criteria he developed to review the operation of EIA worldwide. For this review he has condensed the role of EIA into three goals, to each of which he attaches objectives that also serve as the evaluation criteria for EIA processes as outlined in Table 2.1.

Table 2.1 Evaluation criteria to review operation of EIA

Goal	Objective
To protect the environment from damage	• to ensure that adequate environmental information is available to decision-makers • to ensure that environmental factors are taken into account in decision-making • to coordinate decision-making between agencies • to coordinate policies on environmental aspects of decisions between nation-states • to ensure adequate environmental management over the life of the project.
To improve public participation in government decisions	• to involve the public in all stages of EIA.
Economic efficiency	• to minimise costs and maximise benefits of approval processes.

Source: Hollick 1986

Essentially, the object of the EIA process is to identify the possible risks to the environment (social, economic, biodiversity) that may result from a proposed action. This information is then used to decide whether to proceed with the action, and on what conditions. In most cases the "action" is a specific project or physical works, rather than a policy or plan.

The thinking behind this approach is that if the likely effects of a proposal are known in advance, steps may then be taken to avoid or mitigate for any adverse effects. The EIA process does not eliminate all environmental effects – no matter what we do there will always be some types of effects on

the environment. The critical issues are the magnitude of these effects and the associated risks that need to be adequately assessed. These potential and predicted "risks" are becoming more of the focus of an EIA process. Indeed, various jurisdictions in Australia and around the world are incorporating environmental risk into decision-making frameworks of their EIA legislation and associated regulations or guidelines. This is discussed further in Chapter 6 and 8.5.

Lockie et al suggest that EIA is intended to:

- provide for the integration of environmental management within planning for proposals;
- ensure that proponents assume responsibility for the protection of the environmental values that may be affected by their proposals;
- form a basis for statutory decision-making on whether a proposal meets ecologically sustainable development principles;
- establish relevant environmental management and monitoring conditions; and
- incorporate community and stakeholder views in assessment and decision-making. (2008: 179)

A range of expectations is held for EIA. Broadly speaking, those involved in the business of government and making decisions about the use of resources will probably see EIA as a bridge between their level and the technical and public issues associated with proposals. However, the Canadian Environmental Assessment Agency (2004a) acknowledges that there are others with different, and often conflicting, expectations. In particular:

> Decision-makers see a process that sometimes takes too long, that seems to cost too much, that appears unnecessarily complicated, and that in the end, does not always give them the kind of information they need to make a sound decision. Managers and Practitioners see a process where the results of their work are not always taken into account in the final decisions, and where they do not always have the time and resources to do an adequate job. Members of the public see a process that may exclude them from participating in decisions that affect their lives and communities, or that may provide massive volumes of complex scientific data but few straightforward explanations.

We could add that members of the public look to EIA as one of the few processes available to them to stop proposals that are considered to have an impact on the environment or affect them in some way.

Yet another perspective on the expectations of different people is provided by Anderson and Sadler (1996). They suggested that: to EIA administrators, it is a process that helps safeguard the environment; to politicians, EIA has the potential to "defuse 'explosive' situations" (alternatively, politicians may also see the EIA process as an impediment to development); and to the community, it is often a process which stands or falls on the "right" decision being made at the conclusion of an assessment – or as suggested above, that the community's view is substantiated. Irrespective of the particular

perspectives held, this variety of perceptions and expectations was identified, by Anderson and Sadler, as possibly the crowning success of EIA. Specifically, the range of views has the ability to "remove some of the political heat from environmentally contentious decision-making". Essentially, the EIA process is a mechanism for all stakeholders to become involved in the environmental assessment process, whether it is the community, government or private industry, prior to decisions being made.

Having a range of perspectives may make life interesting, but it confuses the operation of EIA. To clarify the position and to provide some consistency for those closely involved in the EIA process, the objectives of EIA have been identified by the International Association of Impact Assessment (IAIA), one of the key associations of international EIA professionals. In line with the thinking of the 1990s, these objectives, below, built in aspects of sustainability (or sustainable development, see 10.7):

- to ensure that environmental considerations are explicitly addressed and incorporated into the development decision-making process;
- to anticipate and avoid, minimise or offset the adverse significant biophysical, social and other relevant effects of development proposals;
- to protect the productivity and capacity of natural systems and the ecological processes which maintain their functions; and
- to promote development that is sustainable and optimises resource use and management opportunities. (IAIA (2000))

In addition, we can identify that an underlying "agenda" for EIA has been to educate people about the environment. While less prominent, EIA has the objectives of:

- requiring proponents, and decision-makers, to think about the possible environmental impacts that can arise – if proponents recognise the impacts they have the opportunity to revise the plans and designs for the proposal, and so reduce the impacts; and
- informing the public about the proposal, and the potential environmental impacts.

Along with the many objectives of EIA identified above, Abaza, Bisset and Sadler (2004) have identified specific principles of EIA that are based upon the IAIA's objectives and Sadler's (1996) international study of Environmental Assessment effectiveness. These principles are that EIA should:

- be applied as a tool to help achieve sustainable development;
- be integrated into existing development planning and approval processes;
- be applied as a tool to implement environmental management;
- be integrated into the project life-cycle;
- be applied to all proposed actions that are likely to have a significant adverse effect on the environment and human health;

- include an analysis of feasible alternatives to the proposal;
- include opportunities for public participation;
- be undertaken in an interdisciplinary manner; and
- integrate information on social, economic and biophysical impacts. (2004: 42)

In a practical fashion, Spry (1976) suggests that for the developer, the short-term purpose of EIA is to comply with regulations imposed by some authority. For the public, its long-term purpose is to protect the environment. However, Spry proposes that the valuable, long-term uses of EIA relate to the provision of basic information and interpretation of environmental aspects, assisting the developer to decide whether to proceed with the proposal, and defining the monitoring program that would be required during and after the proposal's development.

EIA is also a way of ensuring that environmental factors are considered in decision-making process along with the traditional economic and technical factors. Further, Sadar (1994) points out that EIA requires the scientific (technical) and value issues to be dealt with in a single assessment process. This helps in the proper consideration of all advantages and disadvantages. Environmental considerations may, therefore, be set aside in favour of what are felt to be more important considerations. Alternatively, predicted adverse effects on the environment might lead to strict conditions being imposed to avoid these effects or remedy any adverse effects, or perhaps lead to the complete abandonment of a proposal.

EIA cannot, however, be regarded as a means of introducing an environmental "veto" power into the administrative decision-making process. Decisions which are unsatisfactory from an environmental point of view can still be made, but with full knowledge of the environmental consequences. The final decision about a proposal depends upon the likely risk and severity of the adverse effects, balanced against other expected benefits. However, as outlined further in Chapter 6, some States' EIA legislation enables an "early no".

It is interesting to consider the remarks of Dr Moss Cass when, as Commonwealth Minister for Environment and Conservation, he introduced the Australian Government's legislation for EIA (the former *Environment Protection (Impact of Proposals) Act* (*EP(IP)A*)), and commented:

> It will not grant me the exclusive power of veto over proposals or policies. It will not force developers to abandon environmentally unsound objectives. It will not ensure that the government makes environmentally sensible decisions. It will not give individual citizens the power to stop bad projects or to set conditions for moderate ones.
>
> The legislation will, instead, enable me to gather extensive information on specific proposals. It will force developers to include environmental impact in their planning. It will present the government with comprehensive information about environmental impact as an aid to decision-making. And it will enable the public to argue a case publicly, to have the case

published, and to force governments to justify their decisions. (Fisher (1980: 13))

As can be seen from Dr Cass's conception of EIA, which is the same worldwide, it is an administrative process that identifies the potential environmental effects of undertaking a proposal, and presents these environmental effects alongside the other advantages and disadvantages of the proposal to the decision-makers. The implications of these principles can be seen from Sadler's study to assess the effectiveness of EIA (see 10.1). Sadler (1997b) concluded that:

- only a very small fraction of proposals are halted, permanently or temporarily, as a direct result of EIA at the end of the review process;
- preemption or early withdrawal of unsound proposals has been reported though, not unexpectedly, it has proved difficult to document;
- EIA has been useful in developing support for and confirmation of positive environmentally sound proposals;
- the greening or environmental improvement of proposed activities is frequently seen; and
- particular indirect effects of EIA are both instrumental (such as where policy or institutional adjustments are made as a result of EIA experience) and educational where participation in the EIA process leads to positive changes in environmental attitudes and behaviour.

With regard to the last point, there is considerable advantage to the general community where those people involved with the proposal, as well as decision-makers, are required to think about the environmental effects (and therefore avoid negative effects), and the public can be made aware of the details of the proposal.

Clearly, while from the outset limits have been placed on EIA, there are many potential benefits to be realised from the process. These have been identified by Sadler (1997b) as follows:

- Early withdrawal or preemption of unsound proposals.
- Green check or legitimisation for well-founded proposals.
- Last-recourse stop to proposals found to be environmentally unacceptable.
- Improvement tool for proposals, including –
  - relocation of a project or activity to a more suitable site
  - selection of best practicable environmental option (or equivalent concept)
  - redesign of projects to reduce or avoid environmental impacts
  - changes to or rescheduling of planned activities to accommodate environmental and community concerns

- mitigation of impacts by measures additional to those above, including rehabilitation, impact compensation
- policy clarification and redefinition
- institutional response and adjustments
- organisational learning (EIA as a vehicle for education or value change)
- securing equity for communities and groups affected by proposals
- community development through participation

Similar benefits are discussed by many EIA commentators, who frequently, while using their own words, identify essentially the same benefits. For instance Figure 2.1 *below* outlines Sadar's (1994) list of benefits. Interestingly, he has also identified a number of risks (essentially the inverse of the benefits) associated with not using EIA. While there are some differences in emphasis between the lists of Sadler and Sadar, predominantly there is consistency with the general objectives we have noted above.

Figure 2.1 Potential benefits of applying EIA to proposals, and the risks of not using EIA properly

Benefits include:

- Lower project costs in the long term (fewer changes at advanced stages, lower probability of environmental disasters and associated clean-ups).
- Planned or avoided remedial measures which are implemented in time to minimise adverse impacts.
- Improved future planning of future projects.
- Better protection of the environment and minimised social impacts through the consultative processes.
- Opportunity for the public to learn about environmental effects, express concerns, and provide input to the assessment process.
- Opportunity for the public to influence the decision-making processes.
- Enhanced public confidence in public and private institutions.
- Good public relations fostered.

Risks of not using EIA properly:

- Costly litigation and expensive clean-ups.
- Expensive "surprises" in the future which can result in significant economic losses for proponents.
- Loss of public trust in private and public institutions or with individuals in positions of power.
- Worsening environmental conditions leading to a deterioration in the natural resource base of the economy.
- Consumer backlash against industry and businesses responsible for environmental impacts.

Source: Sadar (1994)

2.3 Current Directions in EIA

Almost four decades of operation of EIA have led to innovation in policy-making and administration, and to reconceptualisation of some of the specifics associated with its implementation. In particular, in Australia, where separate procedures have been operating at State and Commonwealth levels of government, there have been moves to develop a more consistent approach. Ministers of the Australian and New Zealand Environment and Conservation Council (ANZECC) have adopted a national approach that presents a series of "principles" which will guide any amendments to the EIA processes that operate in Australia and New Zealand (Australian and New Zealand Environment and Conservation Council (1991a)). The objectives of this national approach broadly follow historical EIA aims, but in some aspects are more precise. They are:

- to ensure that decisions are taken following timely and sound environmental advice;
- to encourage and provide opportunities for public participation in environmental aspects of proposals before decisions are taken;
- to ensure that proponents of proposals take primary responsibility for protection of the environment relating to their proposals;
- to facilitate environmentally sound proposals by minimising adverse impacts and maximising benefits to the environment;
- to provide a basis for ongoing environmental management such as through the results of monitoring; and
- to promote awareness and education in environmental values.

These points have much in common with the Bruntland report "Our Common Future" produced by the World Commission on Environment and Development (1987), and which flagged the role that EIA could play in ecologically sustainable development (ESD). This challenge has been taken up in the ANZECC approach where EIA is specifically seen as one of the ways of achieving the objectives of ESD. The National Strategy for Ecological Sustainable Development, endorsed by all Australian jurisdictions in 1992, defines the goal of ESD as development that improves the total quality of life, both now and in the future, in a way that maintains the ecological processes on which life depends (Environment Australia (2003a)). Although not defined in the Commonwealth *Environment Protection and Biodiversity Conservation Act 1999* (EPBC Act), the Act (which is discussed in detail in Chapter 6, see 6.3.4 on *page 132*) sets out the principles of ESD which includes the precautionary principle and the principle of inter-generational equity. In some respects the link to ESD provided EIA with a theoretical or philosophical basis that was lacking before the concept of an environmental ethic, which became popularised through the Bruntland report.

The Australian Government has now formalised this link in its strategy for ESD, where two objectives are stated to be the incorporation of ESD into EIA, and increasing the sensitivity of EIA to include cumulative and regional impacts (Commonwealth of Australia (1992)). However, little practical progress is evident. For example, Court et al argue that "to pursue the nationally agreed goal of ecologically sustainable development ... it is essential to incorporate Strategic Environment Assessment (SEA) and Cumulative Impact Assessment (CIA) into the EIA process" (1996: 42). While there has been discussion along these lines in an earlier review of Commonwealth EIA processes (Environment Protection Agency (1994b)), Strategic Environmental Assessment and Cumulative Impact Assessment are still marginal activities. However, as discussed in 3.10, SEA is receiving serious attention worldwide for the assessment of the strategic work of government, such as policy-making. Further, in Australia, the EPBC Act (see 6.3.4) has now formally built SEA into government processes. Initially, much of the SEA had been undertaken for fisheries management within Commonwealth waters, however in the last five years we have seen SEA for a range of projects and plans, including a SEA for the Great Barrier Reef which looks at whether the appropriate planning processes and management are in place to protect the Great Barrier Reef from development. Other SEA's include Melbourne's Urban Growth Boundary in Victoria, Fire Management Policy in South Australia and for large iron ore and gas projects in Western Australia. Further discussion on SEA is found in 3.10.

As well as beginning to set out a theoretical base, the national approach shows a new line of thought for EIA in that "protection and management of the environment, for which EIA is a tool, is the responsibility of everyone" (ANZECC (1991a: 4)). To emphasise this, the approach details principles to be adopted when different groups are involved in EIA. Principles are outlined for assessing authorities, proponents, the public and government. These principles also provide a context for broadening the scope of EIA, from the historical emphasis on projects to include EIA objectives in the consideration of policies, plans and programs. The value of EIA in this context is also seen by Htun (1989), who notes that it "provides the means to identify the technical constraints to plan and implement sustainable development projects" (1989: 22).

EIA has been applied to a wide range of proposals. Some have a direct and obvious impact on the environment and their relation to the EIA process is well documented; for example, Buckley's (1998) consideration of the environmental impacts from mining and petroleum extraction. More recently the application of EIA has been considered for tourism and aid proposals. For instance, the Industries Assistance Commission noted that tourism has been associated with significant overdevelopment and degradation of the environment. Consequently, the commission proposed that tourist and travel projects "should be both subject to an initial Environmental Impact Assessment, and should be monitored once operating has begun" (1989: 54). Such recognition is in line with the concerns expressed in the Bruntland report for assessing the effects of development in developing countries.

Australia's national approach is also important in that it proposes that initiation of the EIA process (ie, the stage at which it is decided that the proposal requires an EIA) should be as early as practicable in the planning of a proposal. Also, the public, proponents and decision-making bodies should be able to refer proposals for consideration of an EIA. Associated with these proposals is a likely increase in the number of EIAs. To cater for this, the national approach is that different levels of assessment (ie, the amount of investigation and public review) has been developed to take account of the characteristics of the proposal, and the importance of the environmental and political contexts associated with the proposal (see 6.3).

As EIA is a process used to help make decisions or plan the use of resources, particularly land, there has been some overlap with the older processes of statutory land-use planning. Hogg (1975) presents a theoretical concept for how EIA and "planning" would be linked, and some EIA processes have been integrated with planning legislation, in New South Wales for example (see 6.5). Armour (1989) discusses how the importance of this link continues to be raised. This theme was further reinforced during the World Commission on Environment and Development's establishment of the concept of ESD. However, after considering the issues surrounding integration of EIA and "planning", in connection with technical and disciplinary aspects, consultation, sociopolitical and organisational aspects, Armour concludes, "the barriers to integration seem insurmountable" (1989: 10). Yet, by the beginning of the 21st century most EIA procedures across Australia had moved into a close alliance with the procedures used for land-use and resource planning (see Chapter 6).

In addition to the formal EIA processes, non-procedural environmental assessment frequently occurs. As Ramsay and Rowe (1995) point out, there are situations where an obligation may be imposed on a public authority to take account of possible effects on the environment without any requirement to follow a formal EIA procedure. This obligation may be implicit, such as where those responsible for the implementation of planning acts, or local government activities, are required through some general directive, or objective, to "give consideration to the protection of the environment". The obligation may also be explicit. For example, the Victorian *Planning and Environment Act 1987* has specifically required that consideration of possible environmental effects be given when assessment of development proposals is undertaken; some other States' planning acts also include similar specifications. Likewise, many State laws require a non-procedural environment assessment to be undertaken when pollution licences and approvals are being considered. As detailed in Chapter 6, environmental assessment not only informs land use planning decisions, but is also an integral part of planning.

While there has been an evolution of the formal EIA procedures, EIA has been breaking out of the formal processes that have been imposed by governments. "Informal" environmental assessment may also be undertaken as part of an organisation's internal planning and project assessment processes; for example, the procedures developed by Melbourne Water (1994b). Increasingly, there are examples of the EIA process being incorporated into the decision-making of private organisations (see 10.4). In parallel we are seeing that rather than EIA being a "stand-alone" process, it is being linked to the tools of environmental management, especially Environmental Management Systems (EMS) (see 10.5) and more and more embedded with risk assessment approaches.

The scope of EIA has also been given consideration. Lang (1979) takes the possibilities of EIA substantially further than identifying impacts. He sees that the process can have an impact on both environmental quality and social equity, and so EIA can take on a role as a dynamic instrument for social change. This happens when an assessment is begun early in the planning process and there can be an examination of the need underlying the proposed activity including an early assessment of the proposed site location(s). Early assessments can help break patterns of high consumption, and force organisations and government to be explicit about their definitions of what is needed and how that need can be met. EIA can also be used to reveal present distributional inequalities and hazard a guess at expected inequalities resulting from some proposal. Generally speaking, however, this greater potential goes unrecognised, or is consciously controlled.

Increasing public participation in EIA can have an effect on decision-making. The comprehensiveness of EIA documentation prepared at an early stage (the environmental impact statement, or EIS) was often called into question. Principally, the concern was that consideration of important environmental factors had not taken place, or that the attention given to the factor environment was insufficient. To overcome these concerns the concept of "scoping" was developed. Taylor (1984) has suggested that scoping represented an effort by the Carter Administration in the United States to influence the implementation of EIA processes by opening up to public review the critical early choices about the coverage of EISs. In effect, this exposes the previously internal debate of technocrats to the ideas and concerns of outsiders, who can confirm or debate the internal analysts' technical and political judgments. In scoping, the emphasis is on identification of the significant issues to be addressed in the EIA process and elimination of those that are insignificant.

Since the inception of EIA, its political dimension has been recognised; specifically in respect of the impact on the allocation of resources (use of land, whether minerals should be mined or rivers dammed) and the involvement of the public (particularly interest groups). A broader role has also been identified by Barlett, who suggests that "more than methodology or substantive focus, what determines the success of impact assessment is the appropriateness and effectiveness in particular circumstances of its implicit policy strategy" (1989: 3). This leads to his contention that the purpose of every impact assessment should be to influence policy.

All this suggests that far from being a simple objective process, the results of which are always interpreted and applied in the same way, EIA shows itself to be part of the wider political process. As a result, it is capable of being used, interpreted and ignored as required. This is not surprising because, as Hrezo and Hrezo (1984) point out, environmental issues are political rather than technical issues. Specifically related to the muted enthusiasm for the adoption of EIA, Court et al (1996) point to a lack of political commitment being evident. This was indicated by a lack of appreciation of its need, the conflict of interest with other political goals, and the lack of funding and resources.

We cannot remove the political dimension of EIA. But this means that the links between community values, public opinion and political decisions (and therefore decisions about the environment) need to be appreciated if strategies for understanding EIA and participating in its processes are to be successful.

2.4 EIA as Practised

Building on the need to have a procedure that provides guidance to the decision-makers, the EIA process begins with recognition that there is potential for environmental impacts of a proposal, and a possible need to assess the potential for environmental impacts. In other words, the EIA process begins with determining whether formal EIA is required (known as "screening"). This is followed by procedure to identify possible impacts and to consider ways to reduce any impacts. Under the formal EIA processes (those undertaken through government laws or processes), once it has been decided that an EIA is required the EIA process normally entails two steps. The first provides the information upon which the examination is based, while the second presents the overall assessment:

1. Preparation of a document which provides information on the existing natural, built, social and economic environment; predictions about the environmental effects; and recommendations which could flow from a decision to proceed with the proposal (or alternative to the proposal). This document is usually called an *environmental impact statement* (EIS).
2. Review of the EIS by the public and government officers to consider the accuracy of the EIS, and in view of the predicted effects recommend whether/how the proposal should proceed. This review is reported to whoever makes a decision about the proposal in some form of *assessment report.*

Hence, EIA = EIS + assessment report.

The combination of these two components has been the more usual way in which EIA has been undertaken. With recent moves to "decentralise" EIA, by incorporating environmental assessment into various levels of decision-making, formal documentation through a public EIS and assessment report may not always be apparent; for example, where assessment is incorporated in internal processes of corporations or through planning processes that are often undertaken by the three levels of Australian governments. Nonetheless, informal EIA (see 2.3) such as that associated with an EMS still requires identification and documentation of potential impacts, plus the reporting of how those impacts would be managed. No matter whether the assessment process is formal or informal, or what terms are used, the two stages will be apparent in some form. Again, these two elements combine to make the overall EIA.

During the evolution of EIA, refinements have been added to the essential two steps. (Figure 2.2 *opposite* illustrates the overall process.) Before the production of the EIS there is now an understanding that environmental considerations may be built into the development of the proposal so that the need for a formal EIA is reduced. This may be achieved through SEA (see 3.10), where broad design direction is provided at the national or regional level. Also, as part of the development phase increasing attention is being given to the consultation of interested parties (stakeholders) to assist in decisions about the need for an EIA, and the content of the EIS.

By themselves the objectives we identified earlier (see 2.2) are fairly broad and somewhat idealistic. However, the IAIA has also developed a number of principles that are intended to more closely direct those who apply EIA. These principles have been divided into two groups (see Figure 2.3 on *page 24*). Basic Principles apply to all stages of EIA and are designed to clearly explain the basic intent of EIA. Operating Principles describe how the Basic Principles should be applied to the main steps and specific activities of the EIA process; these steps have been identified in Figure 2.2. Importantly, the IAIA recognises that in some instances the Principles are interdependent and, in some cases, may conflict.

Whenever we are discussing EISs, it is worth remembering that in some EIA procedures, such as that of the Australian Commonwealth, there are two parts to what becomes the final EIS. The initial document produced by the proponent is called the *draft EIS*, which is made available for public comment. The proponent responds to the public's comments in one of two ways. Either a *supplement to the EIS* is produced, and the draft plus the supplement constitutes the final EIS, or alternatively, the draft is revised and released as the final EIS. Once the final EIS has been prepared, it forms the basis upon which the assessment report, of the relevant government department and minister, is developed. The final EIS is also the document that sets the directions for monitoring the environmental management of the proposal.

Looking at the end of the EIA process, management and monitoring of predicted impacts, resulting from specific proposals, has been absorbed into the process, as has auditing of the EIA process itself. We will look at monitoring and auditing on several occasions in the following sections (especially 10.2), but it is salient to note that IAIA (2000), through their best practice principles highlights these latter stages. Follow up is identified as important "to ensure that the terms and condition of approval are met; to monitor the impacts of development and the effectiveness of mitigation measures; to strengthen future EIA applications and mitigation measures; and, where required, to undertake environmental audit and process evaluation to optimise environmental management".

Figure 2.2 – The Environmental Impact Assessment process

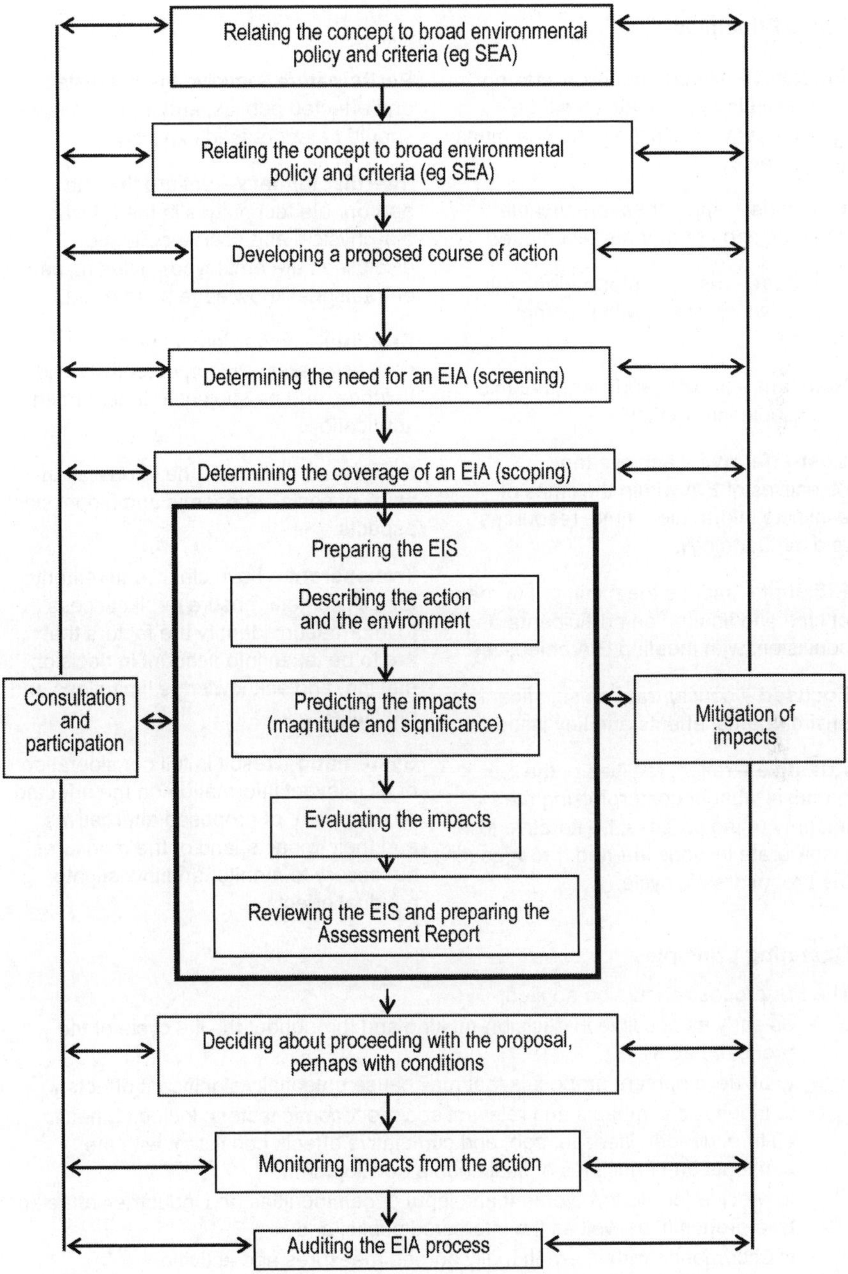

Figure 2.3 Best practice principles for guiding Environmental Impact Assessment

Basic Principles

Purposive – inform decision-making and result in appropriate levels of environmental protection and community well-being.

Rigorous – apply "best practicable" science, and appropriate techniques.

Practical – result in information and outputs which assist with problem solving.

Relevant – provide sufficient, reliable and usable information.

Cost-effective – achieve the objectives of EIA within the limits of available information, time, resources and methodology.

Efficient – impose the minimum burdens of time and finance on participants consistent with meeting EIA objectives.

Focused – concentrate on significant environmental effects and key issues.

Adaptive – reflect realities of the proposal without compromising the integrity of the process; be iterative to incorporate lessons learned throughout the proposal's life cycle.

Participative – involve the interested and affected publics, and their concerns should be addressed explicitly.

Interdisciplinary – ensure that the appropriate techniques in the relevant bio-physical and socio-economic disciplines are employed, including use of traditional knowledge as relevant.

Credible – be carried out with professionalism, rigour, objectivity and balance, and be subject to independent verification.

Integrated – address the interrelationships of social, economic and biophysical aspects.

Transparent – have clear requirements for EIA content; ensure public access to information; identify the factors that are to be taken into account in decision-making; and acknowledge limitations and difficulties.

Systematic – result in full consideration of all relevant information on the affected environment, of proposed alternatives and their impacts, and of the measures necessary to monitor and investigate residual effects.

Operating principles

The EIA process should be applied:

- as early as possible in decision-making and throughout the life cycle of the proposed activity;
- to all development proposals that may cause potentially significant effects;
- to biophysical impacts and relevant socio-economic factors, including health, culture, gender, lifestyle, age, and cumulative effects consistent with the concepts and principles of sustainable development;
- to provide for the involvement and input of communities and industries affected by a proposal, as well as the interested public; and
- in accordance with internationally agreed measures and activities.

Summarised from IAIA (2000)

But the key point is also made that any monitoring should be seen as part of a broad scope of environmental management. Here the proposal is that indicators used to "indicate" the state of the environment associated with the proposal should be designed so they also contribute to local, national and global monitoring of the state of the environment, and ultimately to sustainable development.

The final EIA process outlined in Figure 2.2 (but not of least importance) is public consultation and involvement. Members of the public face many difficulties if they want to be involved in EIAs – lack of time and resources, and limited access to information. Encouraging public involvement in the EIA process ensures a variety of input to the process, and an improved bank of data on which discussions may be based. We will explore public involvement in detail in Chapter 4.

2.5 Issues for EIA

2.5.1 The political perspective

While EIA has been viewed as a technical process, it is inherently a political process. Further, it is not necessarily based on tangible scientific data. As Formby explains, "the Iwasaki case showed the extent to which government can avoid or subvert the Commonwealth *Environmental Protection (Impact of Proposals) Act's* requirements when it so wishes" (1987: 19). Elsewhere, Formby (1989) points out that EIA evolved from the politics surrounding the impacts that development projects were having on the environment. Once established it became dominated by technocratic approaches, which ignore social, political and economic conditions. Most importantly, he notes that the "ultimate purpose of EIA is not just to assess impacts; it is to improve the quality of decisions" (1989: 193). Consequently, the relationships between the political processes and EIA processes need to be understood. For example, Caldwell (1989) suggests that EIA not only forces environmental knowledge into the policy process, but also reveals the inadequacy of the information upon which society and governments propose to act. In this situation, EIA may be seen as a threat to those who avoid change and Formby (1989) reminds us that the politics surrounding EIA are not confined to the politics of parties and elections, but must include the politics of personal and organisational survival.

These points are picked up by O'Riordan's (1985) analysis of the roles of science and risk. Considering why environmental risk issues became political issues, he identifies six explanations: the establishment of a counter-establishment science; post-World War II opportunities for public access to the decision-making processes; environmental martyrdom (such as the close-calls, typified by the accident at the Three Mile nuclear reactor); publicity (through the media); ubiquitousness (effects on "establishment interests"); and culpability (the venting of frustration on a target). As will be apparent from previous discussion, helping to reduce environmental risk is one of the underlying principles of EIA and the integration of a risk assessment approach imbedded in the EIA process is becoming more common.

That EIA can serve political agendas, whether they be the agendas of individuals or groups in power, is illustrated by Richardson (1994). His discussion of the EIA processes for the proposed mine at Coronation Hill, Northern Territory, and the pulp mill at Wesley Vale, Tasmania demonstrates the political factors which influenced discussions about the need for an EIA, and the interpretation of the EIA results. Similar discussions are still occurring regarding various large development projects a decade or so on from Richardson's analysis. Projects such as the Bell Bay Pulp Mill in Tasmania's Tamar Valley and the Channel Deepening project in Victoria's Port Phillip Bay attracted much debate about the EIA process in the last ten years. Flexibility in EIA processes is, according to Leeson, a function of the design of the processes. From examination of several EIAs from the early 1990s in Victoria she sees that:

> The manipulation of the *Environment Effects Act* [1978] by the Victorian Government comes as no surprise. ... The fact that the ... Government is interpreting the *Environment Effects Act* in a manner which it sees fit was the very intention of the government of the day which formulated and passed it. The degree of latitude allowed in the interpretation of the legislation was a deliberate feature on which successive governments have relied. ... Obviously concerned about the implications for economic development, the Government ensured that the Act was weak by maintaining virtually unlimited powers of Ministerial discretion and avenues by which the obligations of the Act could be avoided. Not only is the Act extremely flexible, it does not allow for meaningful review nor does it promote accountability when such discretion is exercised. (1994: 90)

The meaning of "environment" in the context of an EIA process was tested during the Victorian Supreme Court case of *Friends of Mallacoota Inc v Minister for Planning and Minister for Environment and Climate Change* [2010] VSC 222 where the Friends of Mallacoota contested that, amongst a number of matters, the Minister did not follow process as intended under the *Environment Effects Act 1978* when he made his Assessment under the Act and considered social matters (including safety). In this case, Justice Osborn concluded that the Act should be understood as one which embraced a flexible concept facilitating the assessment of potential harm to the environment in the broad sense, despite there being no definition of "environment" in the Act.

A significant political issue is the choice of proposals to which EIA is applied. Leeson discusses some examples of projects that have escaped assessment, while Shrybman (1990) points out that international trade agreements can also be exempt. In particular, the *General Agreement on Tariffs and Trade* (GATT) gave little recognition to the links between trade and environment. To address this omission, he proposed that GATT should be subject to assessment processes to identify the significance of these negotiations. While GATT has been superseded by the World Trade Organization (WTO) which has developed environmental guidelines, Thomas (2007) notes that concerns about the effects of trade continue to be expressed by Non-Government Agencies (NGOs) and researchers.

The community has the opportunity to bring about such changes, but the public's ability to be involved in the EIA process has been limited by the time and resources (such as technical expertise and access to information) available. Martyn, Morris and Downing (1990) point out that at present only the larger conservation groups and professional bodies have the internal expertise to provide comment on proposals. Smaller groups and individuals are unlikely to have access to this type of expertise, or the funds to acquire it. However, a system was developed in Ontario, Canada, to provide assistance to groups, where they requested it through an application process. Without consideration of the difficulties that face members of the public attempting to be involved in EIA, comment on their input is likely to be misleading. Another issue with public participation in the EIA process is that it often occurs late in the process, after a proponent has determined what the development will be and where it is going to be located.

2.5.2 EIA is an extra impediment to development

EIA becomes an extra step in the project development phase *only* if the project is likely to produce some substantial adverse environmental effects. If the project is "properly" planned (ie, one of its objectives is to minimise environmental effects), and potential environmental impacts have been considered during project planning, then the proposal is less likely to require an EIA. However, if the only objective is to show a quick profit, it is likely that an EIA would be involved.

The vast majority of proposals do not need formal EIA. For example, Bates (1983) notes that during the first few years of the operation of the Australian EIA procedures, some 3250 proposals were identified as being sufficiently environmentally significant to come under the procedures, but only 50 EISs were subsequently required. The Commonwealth Department of the Sustainability, Environment, Water, Populations and Communities (DSEWPC) 2011-12 Annual Report (DSEWPC (2012)) states that of the 412 proposals referred to the Commonwealth for a decision under the EPBC Act, 132 were determined to be controlled actions requiring formal environmental assessment under the Act, of which 44 related to mining projects. Even still, at the Commonwealth level, such figures indicate that the majority of proposals do not require formal EIA.

Similarly, in the United States, Hart (1984) discovered that less than 1 per cent of State actions would require preparation of an EIS. He proposed that generally 80-90 per cent of proposals would be exempt from the EIA process. Similarly, delays were infrequent at the next level of the process, where of the 11,000 EISs prepared at federal level between 1970 and 1978 in the USA, only 2 per cent resulted in injunctions and action through the courts. To complement this finding, another United States study by Sewell and Korrick (1984) suggested that factors other than the EIA process were responsible for delaying projects. Such factors include EISs that resulted in litigation (1 per cent of all EISs in the United States). Sander (nd) suggests that the EIA process is justified by the environmentally unwise projects that have been avoided due to the process. However, no quantitative study appears to have been done.

An early 1990s survey of large Australian business by the Bureau of Industry Economics (1990) produced similar findings. The survey found that the cost of EIA was generally not a great concern, but that delays resulting from the EIA process were the most significant cost, being up to 10 per cent of the project cost. Sources of delay were the number of authorities involved and the lack of coordination between them, the lack of uniform standards, and unpredictable changes to the rules of environmental assessment. An example of this concern is given by Short (1991) reporting on a mining proposal in Victoria, where it was claimed that the proposal was delayed by the need to do "costly" environmental impact studies and by public objections through the parallel approvals processes of environmental protection and planning.

Suggestions proposed by the Bureau of Industry Economics (1990) to improve the situation were to set and adhere to strict time limits, to develop consistency between State and Commonwealth procedures, and to better coordinate authorities. In other words, a more formal and organised process was sought. The report importantly noted that there are benefits of EIA which are apparent to industry, and which are in addition to those accruing to the community as a whole: increased discipline in project planning (with savings in cost and time); increased discipline in production processes; and improved environmental standards (through identification and control of pollution sources).

A subsequent survey by the Bureau, discussed by Leeson (1994), indicated that EIA was ranked the fourth most important impediment, behind operating costs, construction costs and cost of capital. This put EIA in front of exchange rate volatility, transport costs, company tax, and overseas trade barriers. These results suggests that while EIA is a concern for organisations, it is by no means the major concern.

This finding is broadly consistent with the international comparison that Wood (1995) has undertaken. He comments that:

> It is, perhaps, a testament to the inherent effectiveness and efficiency of EIA that, despite the marked differences in ... EIA systems, there should be a uniformity of view that the benefits of all the EIA systems outweigh their costs. (1995: 265)

2.5.3 EIA is just an additional cost

Two people are looking at a recently built building. One comments, "They say the EIS cost more than the building". The other responds, "Well it should

Figure 2.4 EIA Costs as % of Project Development Costs by Size of Project (€m)

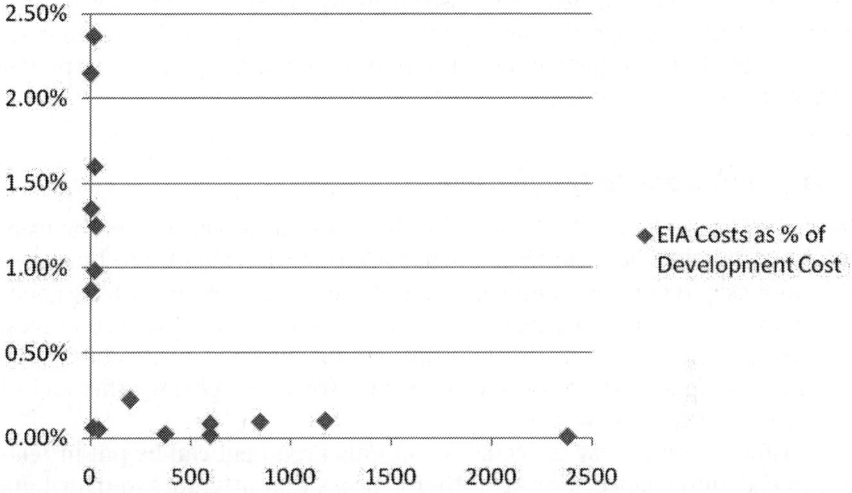

Source: GHK (2010: 23)

have, it was bigger!" The "joke" has many forms, but is basically promoting the suggestion that production of an EIS (the main cost in EIA and that which falls on the proponent) is a significant part of the proposal's cost, leaving aside the costs of government administration of EIA.

When the costs for EIS preparation are looked at, however, a different picture emerges. The United States has the most experience with EISs, and work done by Cook (1981) suggests that EIS production has been a very small proportion of the cost of proposals. Examining a sample of 60 projects, he found that the average cost of EIS preparation was less than 0.2 per cent of the total cost estimated for construction. A similar analysis by Hart (1984) produced figures of 0.5 per cent for (American) State projects and 0.1 per cent at the federal level. Hence, an insignificant cost was attributable to the consideration of environmental effects, particularly when in most EISs much of the data collection and analysis would be required even without preparation of the EIS, for detailed site planning, general government approvals and the like. More recently, research on EIAs carried out in the European Union across a range of projects and sectors suggests that costs for EIA represent on average approximately 0.7% of total project development costs (GHK (2010: 24)). This research found that where the capital cost of the project was the largest, the per cent of EIA costs was the least.

Further, Cook found that preparation of an EIS resulted in major changes in nearly all the proposals (presumably to reduce environmental effects). In almost half the EISs studied, resulting changes brought about decreases in the estimated project costs. In one case, the cost was reduced by more than 50 per cent.

The other costs of EIA are those borne by the State, for review of EISs and maintenance of the EIA process by government bodies. Hart's analysis of American States indicated a wide variation, depending on the particular State's process and workload.

2.5.4 EIA is basically defensive

Because of the lack of environmental factors in project planning in the past, EIA has tended to be used to give what some would consider to be excessive emphasis to environmental issues. Eversley considers that EIA is about preserves, territorialism, exclusion, non-access, and the non-use of resources. He goes on to suggest that "much of this window dressing of EIA lies in the pretence that it is a way of preserving values (or resources) for the masses and for prosperity" (1976: 131).

However, it is easy to see how the same argument can be put in relation to economic aspects, or any other process that attempts to distribute resources. Value judgments are always involved. Further, the paternalistic attitude noted by Eversley occurs with every administrative or legislative arrangement, where those with influence get "their way" as opposed to people without access to information and political processes.

2.5.5 EIA is a good ploy

Formalising EIA in government procedures not only provides the safeguard of having environmental effects considered, but also could provide the government with a scapegoat. For instance, where the environmental effects of a proposal are questioned, the proposal could be referred to an EIA, which produces the recommendations desired; that is, "the environmental effects are too great to proceed with the proposal". Alternatively, the proposal's "environmental effects are recognised, but considering the benefits of the proposal it is best to proceed with conditions". This is not a criticism particular to EIA, however, as any social processes can be selective in what is reviewed which consequently affects the outcome. In a similar sense, some governments have included a "this project is environmentally unacceptable" clause at the beginning of the process so that proponents, the community and government do not waste time and resources on assessing the environmental effects of a project that is already considered to be unacceptable for environmental or policy reasons.

EIA could also be used to divert attention from where the action really is. Merely saying that a proposal would be subject to an assessment is likely to divert the attention of the public, and perhaps environmental groups. Eversley has taken this possibility a step further:

> No doubt in the past we have been careless, spendthrift, and often prone to wreck good landscapes or buildings, whether in pursuit of individual gain, or in the name of social good. But relative to the real problems of our time – the wanton killing of human beings by deliberate acts of war, civil war, and terrorism; the persistent inequality between nations, and within countries between different groups – the kind of problems of pollution, erosion, intrusion, or resource exhaustion which are the subject matter of EIA seem rather to pale into insignificance. (1976: 140)

In the context of diverting attention, Lang talks about the rhetoric involved in EIA stemming from "an astute political awareness of the risk it creates for unintended reform" (1979: 250). To counter this risk, EIA is controlled while a facade of reform is built. Lang comments that strong rhetoric suggests the likelihood of a fear of, and need for, such reform. Hence, the promotion of EIA processes as being the way to overcome problems may indicate a simple commitment to the processes rather than to their outcomes. An example would be if the agenda of an EIA (ie, the scope) is not required to incorporate a discussion of the need for the proposal. In this case, the government could say that an analysis of the proposal has been undertaken, but in reality that analysis would have been constrained to accepting that the proposal would go ahead in some form. "Actions speak louder than words" may be a cliché, but it is highly relevant if EIA is not effective in helping us move towards ecologically sustainable development.

2.5.6 Uncertainty and EIA's inflexibility

After almost four decades of operation, it is not surprising that limitations of EIA have been identified. For example, Dorney (1989) has pointed out that EIA is inherently inflexible. He sees that it works well where there is a high degree of certainty, but works poorly in the reverse situation unless EIA is part of a larger process (such as where EIA is associated with SEA). Inflexibility is also seen more specifically by Sammarco (1990), who suggests that most minor proposals are not subjected to EIA, but the cumulative effect of these minor proposals may be significant. Flexibility in EIA would allow the scope of an assessment to consider cumulative effects, about which there is little certainty.

The issue of uncertainty is seen by Vlachos (1985) as a major concern and poses the question of how impacts can be determined when there are so many complicating dimensions to EIA. He identifies five areas associated with change in the environment:

1. Societal complexity and interdependency (and increased vulnerability).
2. Increasing intensity, severity and duration of impacts.
3. Rate of change (related to timing and "future shock").
4. Distribution of effects, in respect of equity and fair access for the community.
5. A society's demand and ability to cope with change.

2.5.7 Responsibility for undertaking EIA

Usually the proponents of the action prepare the documentation for the EIA; in other words, they prepare the EIS. This has been favoured over suggestions for a special government department being set up to undertake the reporting function for the following reasons:

- The proponents should provide the information on the environmental effects in the same way as providing planning information; that is, at their cost and in their timeframes.
- It is more efficient for the proponents to prepare the EIS as they have most of the base information at hand.
- While preparing the EIS the proponents have the opportunity to modify the proposal to reduce the environmental effects.
- Biases are expected to creep into the EIS no matter who does it; setting up a special department to produce the EIS would not solve this problem.

Saying that the proponents should prepare the EIS usually means that a group of professional experts is called upon. This may have its own problems because it tends to put power into the hands of the technocrats. O'Riordan (1976) notes that the "technocratic mode" of looking at environmental issues has left its legacy in environmental policy-making in a number of ways:

- Optimism over the successful manipulation of techniques to extract and manipulate resources – an optimism shared by most policy makers who bask in the reflected glory of technological success.
- Determination to be "value free" in advice and analysis, leaving the "hard" decisions to the political arena that is already shaped by their advice.
- Disavowal of widespread public participation, especially the input of the lay public opinion, a philosophy shared by politicians equally intent on preserving their role in acting authoritatively on behalf of the public.
- Disquieting "fallibility", the constant evidence of error and misinterpretation and of hunches that do not quite pay off (fallibility is tolerable when it is accepted and accounted for but it is dangerous when those in important places ignore its existence).

Ward (1976) considers that the right people to conduct EIAs are members of the public. Here it is perhaps necessary to distinguish between the two parts of EIA: preparation of the EIS; and assessment of the EIS and the environmental effects, and deciding on a course of action.

Looking at these two parts it is difficult to see an EIS being produced without the input of some recognised knowledgeable people (experts, professionals or at least reputable amateurs), and, provided the possibility of there being some built-in bias is recognised, this is probably acceptable.

Ward's main concern seems to be with the second part of the assessment process, where a decision is to be made. In this case, Ward considers that an assessment produced by officers at any level of government (with the collaboration of the proponent) should not be regarded as an impartial "scientific" evaluation by the people whose interests are affected (ie, the public). For example, in the case of a development with an apparently adverse effect on the local population, the very fact that the assessment is prepared by the local planning department (for instance) will effectively destroy the department's credibility.

The alternative, Ward suggests, is where the "local community, in the form of the multiplicity of interest groups, pressure groups, all those self-appointed and unrepresentative spokesmen of local feeling, should be the formulators of an environmental impact assessment" (1976: 187). Unfortunately, as Ward notes, some people know how to manipulate the system to achieve their aims far better than most other people. The possible solution to this disturbing quirk could be to educate the public for participation; that is, to equip everyone with "that lovely bourgeois know-how". Education in this sense means effective community self-organisation, and we have not yet discovered the best recipe for that; maybe there is none (see Chapter 4 for a discussion on public participation).

Overall, while it would be desirable to remove the bias and interest-group pressures that are part of assessment and decision-making, it is hard to see how this is possible. Whether it is a local community or officers of a government department assessing the EIS, human (social/political) interests will always be an influential factor where environmental issues are involved – and the reader will extract the information from the EIS that is going to legitimate their cause – whether it be for, or against, a particular proposal.

The important thing is to recognise that the interpretations put on aspects of the EIA are just that, and they must be allowed for.

2.5.8 Timing and role of EIA

Some of the issues of inflexibility may be addressed by the move to undertake EIA early in the planning stage, when more attention can be given to researching uncertainties, or when they at least can be recognised and alternatives developed and considered. During the mid-1980s, Hollick (1986), among others, raised the issue when he pointed out that the EIS could be prepared early in the planning process when there was a broad range of options to be considered, but when little detail was available. The alternative is for the EIS preparation to be delayed until more information is available, but when basic options may have been foreclosed.

The point at which an EIA is undertaken is also of concern to Dorney (1989). His position supports that of the proposed Australian national approach for early consideration of environmental effects (see ANZECC (1991a)), and re-emphasises the broad concept of EIA. As he says:

> Unless environment assessment becomes part of an overall environmental management process – which considers all stages of project planning, design, hearings and environmental protection as part of implementation – the results are less than satisfactory. The EIA document can keep the approval agencies happy, but little substantive change in the planning, design and implementation process has been achieved. (1989: 110)

A concern about EIA becoming a policy instrument has been expressed by Kinnaird (1990). Citing examples of large development proposals that had been rejected, he noted that instead of the EIA process acting purely as a project approval mechanism, it had become a tool to resolve policy issues (such as acceptable emission standards), which should have been considered before the proposals were caught up in the process. In essence, Kinnaird points out that in some cases the EIA process has been attempting to take on the role of SEA (discussed in 3.10).

Over the 1990s, a key issue for EIA was its role at the level of policy assessment. From the beginning, EIA has been considered to apply equally well to policies and broad programs as to specific projects (see 5.2). However, the practice of EIA has been to apply it almost exclusively to projects (such as individual dams, roads, mining projects). This has opened up discussion about the need for EIA to have a broader scope of application, and to be applied more strategically, so that policy issues like those identified above could be resolved before specific projects are contemplated. This is the rationale for SEA, which is discussed in detail in section 3.10.

Generally, a changing role for EIA can be seen. The development of EIA in conjunction with interested parties outside government can be a way of gaining support for a proposal. As Stern comments in connection with a particular proposal:

> By using environmental impact assessment as a tool to negotiate approval of the project and reduce the potential for later disputes [the proponent] paved the road for an environmentally sound project that was supported by the company's traditional adversaries. (1991: 87)

This suggests that different objectives for EIA are becoming obvious. While government and conservation groups see EIA as a way to give consideration to the environment, proponents could adapt the process to assist them to gain approval for their proposals. Either way, as demonstrated by Stern's observation, the outcome can still be expected to be a proposal which shows environmental sensitivity.

More broadly, O'Riordan (1990) has expressed concern that while EIA has become an accepted part of proposal development, it continues to be considered a hindrance, rather than an opportunity to produce an optimal outcome. Instead of thinking of EIA only in terms of the documentation (ie, the EIS), he sees EIA as being "a process of strategic bargaining between developer, planning regulator and public interest groups" (1990: 11). In other words, a social process.

In this context the evolving role of EIA has meant that those involved in EIA have responsibilities if their work is to be effective. Sadar (1994) suggests that effectiveness in this situation relies on:

- understanding the real purpose and limitations of EIA practices and processes, and being able to explain these to the public;
- understanding and accepting the multi-disciplinary nature of EIA and learning to work with other professionals and the public;
- maintaining and expanding links with appropriate institutions, organisations and individuals in order to incorporate their expertise into the EIA;
- learning to simplify complex issues by making use of available good quality data; and
- communicating regularly, and effectively, with co-workers, peers and the public.

2.6 Overview

EIA is part of the political scene in the area of "environmental justice" (see 3.6). If it is not related directly to the protection of the environment, it is at least concerned with ensuring that the environment is considered when an action is contemplated. O'Riordan (1976) suggests that if EIA is taken seriously by politicians, public officials and the community, then it should provide a positive contribution to democratic decision-making. Almost 40 years on, and this is still an important objective of EIA. However, if EIA is merely tacked onto the older forms of policy-making, then its function may be considered to be obstructionist, delaying or trivial, and detrimental to environmental preservation.

Over four decades of operation have led to the identification of aspects where EIA is deficient. Some of these problems are addressed by the other forms of assessment (Chapter 3) and consideration of the role of public involvement (Chapter 4). Other issues are directly related to the particular application of EIA in specific situations; the latter chapters discuss issues of application and provide guidance for allowing for the inevitable deficiencies of EIA.

Chapter 3

THE MANY FACES OF IMPACT ASSESSMENT

As well as EIA, a range of impact assessment procedures exists. This chapter presents an introduction to the various forms of impact assessment and looks at the range of impact assessment procedures that have been developed and practiced. Risk analysis, social impact assessment, strategic environmental assessment and technology assessment are covered in more detail as these have occasionally been seen as substitutes for EIA, and have received greater use and discussion. Regulatory impact assessment and health impact assessment are recent and developing areas in Australia, while the potential for climate change assessment is growing. On the other hand, cumulative impact assessment and integrated impact assessment (more recently known as sustainability assessment) have received some theoretical attention, but have found little practical application. Yet the concepts which underlie these broad assessment approaches are embodied in strategic environmental assessment, which is assuming greater prominence in the assessment profession and in practice. In relation to the profession, the International Association for Impact Assessment (IAIA) provides information and a point for discussion of several impact assessment areas.

Each of these impact assessment procedures, except for the strategic, cumulative and the integrated assessments, has a particular focus. With a broad interpretation of that focus (eg, energy), the procedure could be defined to embody environmental issues. However, to date, most applications of the procedures have remained fairly focused, although there has been discussion as to how these procedures can be used alongside EIA. While most applications of impact assessment have emphasised one of the forms of assessment we have been talking about, those involved in development agencies have been using a range of approaches. Roche (1999) has discussed the use of forms of impact assessment, and how they relate to analytical frameworks used by development agencies. He specifically makes the point that impact assessment is not a "one off". Rather, he argues that impact assessment needs to take place at all stages of a project: identification of the problem and design of the project; appraisal of the proposal; implementation; and project evaluation.

3.1 Economic Impact Assessment (EcIA)

EcIA has been pursued in many forms since the mid-1970s. The focus of EcIA has been the economic impact statement, which began in the United States as the inflation impact statement. This was supposed to improve federal agencies' consideration of economic costs and benefits of their proposals, but was typically prepared after a proposal was developed, and consequently became a justification of the proposal. Taylor (1984) also lists a range of impact statements that were spawned by EcIA: inflation impact statement became regulatory analysis; competitive impact statement; arms control impact statement; urban and community impact analysis; and small business impact statement. Expansion of impact assessment was also suggested to include regional economic statements, family impact statements and judicial impact statements. Unlike the other assessment approaches, EcIA usually does not seek to identify ways to reduce adverse impacts. Rather the emphasis is on the identification of economic impacts.

James and Boer (1988) suggest that economic analysis can be used at every stage of EIA. The main contributions of economic analysis include the framework provided for collection and interpretation of information; techniques for obtaining environmental values; the capacity for assessing trade-offs; and identification of decision-making criteria which are routinely used in public policy-making.

EcIA is usually undertaken in the form of benefit-cost analysis (BCA), sometimes cost-benefit analysis (CBA), and cost-effectiveness, which are major techniques of economic decision-making. In his summary of economic and fiscal impact assessment approaches Leistritz (1998) also discusses the role of export base models and input-output models.

Application of BCA in EIA is outlined by Hundloe et al (1990), who consider that BCA can be used in the analysis of need for a proposal, and for providing a way of comparing alternatives. To provide this analytical ability, the analysis must include all social and environmental values. This is where a substantial difficulty arises. While James and Boer claim that "monetary values can be placed on conventional goods and on environmental effects" (1988: 94), and outline some techniques, Hundloe et al rightly perceive that there are difficulties in both measuring "externalities" and in using the dollar as a measure of value of natural ecosystems. Notwithstanding, regulators, project proponents and NGOs are attempting to place monetary values on the environment to assist in decision-making. The Victorian government released its final Biodiversity Conservation Strategy and associated documents in May 2013 that prescribe a substantial change to the way biodiversity values are secured, managed and offset within Melbourne's residential growth corridors. Biodiversity values and offsets are prescribed for a number of matters, for example $7529 per Growling Grass Frog habitat removed and $13,218 per scattered indigenous tree that is considered to be removed (DEPI (May 2013: 10)).

3.2 Energy Analysis and Greenhouse Assessment

Energy analysis became used in the late 1970s to provide a different view of projects from that provided by conventional economic analysis. Evans (1982) notes that energy analysis may be defined as the identification and quantification of energy flows into and out of a system, enabling comparisons to be made between alternative proposals for energy use. This analysis could then become part of an overall assessment combining energy, economic and risk analyses with social impact assessment (for example).

During the 1990s there was a growing concern for the effects of climate change produced by the release of greenhouse gases. Consequently an interest developed in applying broad energy analysis to identify the implications for energy use associated with proposals, and hence the extent of emission of carbon dioxide (CO_2), a major greenhouse gas. An indication of the likely growth in this form of assessment is shown by the discussion associated with the introduction of the Commonwealth's *Environmental Protection and Biodiversity Conservation Act* in 1999. As discussed in 6.3, this Act identified six matters of "national significance" that "trigger" the need for an EIA. The issue of greenhouse, or climate change, was not one of the triggers, however, within months of the proclamation of the Act the Commonwealth was considering options for assessing greenhouse effects (Environment Australia (1999)). Certainly, since a heightened international awareness of climate change in 2006-07, and change in Australian Federal Government politics from a Liberal government to Labor in 2007, climate change has become a focus of discussion.

Even without being identified in legislation like the Commonwealth Act, some instances of greenhouse assessments are evident as part of EIA, particularly those for transport, industrial or energy facility (eg, gas plants) proposals. For example, the EIA for the Scoresby Transport Corridor in Melbourne's east included greenhouse gas emissions as one of the significant environmental issues to be considered (Department of Infrastructure and VicRoads (1998)). The focus was on CO_2 emissions and the study provided estimates of the level of these emissions associated with alternatives that included mixes of public transport along with major road construction. Estimates of emissions were made using models that estimated travel distances, number of vehicles, travel times and hence fuel consumption, which was converted into measures of CO_2. The detailed assessment of greenhouse gas emissions for such large projects has increased in EIAs both in Australia and around the world.

From 1 July 2008, the *National Greenhouse and Energy Reporting Act 2007* (Cth) requires corporations to register and report greenhouse gas emissions for the coming financial year if:

- they emit 25 kilotonnes or more of greenhouse gases (CO_2 equivalent), or produce or consume 100 terajoules or more of energy; or

- their corporate group emits 125 kilotonnes or more greenhouse gases (CO_2 equivalent), or produces or consumes 500 terajoules or more of energy to report (Department of Climate Change).

There were 833 corporations registered to report under the National Greenhouse and Energy Report scheme (NGER) for the 2011-12 financial reporting year, (Clean Energy Regulator). Information collected through the NGER scheme provides the basis for assessing liability under the carbon pricing mechanism that commenced in July 2012. This carbon pricing mechanism applies to Australia's largest polluters, which have to report on, and pay a price for their carbon pollution, thus providing an incentive to reduce emissions.

At a conceptual level, Taplin (1998) has been critical of the "integrated impacts assessment modelling approach", which is epitomised by the linear process used to respond to climate change projections from international models. She proposes an alternative approach that is "stakeholder driven" and provides a more integrated assessment framework to provide policy-makers with useful information about the potential impacts of all the development sectors (such as transport and agriculture) in a region.

Even within the examples provided above, energy analysis, greenhouse assessment and assessment of climate change are usually conducted within an existing EIA process, rather than as a standalone analysis. We have seen a move by local Council and State governments towards a more strategic analysis of the impacts of climate change on a regional scale however, rather than for a particular project. The Future Coasts program in Victoria is one such example.

What has occurred over the past decade is an increase in energy auditing, which is an assessment of the amount of energy a particular product/facility produces and therefore an estimate of the amount of greenhouse gases emitted. These assessments highlight to a company or government agency, how much energy they are using and where they can perhaps make changes in their practices to reduce energy consumption.

3.3 Health Impact Assessment (HIA)

Although it can be argued that the first forms of pollution control and reducing impacts on the environment were from health issues to humans caused by bad air and water quality, generally, issues of public health have been given poor consideration in EIA; for example, see Giroult (1988), and Sadler (1988). According to a survey by Steinemann (2000), even those assessments that covered health impacts did so in a very narrow way, and overlooking broader determinants of health, such as morbidity and mortality risks, and cumulative and intergenerational effects.

To redress this situation, HIA has been developed. To illustrate its practicality, and potential, Sadler (1998) provides checklists of issues and risk groups to be incorporated in an HIA. Direct health effects (eg, exposure to pollutants) and indirect effects (eg, expansion of breeding habitats for disease vectors) are also discussed, along with the cultural- and area-specific issue of defining acceptable risks in public health.

The role of HIA is to assess the health effects, positive or negative, of a project, policy or program on a population. It is concerned with the health of populations, and attempts to predict the future consequences on health of decisions that have not yet been implemented. The IAIA defines HIA as *a combination of procedures, methods and tools by which a policy, program or project may be judged as to its potential effects on the health of a population, and the distribution of these effects within the population (IAIA).*

While there has been interest in HIA in Canada (Health Canada, 2000), it has been pursued to only a limited extent in Australia. In Victoria, amendments to the *Health Act 1988* provided for the preparation of health impact statements (HIS). The Health Department of Victoria (1989) noted in a discussion paper that if an individual or a group provided evidence that a proposed activity was a danger to public health, the (then) Health Department could be requested to inquire into the activity. The inquiry process would involve public consultation and preparation of a discussion paper covering scientific, medical, sociological, economic and other relevant perspectives. An HIS, containing recommendations for government departments, public authorities and individuals, would be prepared, following public consultation, to assess the effects of the proposed activity on public health. Replies to these recommendations would be expected to outline what would be done to comply with the recommendations, or the reasons no action would be taken. The resources of the Health Department would be required for conducting the inquiry and preparing the HIS, and to date no HIS has been prepared. Although not directly termed a health impact assessment, the Victorian Environmental Health Indicator Project (VEHIP) is the first attempt at a State level to introduce a comprehensive environmental health survey in Australia. The VEHIP is an initial attempt at a State level to establish an improved basis for understanding the relationship between environmental factors and health and trends in these factors over time (Department of Human Services (2008)). The VEHIP was launched with the release by the Department of Human Services of a discussion paper in December 2006. The project discussion paper presents a brief background to environmental health indicators and describes the aims and structure of a future environmental health surveillance system for Victoria that aims to provide social and economic information for planning at a local government level. The term "environmental health" describes those aspects of human physical and mental health that are directly affected by our natural and built environment which includes biological, chemical and physical media (Kingsland (2006)).

A substantial boost to the promotion of HIA has been given by Tasmania's *Environmental Management and Pollution Control Act 1994*. This Act empowers the Director of Public Health to require that an EIA include an assessment of the impact of a proposal on public health; hence HIA is fully integrated with EIA in Tasmania. As indicated by the following "triggers" for the preparation of an HIA (or health assessment), a broad concept of health is taken:

- The possibility of substantial change to the demographic or geographic structure of the community.
- Potential exposure of individuals to hazardous products or processes, including substances that are clinical or infectious.
- Changes to the environment that may impact on disease vectors or parasites.
- The potential to render recreational facilities or water unsafe.
- Potential impact on land productivity for horticultural and/or pastoral activities.
- Impact on the microbiological or chemical safety of food chains and food supplies.
- Substantial increase in the demands on public utilities.
- Increased traffic flow with increased risk of injury or significant increase in the release of pollutants.
- Generation of a high level of public interest and/or concern about public health issues.
- Identified ecosystems that are vulnerable, and damage to which may cause health effects.
- Potential exposure of the public to contaminants.
- Potential impacts on the incidence of illness or infection in the community, especially in relation to vulnerable populations such as children and the aged. (enHealth Council (2000))

The specifics of an HIA have been directed by guidelines based on the National Framework for Environmental and Health Impact Assessment.

Before a national interest in HIA developed, in the early 1990s the concept of environmental health impact assessment (EHIA) was introduced to explicitly and more effectively incorporate HIA within EIA. Ewan et al (1992) report that EHIA does not imply a separate administrative process; rather, it is a mechanism through which community health issues can be assured consideration in the assessment of proposals. More particularly:

> Environmental health has broadened its sphere of concern beyond the impacts from environmental pollution (chemical, physical and biological), to include the promotion of health through improved housing, living, working and public environments. This broadened view ... implies the contribution of sectors other than health to the EHIA process. (1992: 10)

To guide the preparation of EHIA, Ewan et al provide a comprehensive list of the principles, tasks and responsibilities for the practice of EHIA (ie, HIA in EIA). They also discuss a framework for the practice of EHIA in 21 detailed points. Further assistance can be found in the guidelines that have been produced for a range of situations and listed by Roe et al (1995).

The National Health and Medical Research Council promotes the concept of EHIA. It also suggests that health assessment should be conducted within EIA, but particularly that "an ecological approach which assesses the impacts of development on human habitats is necessary for the protection of human health, both community and occupational ..." (1994: xi). Like EIA, the purpose of EHIA is to provide the decision-maker with information, in this case with emphasis on the environmental health effects of the proposal and on the possibilities for preventing or mitigating any effects. Proponents are expected to:

- incorporate occupational and public health factors fully into the planning of any proposal;
- make and implement a commitment to avoid public and worker health impacts, or where this is not possible, to ameliorate impacts; and
- implement and review corporate occupational and public health policies and strategies.

The National Health and Medical Research Council (1994) recognises that the successful implementation of EHIA requires the collaboration of public health authorities, communities and proponents. Most particularly, EHIA should not shift the burden for public health maintenance onto the proponents of projects. Rather, "more cost-effective planning and implementation of proposals and sharing, collating, and monitoring of data on health outcomes will result if proponents and health agencies work together ..." (1994: xiii). Examples of areas where EHIA could be considered include proposals involving energy-source development; urban development, planning and redevelopment; water and air quality; contingency and disaster planning; and cleaner production.

3.4 Regulatory Impact Assessment (RIA)

A relative newcomer to the range of areas subjected to assessment, RIA has been adopted by some governments in Australia (eg, South Australia and Victoria). Its role is to assess the impact of government policies and law on society. The scope of RIA covers the effects of primary legislation and subordinate legislation (regulations, proclamations, rules, by-laws).

According to the Victorian Office of Regulation Review, "The Regulation Impact Statement (RIS) is a document prepared by the department, agency, statutory authority or board responsible for a regulatory proposal following consultation with affected parties, formalising and evidencing some of the steps that must be taken in good policy formulation. It requires an assessment of the costs and benefits of each option, followed by a recommendation supporting the most effective and efficient option. It must be incorporated into the assessment process used by all areas of government responsible for reviewing and reforming regulations" (1987: A1).

The context for RIA has been the need to avoid an accumulation of regulations that could detract from economic performance. RIA relies on the RIS to report the effects. Essentially the RIS provides an assessment of the costs and benefits of the proposed regulation, it identifies all feasible alternatives to the regulation and analyses the merits of the alternatives relative to the regulation proposed. The Office of Regulation Review (part of the Commonwealth's Productivity Commission) sees that the RIS is a means of effectively providing this information to the public and allowing for consultation on the matter. The Office of Regulation Review publication, *A Guide to Regulation* (1999), is a good source for undertaking a RIS and can be downloaded at <http://www.pc.gov.au/office-regulation-review/guide/regulationguide2>.

South Australia provides an illustration of the application of RIA (Attorney-General (SA) (nd)). A RIS is required when the responsible minister and the Attorney-General agree that the impact of proposed regulation, or deregulation, could impose an appreciable burden, cost or disadvantage on any sector of the community. Once required, the RIS should provide detail about the issues, initially raised in the "green paper" or draft bill. A typical RIS would include objectives, outline (context of the regulation), perceived impacts, cost to industry, cost to government, alternatives to the regulation, reasons alternatives are not appropriate and an outline of the consultation undertaken.

In Victoria, RIA has been undertaken for a variety of types of regulation. An instance of a fairly specific RIA is the revision of the Victorian Fisheries Regulations, which consolidated previously diverse sets of regulations under the one Act, and introduced some new areas of regulation (Department of Natural Resources and Environment (1997)). For this proposal the RIS contained material relating to:

- the background to the proposal;
- objectives of the proposed regulations;
- nature and extent of the problem associated with the previous arrangements;
- authorising power and incidence of the proposed regulations;
- overview of the proposed regulations;
- regulations arising from preservation of the old situation;
- impacts of the proposed regulations;
- cost-benefit analysis;
- comparison of costs and benefits;
- identifications of practical alternatives;
- costs and benefits of alternatives; and
- benefits of regulation versus alternatives.

At a broader level than specific regulations, during the past 20 years there have been several instances of the preparation of a RIA in conjunction with the preparation of State Environment Protection Policies (SEPP). In essence they are RIAs because they relate to changes in the schedules associated with the SEPPs, and which derive from Victoria's *Environment Protection Act*. But to confuse the matter somewhat, these assessments are called Policy Impact Assessments. Broadly, they have focused on the changes associated with the introduction of the proposed SEPP. As an example, the Environment Protection Authority (Victoria), when assessing proposed developments in Port Phillip Bay, sought to provide through these Assessments an understanding of the:

- philosophy and thrust of the draft Schedule;
- key existing and emerging environmental threats to the (Port Phillip) Bay;
- value and importance of the identified beneficial uses to the community and the potential impact of not protecting those beneficial uses adequately;
- proposed changes to water quality objectives and indicators; and
- likely implications of key management actions which will need to take place in order to achieve the Schedule's objectives. (Environment Protection Authority (Vic), 1995: 7)

To do this, the Assessment provides a background to the policy revision, a description of the proposed policy revision, and a summary of policy impacts, which, in the case of Port Phillip Bay, focused on the uses of the Bay.

Examples of RIA are also found in Tasmania. In the examination of the *Environmental Management and Pollution Control Act 1994*, the key issues were: community standards, restrictions on competition, impacts on business, and public consultation (Department of Primary Industries, Water and Environment, 1999).

3.5 Risk Analysis (RA)

EIA involves the estimation of the effects of a proposed change on the environment (including social and economic matters) and the importance of those effects – broadly this equates to an assessment of the risks that may be imposed on the environment. A similar procedure is that of risk analysis (RA), which attempts to identify the hazards involved in certain actions, estimate the associated risks and consider how acceptable the risk may be to the community. Hence, aspects of the RA process could be incorporated in EIA.

In RA the terms "hazard", "risk" and "analysis" have specific meanings (both Conrad (1982) and Kates (1978) provide examples). In essence, a hazard is something that has the potential to lead to a situation that we do not like – so there is a hazard for pedestrians crossing the road, as they may be hit by a vehicle. The risk is the chance (probability, likelihood) of being injured by a vehicle while crossing the road. Pursuing this example, there is greater risk of injury for a pedestrian crossing a freeway than crossing a residential (local) street, since there is more traffic, which is moving faster. Analysis is

Figure 3.1 – Risk framework

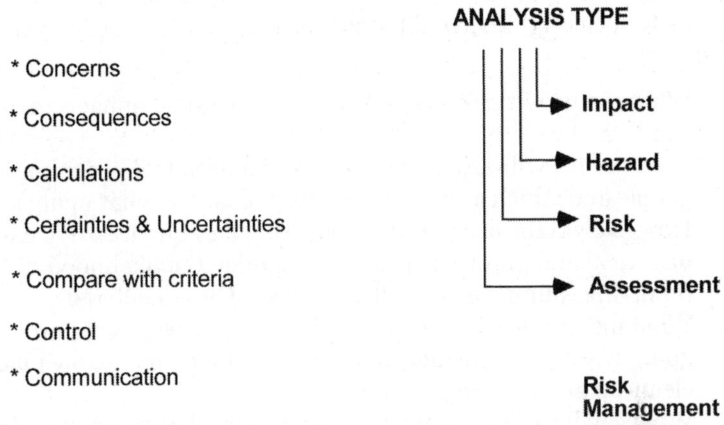

Source: Beer and Ziolkowski (1996)

the overall study of the identification of a hazard through to planning how to manage the hazard, and its associated risk. The link between the terms and stages are illustrated in the risk framework presented in Figure 3.1 *above*.

The principal element is the analysis, or assessment. This is the process whereby the hazards that have been identified are quantified in order to provide a value for the level of risk. Effectively, the outcome of the "analysis" gives the base upon which to make a decision about the action, as does EIA. After this, stage risk management is designed to remove or reduce the risks to acceptable levels.

Beer and Ziolkowski (1996) also identify "levels" of risk assessment. They comment that strategic risk assessment "refers to the use of risk assessment methods to determine corporate activities such as setting environmental priorities, allocating resources or making informed decisions" (1996: 8). Tactical risk assessment, however, refers to quantitative methods used to determine the risk to humans from existing activity, or from planned development; the risks associated with an industrial accident would be assessed as "tactical". The distinctions made between these two levels of risk assessment are similar to those drawn between Strategic Environment Assessment (see 3.10), which considers broad policy and planning matters, and the more usual use of EIA which focuses on specific proposals or projects.

However, while RA can produce information on one aspect of a proposal, EIA is capable of bringing together information on a variety of aspects. In this context, RA could be used as part of EIA to provide information that would be used with information from other sources to contribute to an overall decision. Certainly, there has been a move towards the use of RA in the EIA process, with more recent examples of EISs having chapters or sections on "risk" and "risk management measures".

Broadly speaking, Roe et al (1995) see that any RA needs to consider five questions:

- What can go wrong? – eg, what are the possible impacts on human health and welfare.
- How severe will any adverse consequences be? – eg, how many people and which areas may be affected and at what financial cost.
- How likely is the occurrence of adverse consequences? – what is the historical and empirical evidence to judge the likelihood of failure occurring (human error will also need to be considered).
- What measures will need to be taken in the event of the procedures going wrong? – consideration could include emergency planning, clean-up and recovery planning.
- What can be done and at what cost to reduce unacceptable risk and damage? – such as requiring the risk of failure to be carried by the proponent.

As indicated, the basis of RA, and risk management broadly, is to eliminate or contain risk. This is consistent with Wildavsky's (1988) observations that the underlying emphasis of most public policy is focused on reducing risk, and establishing certainty. He comments that society generally considers issues associated with change, such as the introduction of new technology or the implications of greenhouse gas emission, from either of two approaches. We can seek a situation of certainty as with "trial without error", where we do not proceed unless we are sure that problems will not arise. Or, we can proceed on a "trial and error" basis, where we expect that there will be problems and are prepared to attend to these as they occur.

Wildavsky comments that "trial without error", for example, is generally favoured by planners and policy-makers because it offers safety to the public and is politically acceptable. "Proponents of trial and error can speak only of safety later, a safety, moreover, not for this or that specific group, but – because they cannot predict or control what will happen ... for society in general. You'll be better off in the 'by-and-by' has never been noted as a politically potent appeal. The benefits lost because of rejected opportunities are seen as inferior political goods." (1988: 36)

As epitomised in *Searching for Safety*, the title of Wildavsky's book, his concern is to examine the possibilities for reducing risks. Drawing on examples from many areas of society, he argues that the preferable approach is to be much more prepared to follow the "trial and error" approach. He argues that rather than trying to achieve very low levels of risk for everyone, at associated high costs, it may be preferable, and cheaper, to "put out the bush-fires" when they develop. This means developing resilience to respond to outcomes, which may be unexpected, or may be anticipated in some form. Consequently, we would need to build sensitive monitoring programs and create the mechanisms to be able to respond to unwanted situations; as we do in having plans for evacuating flood prone areas, for example.

RA can be used to provide input to EIA about a number of situations (eg, see Ramsay (1984)). RA does not have to be confined to assessing the acceptability, in terms of safety, of, for example, siting a chemical plant adjacent to a residential area. Such an analysis could also consider the acceptability of noise intrusion or air emissions, or some other aspect. There are examples of EISs that include a risk analysis of the impact of climate change (and predicted sea level rises), which demonstrates that RA is becoming more of an essential element in the EIA process than it had previously.

RA could also be used to consider the acceptability of actions that could result in other environmental effects; for instance, whether the effects of filling a swamp were acceptable in respect of the resident bird population, or the acceptability of effects of forestry operations on erosion of soil, or whether the effects of freeway construction on the social patterns within a community were tolerable.

It is apparent that RA is becoming a useful way of collecting and presenting information for EIA, and although in the past it was seldom used or incorporated in EIA in Australia, it now is an essential component of the EIS studies and in decision-making. Whether it is being undertaken effectively is a topic for future research.

The following case study illustrates the use of a risk assessment approach in EIA.

3.5.1 Case Study: Shell Australia's Darwin Joint Terminal project

In the Public Environment Report (PER) for Shell Australia's Darwin Joint Terminal project, a hazard and risk assessment was undertaken as part of the PER process to identify hazards and risks in the operation of the Darwin Joint Terminal. This is an approach that categorises risks from threats to people, assets and the environment. A scale of consequences from "0 to 5" was used to indicate increasing severity (Sinclair Knight Merz (1999: 33)). The consequences are potential outcomes that can develop from release of a hazard.

The Shell Risk Assessment Matrix used in the PER for the Darwin Joint Terminal project is presented in Figure 3.2 (*see over page*).

The PER explains the RA approach as determining the likelihood of potential consequences for the assessment on the basis of historical evidence or experience with the Terminal facilities of Shell Australia, and the Australian oil industry. All relevant aspects of the proposed operation have been reviewed, including product delivery at the berth and to the terminal, storage of bulk and packed product, product loading and the drainage and interceptor system. A realistic qualitative assessment has been made with respect to people, assets and the environment. For this project proposal, the hazard and risk assessment determined that the risk to people, assets and the

Figure 3.2 Shell Risk Assessment Matrix

CONSEQUENCE				INCREASING LIKELIHOOD				
				A	B	C	D	E
SEVERITY	People	Assets	Environment	Never heard of in Industry	Heard of in Industry	Incident has occurred in our company (Coy.)	Happens several times per year in our Coy.	Happens several times per year in a location
0	No health effect/injury	No damage	No Effect					
1	Slight health effect/injury	Slight damage	Slight effect			Increasing Risk		
2	Minor health effect/injury	Minor damage	Minor effect					
3	Major health effect/injury	Localised damage	Localised effect	LOW RISK				
4	PTD or 1-3 fatalities	Major damage	Major effect		MEDIUM RISK			
5	Multiple fatalities	Extensive damage	Massive effect			HIGH RISK		

Source: Sinclair Knight Merz (1999: 33)

environment is "Low" or "Medium" for all hazards identified (Sinclair Knight Merz (1999: 33-34)).

Application of this approach can be seen below, where three hazards, as detailed in Appendix A, of the PER are presented (Sinclair Knight Merz (1999: 54)). More detail can be found in the PER.

As with all procedures, RA has limitations. In particular, there is the difficulty of obtaining sufficient data to determine hazard and risk; allowing for the inevitable biases that occur in the compilation of this data, and the evaluation of risk. There are also methodological and ethical problems of evaluating (costing) risks, as noted by Shrader-Frechette (1985). However, RA is not the only approach that exhibits these types of limitations (see 8.2).

RA has been developing over a number of years, and Brownlea (1988) provides an overview of RA and its general applications. Of particular relevance is his summary of the problems in applying RA to ecological and environmental situations. Nonetheless, guidelines have been developed to address these problems. In reviewing the key (mainly United States) guidelines, Gabocy and Ross have clarified the concepts of ecological risk as being: "the evaluation of the probability and resulting adverse effects on the non-human population or the ecological system in a particular region or area from an environmental hazard or stressor" (1998: 193).

Expanding on the human and scientific emphasis of risk, Brown and Campbell (1990) note that, historically, RA is derived from the engineering sciences, and therefore focuses on the physical and the measurable. For example, indices of risk are usually based on mortality and morbidity.

Figure 3.3 Shell Public Environment Report Hazard and Risk Assessment

Potential Hazard	Potential Consequence(s)	Risk Assessment Matrix Score			Management Control(s) to Minimise Risk
		People	Assets	Environ.	
PRODUCT DELIVERY					
Fire or explosion of tanker at Wharf during delivery of product to Terminal	• Radiant heat or spread of fire- fire or explosion at land backed Wharf facilities.	A4 Low	A5 Med.	A4 Low	• Fire protection system- as installed by NT Govt during Wharf construction (ie. fire fighting equipment, fire alarms). • Tanker fire protection system- fire water system, fire fighting equipment. • Operating procedures- tanker discharge safety procedures, safe work permit system for JUT controlled work, Wharf operating and safety procedures, restricted Wharf access during tanker discharge, security controls to enforce ban on ignition sources at Wharf (ie. matches, lighters, mobile phones). • Emergency response plans- harbour oil spill response.
	• Equipment failure due to fire or explosion- loss of containment- impact to site soil, groundwater or harbour.	A1 Low	A4 Low	A4 Low	
	• Injury to Wharf or tanker discharge staff.	A5 Med.	-	-	
Corrosion of Wharf-Terminal delivery line	• Corrosion hole- loss of containment- hydrocarbon impact to off-site soil, groundwater or harbour.	C1 Low	C4 Med.	C3 Med.	• Maintenance- inspections as per AS1940-1993 and regulatory requirements, recoating of paint surface, pipeline pressure testing, cathodic protection systems testing. • Leak detection systems- stock control reconciliation after product receipts. • Pipeline design- design and construction standards. • Anti corrosion controls- painting pipe surfaces, cathodic protection systems, wrapping, pipeline trench backfill material. • Emergency response plans- harbour oil spill response.
Physical damage of Wharf-Terminal delivery line	• Loss of containment- hydrocarbon impact to off-site soil, groundwater or harbour.	B1 Low	B4 Med.	B3 Low	• Pipeline design- protection to design against collision with vehicular or shipping traffic, pipeline warning signage, notification to JUT operator prior to site works or excavation in pipeline vicinity.

Source: Sinclair Knight Merz (1999: 54)

Expanding on the human and scientific emphasis of risk, Brown and Campbell (1990) note that, historically, RA is derived from the engineering sciences, and therefore focuses on the physical and the measurable. For example, indices of risk are usually based on mortality and morbidity. These measures ignore other harmful outcomes which are hard to quantify, such as public anxiety or impaired aesthetics. Other problems with RA as practised are seen to be:

- a focus on vulnerable individuals at risk, to the extent that other groups may be excluded;
- the difficulty of the public revising the conception that accidents happen, but not to oneself;
- whether groups or individuals are affected and how the analysis reflects this;
- how to treat catastrophic potential situations (injuries and deaths concentrated in one accident rather than over time);
- whether the risk is voluntary or has been imposed (the latter is usually less acceptable); and
- underlying values of the individual preparing or acting on the RA (ie, whether the development is seen as being useful).

These types of problems have led Kannegieter to suggest that RA is "so fraught with uncertainties and so vulnerable to economic and political manipulation that it should be replaced with a more qualitative approach with greater community participation" (1991: 22). One of the main shortcomings he sees with traditional RA is that it encourages conservative judgments about acceptable levels of risk to be made at the point at which risks are assessed (ideally through dispassionate scientific inquiry) rather than when the risks are to be managed (by social processes often after the proposal has commenced). He proposes that a method to describe an accident accurately is needed. This approach would have parameters which would encompass the environment affected, nature of the effects, species at risk, area affected, duration of the effect, and cost and efficiency of the clean-up. By giving a score to each parameter and choosing suitable frequencies for "accidents" of particular severities, information would be available against which proposals could be judged.

Stern and Fineberg (1996) also argue for the evolution of RA to a broader based process. Without proposing any particular assessment approach, they emphasise the need for an "analysis-deliberation" process based on the collection and examination of information from a range of groups in the community, and on several feedback stages. In summary, information from public officials, scientists/technicians and interested and affected parties would all be included in the stages of problem formulation, process design, selection of options and the synthesis of the data. Throughout these stages feedback to previous stages, and resultant learning, would be used before a decision would be taken about a proposal. Feedback and learning, or monitoring, would also be included in implementation and evaluation stages of the proposal. Effectively, this process would mean that RA would be embodied in proposal planning, and that there would be continuous consideration of risks, rather than an isolated RA being used as one of the many independent inputs to the final decision.

RA is being used more and more as a tool within EIA to assist with identifying key risks and residual risks to be assessed as part of the EIA and the use of the standard *AS/NZS ISO 31000: 2009 Risk management – Principles and guidelines* is increasing (see 8.5). In some instances where adequate data is available, RA may become a more widely used tool. Alternatively, if Stern and Fineberg's (1996) proposals are taken up, we could see considerations of risk incorporated in all stages of the planning for a proposal. To a degree we can see this developing since risk assessment and analysis are at the heart of environmental management generally, and EIA. It provides a framework for deciding which are the important (ie, significant) impacts, and therefore which will be included in the study, and which will be emphasised in management plans.

3.6 Social Impact Assessment (SIA)

As with land-use planning, the effects of proposals on social factors have sometimes been considered separately from other effects. In the "early days" of EIA, when most emphasis was given to effects on the physical environment (flora, fauna, soil, air, water), an SIA was occasionally undertaken to assess social effects of actions. In these cases emphasis was given to identifying effects on social and public services (eg, police, medical, educational requirements) and/or on groups within the community (eg, residents, shoppers, shopkeepers, cultural groups). Finsterbusch (1980) provides guidance for identifying and assessing these aspects. Burdge (1988) presents the history of SIA, while Formby (1986) and Rickson et al (1988) provide a clear overview of the intentions and general operation of SIA.

Comparing the various definitions of SIA, including other forms of social assessment such as technology assessment, Soderstrom concluded that:

> Social impact ... is the change in the activity, interaction, or sentiment of a unit ... as it responds to the changes on it from the surrounding environment and the resultant changes which occur due to the interdependent relationships of the system ... [A] project will alter one or more of these elements in the units to differing degrees, and those changes will in turn alter other elements and units. (1981: 11)

Specifically, Bowles (1981) proposes that SIA has two basic objectives:

- to anticipate the future consequences of changes associated with proposals; and
- to help develop policies that will guide proposals so that desirable changes are emphasised, and undesirable effects are minimised.

Figure 3.4 Model for SIA

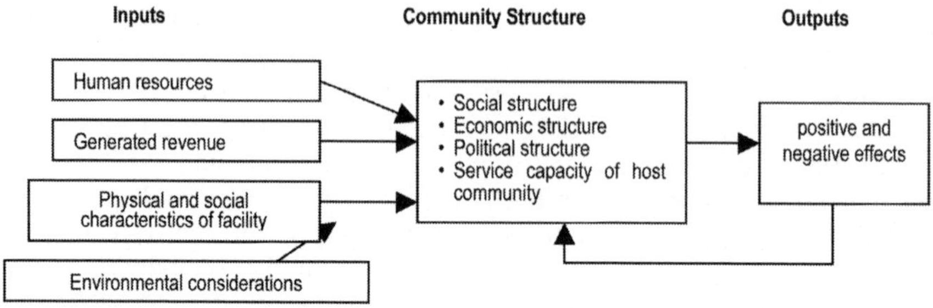

Source: Based on Soderstrom (1981)

A model for assessing this type of change, or impact, is shown in Figure 3.2. An implicit element of the model, and the discussion below, is that once negative effects are identified, efforts would be made to redesign the proposal to reduce unwanted impacts. The basis for assessment is the social indicators that describe the characteristics of a particular social group, and which are usually collected at different times to form a time series. Soderstrom provides detail on how theseindicators are identified, defined and used.

An important aspect of the model (Figure 3.4) is the feedback loop where social groups are brought into the assessment. This is the part of SIA where Peterson and Gemmell (1981) have identified three challenges for assessment:

1. Clarifying, for each person or group involved, all of the factual issues that are of concern.
2. Improving the "player's" skills and opportunities to play the game.
3. Helping to identify which issues and disagreements are ideological or value-related (and hence not resolvable through known or knowable facts).

The authors also raise a number of criticisms of SIA, particularly the need for clarification as to whether an SIA should restrict itself to providing information or become involved in evaluation (which presupposes that someone has obtained the criteria with which to make value judgments, eg, that some effect is undesirable). Other concerns are whether SIAs are to inform, confuse or divert the public (is the assessment undertaken seriously, or only to satisfy the bureaucratic aims?). Some areas of improvement are also suggested.

These criticisms of SIA raise the issue of the primary direction for SIA. Whereas SIA was originally conceived as a technical approach, essentially a means of only identifying social impacts, a political perspective has also evolved as discussed by Lang and Amour (1981) and illustrated in Table 3.1.

Goodrich, Taylor and Bryon (1987) have proposed that SIA be built on both critical theory and environmental sociology so that this broader political

Table 3.1 Contrast between technical and political approaches

	Technical approach	Political approach
Focus	Improved public decisions via improved social impact studies	Improved public decisions via improved socio-political processes
Key assumption	Better information inputs lead to better decisions	Open participative process leads to better decisions
Faith in	Rationality and processed knowledge Science/scientific method etc	Innate wisdom of the people Participation, pluralism etc
Reacts against	Overlooked social issues (the result of uniformed, arbitrary, narrow, short-range political decision-making)	The technical approach and rule by experts (technocrats); basic problems in the political system

Source: Based on Lang & Armour (1981)

perspective to assessment is appropriate. Craig (1990) compares the "technical" and "political" approaches to SIA (the product and process approaches) and points out that it is the political approach which emphasises community involvement and development. Also, in consideration of the role of SIA, the author proposes a wide role so as to take into account broad planning and policy issues, as well as the traditional project appraisal, where SIA is usually seen as an extension of EIA.

Gale (1983) provides some additional "teeth" to the political perspective by proposing that SIA provide "impacted populations" with the opportunity to gain insight into their current social situation, plus a minimal understanding of historical patterns and (for well-done assessments) an idea of likely alternative futures. Further, he suggests that SIA should raise the consciousness of the impacted populations so that these groups can more effectively communicate their values and preferences to decision-makers. Communication of values enables the actual assessment in SIA in that it becomes possible to evaluate impacts (desirable or undesirable) with respect to the values of the people affected, instead just of those preparing the SIA.

To determine values and preferences it is necessary to contact and involve the public. Garcia and Daneke suggest that a "baseline SIA" can be used to identify issues and the impacted populations. "Thus, SIA can provide an agenda to meaningful public involvement" (1983: 161). They also stress that SIA and public participation go together. This is the case whether the SIA has a technical perspective (where public participation provides information which is fed into the decision-making process) or a political perspective (when the public's consciousness is raised, which may lead to expansion in the scope of the SIA process).

With the broader concept of EIA, in relation to the definition of environment to encompass physical and social aspects, it would be expected that

EIA would embody those issues previously considered under an SIA. Further, there is a synergistic effect (the outcome is more than just the sum of the parts) where consideration of the physical environment plus the social environment produces a more comprehensive assessment of potential changes. EIA is now able to take into account the individual physical and social factors, and the relationships between them; for example, retraining schemes needed because of redundancies in a timber industry from the cessation of logging due to creation of national parks demanded by a community majority). Also, EIA encompasses the involvement of the public in making comment on proposals (and their environmental effects) and making decisions about the proposals, so that consideration of most of the issues of concern in SIA can be handled by EIA; in theory at least. SIA is usually embedded somewhere in the EIA process these days and more so when a proposed project is in close proximity to residential areas or communities. A good EIA process will have incorporated SIA from the beginning (project concept stage) and engaged with the community early regarding the development proposal. Chapter 4 discusses the role of community engagement and participation in more detail.

Under the banner of EIA, SIA can be undertaken to great effect because physical and social systems can be considered together, and the public can be provided with the opportunity to be involved in considering the many implications of these interactions. In this context, Lang (1979) has commented on the potential of EIA to involve issues of social equity and reform.

Another way of considering SIA is outlined by Wildman and Baker (1985), and Wolfe (1983), where SIA is seen to be the process by which the physical and social environmental issues are identified and assessed. In this case the public has a large input to the study right from the beginning and the EIA is subsumed into the overall SIA process.

As a consequence of the development of SIA, and the issues it attempts to address, there are many similarities with EIA as illustrated by Lang and Amour (1981) (see Table 3.2 *opposite*).

The methodologies for undertaking SIA are presented by Wildman (1988), who also discusses the need to include SIA in the determination of social policy. Providing more detail, Burdge (1989) lists 26 variables that are relevant to SIA. Although much has been written on the theory of SIA, compared with EIA there are few examples of its use. Canada, the United States and New Zealand have a number of cases where SIA has been adopted, but Australian SIA studies are hard to find, although there are many studies where elements of SIA are evident (see Armstrong (1982) and Weston & Crawshaw (1984)). Otherwise, Carley and Bustfeld (1984) provide a review of relevant literature that has practical applications for SIA. In addition to these and the other authors cited, Finsterbusch, Llewellyn and Wolfe (1983) have assembled the components of SIA methodology to provide guidance for preparation of an SIA.

Rickson, Western and Burdge (1990) see the constraints on the use of SIA to be generally bureaucratic rigidity and disciplinary inertia. In developing countries this is compounded by the Western idea of public participation, which is often incompatible with local culture and traditions, while rancorous conflict, extreme poverty and ignorance (by planners responsible for social issues and the methods of SIA) may also be involved.

Table 3.2 Similarities between methodological frameworks of EIA and SIA

Methodological framework for EIA	Methodological framework for SIA
Identify the full range of environmental features and processes likely to be affected by the proposed project or its alternatives; establish the "before" condition and current trends	*Profile* the existing social conditions in the area likely to be affected by the proposed project or its alternatives; establish the "before" conditions, current trends and concerns
Predict the magnitude, spatial dimensions and probability of potential environmental modifications, drawing attention to direct and indirect effects and primary and secondary effects, in the immediate area and beyond	*Project* the social changes that are likely to occur, drawing attention to the distribution of expected changes among the people affected by the project and its alternatives
Assess the relative importance of the predicted effects, taking into account the current condition and the future condition that would result in any case, as well as possible mitigative measures	*Assess* the relative importance of the expected changes for each of the groups affected, taking into account social conditions they now experience, future social conditions in any case, and the future social condition they would prefer, also taking into account possible mitigative measures
Evaluate the overall acceptability or "impact" of the proposed project and each of its alternatives	*Evaluate* the overall acceptability or "impact" of the proposed project and each of its alternatives

Source: Based on Lang & Armour (1981)

In New Zealand, SIA has had some status and acceptance within government authorities. The Town and Country Planning Directorate (nd) provides an overview of the operation of SIA and outlines its role, the process to be followed and issues to be considered. These "guidelines" also discuss the legislative basis for SIA in New Zealand, noting that there is no specific legislation, but that SIA has become expected through the interpretation and operation of the planning legislation and the environmental procedures.

Nonetheless, a high profile Australian example of the application of SIA has been presented by the Olympic Games Social Impact Assessment Steering Committee (1989). While the SIA was focused on a specific idea, that of holding the 1996 Olympic Games in Melbourne, the SIA was begun in the early stages of consideration of the concept of the bid for the Games. As a result, this example has elements of Craig's (1990) political approach to SIA in that it was undertaken early enough to enable emphasis on the development of issues of concern, followed by a preliminary assessment of impacts.

This can be contrasted with the "technical" SIA prepared for a late 1980s road proposal. In this example reported by VicRoads (1989), the social assessment is described in terms of a community consultation process to consider identified alternative locations:

> [Study staff went] into the community to listen to people's views and to understand their concerns. Views expressed and issues identified ... were documented and fed into the study process ... This allowed the community to influence the direction and outcome of the study. (1989: 5)

While it is explained that there was strong support for the concept of the road, the report does not elaborate on how the community was able to influence the study.

A variation of SIA has developed to give particular attention to people who may be disadvantaged by environmental inequities. Environmental Justice Impact Assessment is intended to identify people at risk of injustice and to review ways of reducing this risk. Wilkinson (1998) notes that environmental justice is defined as fair treatment such that no group of people bears a disproportionate share of human health or environmental impacts; for example, people living down-wind of a power station and being exposed to its emissions. The impetus for this form of assessment comes from an Executive Order by the US President in 1994, which effectively required agencies to incorporate environmental justice into their EIA processes under NEPA (see 5.2). In summary, the assessment process has three key elements:

- Demographic assessment – including the identification of minority and low income groups, and other sensitive groups depending on the proposal (if air emissions were an issue then asthma suffers would need to be considered).
- Impact assessment – this stage includes consideration of: human health, environmental and socioeconomic effects; disproportionately high and adverse effects; multiple and cumulative effects; normal operations and accident scenarios; avoiding and reducing impacts through mitigation and alternatives; integrating impact and demographic data.
- Community involvement to ensure that affected communities are consulted throughout the assessment process.

Environmental Justice Impact Assessment has much in common with SIA, but there is an emphasis on identifying the distribution of impacts across specific groups within the community. The US EPA defines Environmental Justice as the fair treatment for people of all races, cultures, and incomes, regarding the development of environmental laws, regulations, and policies (US EPA (2008)). Since the 1990s attention to the impact of environmental pollution on particular segments of our society has been steadily growing, and the US EPA has strengthened its commitment to addressing environmental justice issues by releasing the Environmental Justice Strategy in the mid-1990s, which focused on the EPA's efforts in addressing environmental justice concerns.

3.7 Species Impact Assessment (SIS)

The development of international agreements and treaties related to biodiversity, such as the Convention on Biological Diversity (1992), has been growing over the past two decades. Australia's response has been through the 1996 National Strategy for the Conservation of Biological Diversity and the introduction of the *Environment Protection and Biodiversity Conservation Act 1999* (EPBC Act) (discussed in Chapter 6). The combination of these influences has led to the specific need to consider the short and long-term effects of proposals on flora and fauna.

The response in New South Wales has been the development of Species Impact Statements (SIS). The NSW Department of Planning indicates that under the *Environmental Planning and Assessment Act* an SIS must be prepared if a proposal is:

- on land that contains a "critical habitat"; or
- likely to significantly affect threatened species, populations or ecological communities or their habitats.

The form and content of the Statement is set out in detail, but has many similarities to an EIS. Also, an SIS may be prepared as a separate document, or it may be incorporated into an EIS for the proposal. In any case the SIS goes to public exhibition. As mentioned further in 6.5, the Commonwealth has accredited the NSW SIS process under the EPBC Act. This means that if a development has the potential to impact on Commonwealth listed flora and fauna species, and is determined to be a "controlled action" by the Commonwealth, the NSW SIS process can assess the potential impacts – thereby avoiding duplicating the Commonwealth assessment.

A related form of assessment is Biodiversity Assessment. Brooke (1998) reports that EIA and associated forms of assessment do a poor job of identifying the impacts on biodiversity, and hence do not assist governments to deliver their obligations under the Convention on Biological Diversity. Rather than proposing the establishment of yet another form of impact assessment, Brooke argues that the assessment of impacts on biodiversity should be better integrated with EIA. In particular, EIA would need to demonstrate a more positive approach to biodiversity, so as to consider impacts on whole ecosystems, to identify fragmentation of habitats or isolation of species, and to more clearly identify enhancement opportunities. At its sixth meeting in 2002, the International Convention on Biological Diversity adopted guidelines for incorporating biodiversity-related issues into impact assessment processes. The Convention has been addressing the role of biodiversity in various fields of impact assessment, including the experience of biodiversity in strategic environmental assessment (SEA) and biodiversity in sustainability impact assessment of trade agreements. Most recently, the Convention has prepared voluntary guidelines to assist parties in incorporating biodiversity-related issues into environmental impact assessment and strategic environmental assessment legislation and procedures (Convention on Biological Diversity (2007)). The guidelines provide details on whether, when and how to consider biodiversity in both project and strategic level impact assessments. They were prepared in collaboration with the International Association for Impact Assessment (IAIA), the Netherlands Commission for Environmental Assessment (NCEA). (Relevant organisations and a link to these guidelines can be found on the Convention's website <www.cbd.int/programmes/cross-cutting/impact>. The voluntary guidelines were endorsed by the 8th meeting of the Conference of the parties to the Convention in March 2006.

In the context of protecting RAMSAR wetlands, Bagri and Vorhies (1999) point out that there are many EIA guidelines available that include directions for including wetlands and water resources in EIA. However, to improve the attention and priority given to wetlands, they suggest that the economic values of wetlands need to be integrated into EIA to influence the decision-making process.

Also with a focus on identifying the impacts that ecosystems may experience, Treweek (1995) introduces, and presents a clear argument for, Ecological Impact Assessment; an approach that appears to overlap with ecological risk analysis (see 3.5). Treweek and Hankard (1998) define Ecological Impact Assessment (abbreviated to EIA, adding to the confusion in assessment approaches) as "a formal process of identifying quantifying and evaluating the potential impacts of defined actions on ecosystems" (1998: 263). They note that ecological impacts often "fall through the net" in project level Environmental Impact Assessment. However, the broader scope and longer time frames of SEA offer good opportunities for the development of ecological assessment.

3.8 Technology Assessment (TA)

> Technology assessment is an open-ended search, through a variety of qualitative and quantitative techniques, for potential impacts on the economy, the environment, the polity, and social behaviour and institutions... It maps the uncertainties [of government] interventions, and lays before decision-makers an ordered set of alternatives or complimentary policy options together with their long-range implications. (Coates & Coates, 1989: 17)

TA is more like EIA and SIA in that it has a broad scope and, unlike RA and energy analysis, the assessment methodologies take the results of analysis further to provide a discussion of what the analysis means. With TA the approach is to look at a comprehensive study of the possible effects on society resulting from the introduction of or change in some technology. More particularly the characteristics of TA have been identified by Barbour (1980) as:

- the attempt to anticipate consequences of actions;
- consideration of a wide range of impacts;
- identification of a range of diverse stakeholders and assessment of how they would be impacted upon; and
- analysis of policy effects.

Similarly, Street sees that TA is essentially "the act of describing, analysing and forecasting likely effects of technological change, the main purpose of which is to provide an input to technological policy making" (1997: 145). He amplifies this by commenting that the elements of forecasting include technology forecasting, social, economic and political forecasting, and forecasting or analysing relationships between new technology and the "world" of the future. These forecasts, he suggests, cannot state what the future holds, but are useful in indicating the possibilities and opportunities that exist for shaping the future.

As noted by Armstrong and Harman (1980), the thrust of TA derives from the assumptions that the implementation of new technology, or changes to an old one, should be a conscious social choice. In particular, it is not generally technology in and of itself that is inherently harmful, but rather its management. These authors outline a methodology for assessment, comment on the understanding of social values and uncertainty, and indicate its practical application. Borouse, Chen and Christakis (1980) devote considerable effort to the discussion of TA application, and give particular consideration to the institutional role for TA (in industry and government) for different countries. TA in developing countries is the main focus of Srinivasan (1982), who discusses both the philosophy of TA and specific applications.

Legitimisation of the concept of TA came in 1972 with the passing of the *Technology Assessment Act* (US), which set up the Office of Technology Assessment in the United States; this Office was closed in the mid-1990s. Countries within the European Community (EU) have employed TA over a number of years. Becker (1991) reports that, through this practice, TA has evolved from a "rational" approach to decision-making, where there was conviction that future trends could be predicted, to a decision-making process based on consultation, and where political and social considerations are involved.

TA has also been discussed in Australia. The Senate Standing Committee on Science, Technology and the Environment (1987) concluded that TA could concentrate on how to assist new technology, while socioeconomic impacts would not be considered; this was termed "technology promotion". In the same report the Australian Council of Trade Unions was noted as having a policy on the implementation of a *Technological Change (Impact of Proposals) Act* (having a scope similar to the Commonwealth's former EIA legislation – the *Environment Protection (Impact of Proposals) Act 1974*) and which would promote the use of a technology impact statement focused on employment and the workplace. After considering the potential for TA, the committee proposed that a Technological Change Committee be established as part of the Australian Science and Technology Council, to advise the Prime Minister about the effects of technological change on socioeconomic and technical areas. Four decades later TA has yet to be formally implemented.

Activity in the United States and proposals for Australia appear to typify the concern expressed by Kay (1989) that most discussion about TA is itself technological in that it is primarily concerned with practical issues of methodology and policy formation. He concedes, however, that a broader approach is made difficult because of:

- Epistemological issues, in that while TA is thought of as applied science, it can include social technologies (eg, economics, astrology).
- Ontological issues, such that "identical technologies may produce completely dissimilar outcomes when introduced into different cultural settings ... [and] technological change can take place even if the technology itself remains unchanged" (1989: 128). That is, different groups use technology differently.

Nonetheless, the Organisation for Economic Co-operation and Development (OECD) (1988) sees a role for TA. When investigating the effects of technology, the OECD pointed out that long-term socio-economic strategies will have to take into account that, for the technical and economic potential of technologies to be realised, related social and institutional changes at all levels of society will have to be anticipated and assessed. A key input to these strategies would be the report's recommendation that TA be developed in various forms. Perhaps in response, Strohmann (1988) reports the development of the International Environmental Technology Center (IETC) in the United Nations Environment Programme. Through this Center the emphasis is on the assessment of technology for its appropriateness to the social and environmental situation to which it will be introduced in "developing" countries. Specifically, the IETC promotes and implements environmentally sound technologies (ESTs), including management systems, for disaster prevention, production and consumption, and water and sanitation. Since 2004, IETC has been focusing on disaster management, sustainable production and consumption, and water and sanitation.

3.9 Cumulative Impact Assessment (CIA)

While much experience has been gained with the application of the different forms of impact assessment, especially EIA, according to Vlachos a fundamental problem remains: being able to specify methodologies "for carrying out a sweeping assessment mandate and for accounting for the far-reaching, aggregate effects of projects and technologies" (1985: 60). As a consequence, he proposes development of cumulative impact assessment (CIA).

The central concept of CIA (or Cumulative Effects Assessment, as it may be referred to) is explained as the development of an approach that recognises that although individual actions may have an insignificant effect by themselves, the aggregate of these effects may have a significant effect. Individual impacts cannot be added, Vlachos notes, as the cumulative effect would depend on whether the actions were repetitive, continuous or delayed. Cumulative impacts are those impacts that result from incremental changes caused by other past, present or reasonably foreseeable actions together with the existing project (European Commission (1999: iii)).

While Erickson (1994) talks of cumulative impacts that are simply additive and those that are synergistic, Vlachos argues that the whole is more than the sum of the parts; that is cumulative impacts are synergistic, where interactions, combinations and new patterns of connection are important considerations. Irrespective of terminology, both authors indicate that CIA should be based on a holistic perspective, emphasising dynamic change. A similar position is taken by Buckley:

> Cumulative impacts range from simple additions to prior impacts of similar types, to complex interactions of environmental stresses due to the multiple impacts of many different types of development. ... Cumulative effects, therefore, are the rule rather than the exception ... (1998: 95)

To elaborate on this loose definition of cumulative environmental impacts, Buckley (1998) includes the following examples of impacts:

- Small additional impacts of new development in an area with prior impacts.
- Total aggregate impacts of multiple developments in a defined area, or on a specific ecosystem.
- Overall impacts of many similar concurrent developments.
- Interactive impacts from nearby developments of different types.
- Interactions between impacts from diffuse and point sources.
- Increases in impacts over time, from an expansion in activity or from decreasing quality of outputs (eg, increasing emissions from deteriorating equipment).
- Net impacts of multiple developments on particular environmental parameters (eg, water quality).
- Joint effects of multiple stresses on plant and animal populations (eg, through habitat clearance).

The sources of cumulative environmental changes and their impacts are discussed by Sadar (1994) and illustrated in Table 3.3 *over page*.

CIA would not use the deterministic methods employed in the like of EIA and TA. Rather, Vlachos (1985) sees that the methods would come from the categories of intuitive/holistic (eg, brainstorming), metaphors and scenarios (eg, alternative futures), extrapolative/time series (eg, probabilistic analysis), models/matrices (eg, systems analysis), and decision-trees (eg, judgment theory).

While few impact assessors have gained confidence and competency in the methods commonly used with EIA and like processes, these "holistic" methods are as yet not within the grasp of the majority of assessors. As found from a survey of CIA practitioners by Cooper and Canter:

> Although additive impacts were considered to be relatively easy to identify, synergistic and interactive impacts were not well understood, and they were difficult to accurately predict. Identification of these latter two types of impacts typically requires the use of complex and expensive models. (1997: 27)

Table 3.3 Sources of Cumulative Environmental Change

Issue Type	Main Characteristics	Examples
Time crowding	frequent and repetitive impacts on a single environmental medium	wastes sequentially discharged into lakes, rivers or watersheds
Space crowding	high density of impacts on a single environmental medium	habitat fragmentation in forests, estuaries
Compounding effects	synergistic effects due to multiple sources on a single environmental medium	downstream effects of several projects in a single watershed
Time lags	long delays in experiencing impacts	carcinogenic effects
Space lags	impacts resulting some distance from their sources	gaseous emissions into the atmosphere
Triggers and thresholds	impacts to biological systems that fundamentally change system behaviour	effects in changes in forest age on forest fauna
Indirect	secondary and tertiary impacts resulting from a primary activity	roads to resources which open up wilderness areas

Source: Sadar (1994)

The limited understanding and use of CIA was also identified by Canter and Kamath. From examination of a small number of cases they found no consistency in the methods used to investigate cumulative impacts, and "a general lack of detailed incorporation of cumulative impact concerns in the overall EIA process" (1995: 337). On the basis of their work they have developed a generic checklist, to be part of scoping, for incorporating consideration of cumulative impacts.

Also, Vlachos (1985) notes that with CIA a difficulty is that while people have a feeling that several small actions could cause significant problems in the future, there is little agreement as to what "significant" might mean. To date, the complexity and challenge which undertaking a CIA represents have meant that, compared to EIA activity, few CIAs have been reported.

However, Court et al (1996) see that CIA forms an integral relationship with ESD, SEA and EIA (see Figure 3.5 *over page*), so that a greater emphasis may be placed on CIA in future. Likewise Anderson and Sadler (1996) identified the complexities associated with identifying cumulative effects. It was generally agreed that there was no single answer to the problem of these effects, and that a range of mechanisms should be brought to bear, including:

- auditing;
- linking EIA more closely with land use planning;
- mitigation measures;
- incremental changes to policy planning;
- regional planning; and
- SEA for policies, plans and programs.

Against this, Erickson (1994) points out that there are "bureaucratic" constraints on the assessment of cumulative impacts. Most especially, EIA is typically undertaken "within the strictly limited jurisdictional authority of specific governmental agencies having operational (as opposed to environmental policy) responsibility" (1994: 236). So with limited operational scope, most agencies have had little incentive to look past the specific project impacts. But with increasing interest in the application of SEA there should be encouragement to take a broader perspective, so that the relationships indicated in Figure 3.5 develop and CIA becomes a more usual occurrence.

The need for approaches for CIA has been given increased emphasis by the inclusion of requirements for CIA in the *Canadian Environment Assessment Act 1992*. The Canadian Federal Environmental Assessment Review Office (1993) points out that "cumulative environmental effects, and a determination of the significance of such effects, are a key component of every environmental assessment conducted under the CEAA ... (where) every screening or comprehensive study of a project and every mediation or assessment by a review panel shall include a consideration of ... any cumulative environmental effects that are likely to result from the project in combination with other projects or activities that have been or will be carried out ...". To support these assessments the Federal Environmental Assessment Review Office provides a list of supporting studies and references. (Likewise, Canadian Environmental Assessment Agency (1996) outlines a comprehensive list of CIA studies undertaken worldwide.)

Since this time many have worked on the methods for developing the utility of CIA to better protect environmental matters. Some of this development is noted by King and Pushchak (2008) who report experiences with CIA in the environmental review of aquaculture projects in Canada. From this review their conclusions echoed the belief that a broader approach to planning these projects, through CIA, was needed, and that "many sites would not have been approved if siting decisions had been made within a regional management framework and cumulative impact criteria were considered in the site evaluation process" (2008: 572). The key issue here was the effect on

Figure 3.5 The Relationship of SEA, CIA and EIA to ESD

ECOLOGICALLY SUSTAINABLE DEVELOPMENT

Fundamental change to decision-making
Range of tools to implement SEA, CIA, SER, EIA

↓

STRATEGIC ENVIRONMENT ASSESSMENT

Impact assessment of Policies, Plans and Programs, being:

- sectoral
- regional
- indirect

This is a government function

↓

CUMULATIVE IMPACT ASSESSMENT

Carrying Capacity and Limits of Acceptable Change coming from:

- databases and predictive tools
- landscape units
- trigger mechanisms

This is a proponent of government function

↓

ENVIRONMENTAL IMPACT ASSESSMENT

Existing project specific assessment
This is a proponent function for major projects only

After Court et al (1996)

regional ecosystems that were incrementally affected by individual projects. In May 2013, the Canadian Environmental Assessment Agency released its *Operational Policy Statement for Assessing Cumulative Environmental Effects* under the *Canadian Environmental Assessment Act 2012* (CEAA). The CEAA 2012 specifies in s 19(1)(a) that a project must take into account cumulative effects that are likely to result from the project in combination with other physical activities that have been or will be carried out.

We could expect better of our professionals charged with making these sorts of decisions. However, as should be clear from the material in this book, we still have a lot to learn about assessment processes. In regard to CIA, or increasing called Cumulative Effects Assessment (CEA), an important consideration in the limited success, to date, of CIA has been the difficulty of selecting the appropriate scale of the assessment. From a review of literature and case studies Thérivel and Ross (2007: 365) concluded that:

> [S]cale issues are poorly considered at both (project and strategic) levels, with particular problems being unclear or non-existing cumulative effects scoping methodologies; poor consideration of past or likely future human

> activities beyond the plan or project in question; attempts to apportion "blame" for cumulative effects; and, at the plan level, limited management of cumulative effects caused particularly by the absence of consent regimes. Scale issues are important in most of these problems.

Frequently a regional scale needs to be set, but a national approach can be appropriate. An example of a CIA, provided by Sadler (1997a) relates to the work undertaken by the 1993 Coastal Zone Inquiry in Australia. He points out that Inquiry was based on a "broad-brush" assessment of cumulative effects, that is, linking patterns of growth to region-wide environmental change. Ultimately it recommended a National Coastal Action Program that focused on the following issues that he identified as critical for addressing cumulative effects:

- Adopting a long-term, holistic perspective (over short-term expediency).
- Greater community and industry involvement in decision-making.
- Use of innovative tools and measures to assist integrated Coastal Zone management (eg, economic instruments to fully implement the user-pays principle, strengthening the integrity and reliability of the EIA process).

From the investigations associated with the International Study of the Effectiveness of Environmental Assessment presented by Sadler (1997a), lessons have been drawn for the operation of CIA, especially regarding its relationship to SEA.

More recently, an example of incorporating CIA into EIA occurred as part of the Ryan Corner Wind Farm development proposal in south-west Victoria.

3.9.1 Case Study: Ryan Corner Wind Farm – Victoria

The Ryan Corner Wind Farm project proposed by Union Fenosa WindAustralia, involves development of a wind energy facility of up to 60 wind turbine generators, each with a capacity of two megawatts (MW). The overall height of the wind turbines would be approximately 123 metres. The proposed site is about 10 kilometres north-west of the coastal township of Port Fairy in south west Victoria. The wind farm would be developed on a number of rural landholdings, together of approximately 3000 hectares in area. At its closest point, the site is six kilometres from the coastline.

As part of the Environment Effects Statement (EES) (equivalent to an EIS), the proponent needed to assess the cumulative impacts in combination with the nearby Codrington and Yambuk wind farms and other proposed wind farms in the region, on environmental assets. These assets are described in the Assessment Guidelines as "the components of physical and ecological systems, human communities and economic activity that would be affected by the Ryan Corner proposal" (DSE (2005)).

The assessment of cumulative impacts included the following:

- Identify each environmental asset likely to be subject to cumulative impacts.
- Define the geographic extent of investigation required for each environmental asset that is likely to be subject to cumulative impacts. This will vary according to the nature of the asset (eg, a viewshed for landscape assessment, noise contour for noise impacts, the range of a particular bird or bat species).
- Identify existing wind farms (particularly Codrington and Yambuk) and reasonably foreseeable wind farm proposals which could give rise to cumulative impacts on each environmental asset.
- Assess cumulative impacts in a manner that is proportionate to the level of cumulative risk to the asset.

The EES also needed to assess the cumulative impacts on landscape and visual amenity where the Ryan Corner and one or more other wind farms would be simultaneously visible. For recognised tourist routes, the cumulative landscape impact should also take account of wind farms viewed sequentially, and not necessarily simultaneously, along such routes. The assessment of cumulative impacts should specifically address potentially significant impacts on protected birds and bats.

In regards to cumulative ecological impact assessment, the EES concentrated on threatened species that have a moderate or higher likelihood of occurrence at the Ryan Corner site and compared this information with available data for threatened species at the other nearby wind farms, being the Macarthur, Hawkesdale, Woolsthorpe, Yambuk and Drysdale wind farms. It was determined that the Wedge-tailed Eagle, and possibly the Southern Bent-wing Bat and Great Egret were the only species that may occur at all of the wind farm locations and that the potential impacts on each of these species was likely to differ across the various sites.

The EES considered that with the number of wind farms proposed in the region, there would be significant economic and social benefits arise from the wind farms, including the potential to employ over 400 people on site during construction (up to 40 during operation), and potential to contribute $4-5 million per annum into the local economy through wages, the purchase of goods and services, licences and rates.

In regard to the cumulative impacts on the landscape, the EES concluded that the major cumulative impact would be on viewers using the Princes Highway, as both wind farms would be visible. This impact was considered to be low because of the low visual sensitivity of the surrounding landscape.

The government's assessment outcome of the Ryan Corner Wind Farm was released in May 2008. The Minister for Planning supported the Inquiry's recommendation that the environmental effects of the Ryan Corner Wind Farm can be managed and a permit should be issued (Minister for Planning

(2008: 19)). As of mid-2013, Union Fenosa is preparing for construction of the Ryan Corner Wind Farm and has commenced early site works (Union Fenosa website).

3.10 Strategic Environmental Assessment (SEA)

Worldwide there has been a movement to take environmental issues into account early in the planning process. The *National Approach to Environmental Impact Assessment in Australia* notes that environmental factors need to be incorporated fully into the planning for proposals, meaning that these factors have to be part of the planning rather than thought about afterwards (Australian and New Zealand Environment and Conservation Council (1991a)).

In agreement is Wood (1993) who points out that, from the outset, the US *National Environmental Policy Act 1970* provided for the early consideration of environmental factors in plans and policies of government. Now, he says, taking the environment into account earlier in the planning process is more widely accepted. This concept has become known as strategic environment assessment (SEA) because of its focus on strategic planning, or the "broader picture", compared with the traditional way in which EIA has been applied to specific projects. SEA is the environmental assessment of policies, plans and programs. It serves to strengthen accountability and provide greater public confidence that government decisions are being made in full awareness of the potential environmental impact. The Canadian Environment Assessment Agency promotes SEA as a tool that contributes to informed decisions in support of sustainable development by incorporating environmental considerations into the development of public policies and strategic decisions (CEAA (2010)).

The integration of SEA has aided in the use of sustainability principles in decision-making. Recognising SEA's potential role in addressing concerns about cumulative impacts and sustainability, Partidario (in Stinchcombe and Gibson) offers the following definition:

> SEA is a systematic ongoing process for evaluating, at the earliest appropriate stage of publicly accountable decision making, the environmental quality, and consequences of alternative visions and development intentions incorporated in policy, planning, program initiatives, ensuring full integration of relevant biophysical, economic, social and political considerations. (2001: 345)

Although there is a close relationship between SEA and other forms of assessment such as EIA, a major criticism of EIAs has been that they have been site-specific and have not considered the cumulative effects of development. SEA, however, allows consideration of environmental impacts across a larger geographic area, and over a period of development. Importantly SEA is not necessarily about the impacts of a particular development, but focuses on the impacts of proposed policies, programs or plans, eg, Forest Plans. Sadler and

Verheem (1996) also provide a useful definition of SEA: SEA is a systematic process for evaluating the environmental consequences of proposed policy, plan or program initiatives in order to ensure environmental factors are fully included and appropriately addressed at the earliest appropriate stage of decision-making on par with economic and social considerations.

The differences between EIA and SEA have been succinctly identified by Arce and Gullón (2000) (see Table 3.4).

Table 3.4 Differences between EIA and SEA

EIA	SEA
Typically begins when considerable thought and planning has been given to a proposal	Typically begins at the earliest stage of publicly accountable decision-making
Focuses on the better execution of specific actions	Focuses on the previous conditions in which specific action are inserted
Reactive approach	Proactive approach
Scope is site specific with an emphasis on a short time span	Scope is “global”, to broaden the spatial and temporal range of the assessment
Use is for the assessment of specific projects	Use is for the assessment of programs, plans and policies

Typically, SEA would be used at a national or regional level to assess broad policies (such as greenhouse policy), plans (eg, strategic land-use plans), and programs (like that for public transport expansion). On the other hand, more local projects, like a road link, would be assessed through the EIA process.

As a result of the review of environment assessment procedures, Sadler (1997a) concluded that SEA and EIA can and should be tiered or vertically integrated. This assists the opportunities for investigating environmental considerations at the appropriate level of decision-making. He noted that this vertical integration is most successful where SEA is applied to policies, plans, and programs that initiate projects, and that several benefits of the use of SEA have been proposed. This study also provided the opportunity to consider the relationship of SEA to CIA. Specifically, Sadler noted that there are three elements that provide a point of entry for SEA to address cumulative effects, that is:

- Sectoral or programmatic types of decision-making activities can focus on sources, or the activities that lead to cumulative effects.
- Regional plans shift the attention toward effects and the sensitivities and capacities of the receiving environment, as indicated by keystone or indicator species.
- Policy appraisals may benefit from taking a synoptic, process perspective of relationships and consequences.

Strong advocates for SEA, Arce and Gullón argue that SEA makes a fundamentally important contribution to sustainability (or sustainable development). They comment that:

> SEAs increase the possibility of analysing and proposing alternative solutions and incorporating sustainability criteria throughout the planning process, as they carry the principles of *sustainability* down from policies to individual projects. The contribution of SEA towards sustainability stems from several points: (1) SEA ensures the consideration of environmental issues from the beginning of the decision-making process; (2) provides a framework for a chain of actions; (3) contributes to integrated policy making, planning and programming; and (4) can detect potential environmental impacts at an early stage, even before the projects are designed. At the level of planning, programming and policy design, many decisions are taken. By developing SEAs, these decisions can be integrated into a systematic process of environmental assessment, which subsequently influences the following project designs. SEA provides a sound and holistic framework for planning and strategic decision-making. (2000: 394-395)

SEA has been incorporated into a number of EIA procedures (eg, in California, Canada, New Zealand, France, and Australia). In respect of the EU, Wood (1993) suggests that "there is growing interest in the possibilities of using some form of SEA as an integrative instrument in promoting sustainable development" (1993: 24). Thérivel (1993) sees that this link is particularly emphasised by the Dutch system, which outlines criteria for sustainable development which are to be met through the use of SEA.

The European Commission first put forward a proposal to introduce SEA through the impressive, but perhaps vague, title of Proposal for a Directive on the Assessment of the Effects of Certain Plans and Programs on the Environment. Feldmann comments that implementation of the proposal could rely on "awareness raising for most individuals and levels involved ... to bridge borders, which are mainly administrative and political" (1998: 13). After consultations and amendments, the Environment Ministers of the European Parliament reached a common position in 2000. The SEA Directive 2001/42/EC was formally adopted in May 2001. The purpose of the SEA Directive is to ensure that environmental consequences of certain plans and programs are identified and assessed during their preparation and before their adoption (European Commission (2004)). The European Commission believes that SEA will contribute to a more transparent planning and policy process and ultimately achieve the goal of sustainable development. A review of the effectiveness of the SEA Directive was undertaken in 2009 and it found that the use of the SEA Directive by member states was still in its infancy. However, overall the study concluded that the SEA Directive *contributed to the systematic and structured consideration of environmental concerns in planning processes and better integration of environmental considerations upstream* (CEC 2009: 11)).

During the 1980s increasing attention was given to SEA, especially its application to policies, plans and programs. Thérivel (1993) provides a comprehensive review of the European and United States situations, and discusses the experience which has been accumulated, including:

- SEAs have been typically sectoral (eg, for water supply, agriculture, energy), regional (eg, metropolitan plans, community plans, redevelopment plans) or indirect (eg, science and technology, financial policies, enforcement policies).
- The impacts considered have been traditional (eg, air, water, soils), sustainability-related (eg, unique natural features, energy, non-renewable resources) or policy-related (eg, safety and risk, climatic hazards, social conditions).
- Public involvement is complicated by the breadth of issues considered, arranging meetings of the different groups involved, the appropriate level of the decision-maker and the lack of a specific deadline for a decision.

For Bagri and Vorhies (1999), SEA holds great potential as a means of identifying the role of wetlands, especially wetlands that are recognised under the RAMSAR convention. They see that SEA could be a tool for the legal and institutional review of proposals that affect wetlands, and could set the right incentives for ensuring that biodiversity is adequately considered in policy and planning. However, they also identify a number of hurdles for the integration of wetlands in current SEA processes.

The extent of activity clearly suggests that internationally SEA is becoming an important element of planning in a broad sense. Wood's (1995) review of international EIA systems indicates that while few countries specifically require SEA, there are several where SEA is covered by the breadth of the EIA provisions, and identifies instances in both groups of countries where SEA is practiced. This leads him to the observation that, around the world, SEA practice is developing quickly. In New Zealand, for example, the *Resource Management Act 1991* provides for both project and policy based environmental assessments. Based on this experience, the International Study of the Effectiveness of Environmental Assessment identified a series of steps that represent good practice for SEA:

- Screen policies, plans and programs to trigger SEA and identify likely scope of review that is needed.
- Use a scoping process to identify key issues and alternatives, clarify objectives and to develop terms of reference for the SEA.
- Elaborate and compare alternatives, including no action options to clarify implications and trade-offs.
- Undertake an impact analysis or policy appraisal to examine effects (issues), evaluate alternatives, and identify mitigation and follow up measures.
- Document the findings of the SEA with supporting advice on terms and conditions for implementation.

- Check the quality of the SEA report to ensure it is clear and concise, and the information is sufficient and relevant to the decision being taken.
- Establish necessary follow up measures, eg, for monitoring effects, checking implementation, and tracking any arrangements for subsidiary level assessment. (Sadler (1997a))

From the investigations associated with the International Study presented by Sadler (1997a) lessons have been drawn for the operation of SEA. Interestingly, the majority of these relate more to the political nature of SEA (such as where it fits in the planning and decision-making processes) rather than focusing on the technical issues (such as the content). Unrelated to this study, Brown and Thérivel have discussed the methodology of SEA and, in what can be seen as an improvement to past practice, argue that SEA has a clear role as a policy, plan, program formulation tool. In particular, "it is at the stage of PPP formulation, rather than of appraisal of an already formulated PPP (for instance green paper stage, review, public consultation) that SEA can be most effective" (2000: 187).

A move on the technical side is reported by Clement (2000) who notes that Strategic Sustainability Assessment (SSA) has been explored to support the application of SEA. SSA is intended to translate sustainability priorities into measurable indicators, with the objective of identifying the range of indicators that may be relevant within the context of sustainability. The maximum value of an indicator would be the point beyond which sustainability could no longer be ensured. Clement comments that "for a specific policy context, SSA would be used to define targets for environmental indicators and advise on the appropriate combinations of indicators ..." (2000: 27). However, no application of the assessment approach was reported.

In effect, SEA is equivalent to the environmental planning which was undertaken by planning authorities in the 1970s, and the many regional planning activities of the past three decades such as the assessments undertaken by the Victorian Land Conservation Council (now replaced by the Victorian Environmental Assessment Council (VEAC)). Although not a precise example of SEA, independent agencies such as the VEAC had some similarities. So we have had a reasonable history of strategic planning, but as Dover (2002) suggests, the dominant political fashion of the past two decades has not been kind to independent, informed, inclusive institutions and processes that persist and unsettle the status quo; and the questions that an SEA could raise would often be unsettling.

However, Conacher (1994) points out that these earlier regional planning examples have been less than ideal in integrating environmental assessment and resource management. To achieve integration, Elling (2000) has proposed a five-phase process, which has much in common with Sadler's (1997a) process (above):

- Phase I – scoping of the existing plan, and identification of environmental criteria to assess the extent to which the plan includes environmental protection.
- Phase II – environmental assessment of the existing plan to contribute to the identification of the scope of a revised plan.
- Phase III – public contribution of ideas for the scope of the revised plan, definition of planning objectives and strategic guidelines for the environmental assessment of the plan.
- Phase IV – drafting of the revised plan and its public review.
- Phase V – finalisation of the plan and its adoption.

In many respects, the process by the international community during the late 1990s and into the 21st century, to develop and agree on a greenhouse policy, illustrates these phases. At a more regional level, the integration of spatial and environmental planning is the basis of the environmental plans developed in conjunction with the EIA process in New South Wales, and the integrated development approaches increasingly being favoured around Australia (see Chapter 6), and in New Zealand (see 5.5). SEA, or environmental planning, provides the general direction for the use of resources in a region. If a proposed project fits with those "allowed uses", and if it requires an EIA, then the EIA would not need to cover the regional environmental issues or look into alternative sites. Instead the EIA could concentrate on details and alternatives on the site (eg, landscaping, and types of pollution-abatement equipment). This is similar to the idea of a two-stage process for EIA. In this process an initial broad-ranging study outlines the types of activities and their extent appropriate in a region, then specific proposals which accord with the study are accepted in principle, while only the detailed site-specific concerns are examined.

Consideration of these points by the administrators of the EIA process in Western Australia has led to the introduction of SEA in the environmental assessment process. Informal SEA occurs in Western Australia under s 16 of the *Environmental Protection Act 1986*, which allows the EPA to advise the Minister on various environmental matters, such as water and wastewater management strategies, planning policies, transport and infrastructure proposals. The Minister for the Environment recognised that the Act did not allow for the formal assessment of strategic programs. Consequently, amendments to the Act occurred in late 2003 which enables the assessment of policies, plans or programs (such as energy supply options), but where the right would be retained to assess the detail of subsequent individual proposals. The amendments used the existing provisions for the assessment of proposals, strengthening them to incorporate SEA. Provided issues had been adequately considered in a strategic assessment, they would not need to be considered at the level of the individual project. There are also opportunities to undertake an SEA for a class of projects, a code of practice, or a policy. Following this, proposals that demonstrated compliance with the class, code or policy would not need to be assessed individually. The limitation to the

WA system, as we can see in other jurisdictions, is the voluntary nature of proponents being able to initiate strategic assessment. The EPA cannot "call in" a strategic proposal for assessment nor can third parties request referral of strategic proposals to the EPA (Malcolm (2002)).

Overall, the notion of SEA, or whatever it may be termed, provides opportunities for ensuring that the broadest concepts of environment are factors in decisions which shape development. In this respect, SEA contributes to sustainable development. This link was emphasised by consultants who advised on aspects of the 1993-1995 review of the Commonwealth's *Environmental Protection (Impact of Proposals) Act* (EP(IP) Act). More particularly, their report recommended the adoption of SEA, incorporating cumulative impact assessment, as the ultimate way of achieving ESD (Environment Protection Agency (1994a)). When the Commonwealth replaced the EP(IP) Act with the *Environment Protection and Biodiversity Conservation Act 1999* (EPBC Act) (see 6.3.4) the opportunity was taken to build SEA into the Commonwealth's processes. Although the Act itself does not define strategic assessment, s 146 of the EPBC Act provides for SEA of actions arising from policies, programs and plans generally. In particular, ss 147-154 provide for SEA of Commonwealth managed fisheries and many draft strategic EISs have been prepared under this provision (these can be viewed at the Department of the Environment, Water, Heritage and the Arts website <http://www.environment.gov.au/epbc/assessments/strategic.html>.

While there have been some moves to develop SEA in Australia, Buckley (1997) considers it to be the most important issue for EIA, and has proposed that it be applied to the assessment of government policies and associated legislative proposals. To ensure that this happens he proposes that formal and enforceable mechanisms to trigger SEA of policies have to be developed and applied (Buckley (1997; 2000)). Arguments in support of the application of SEA to legislative proposals are provided by Marsden (1997). The process, and trigger to initiate a SEA, he suggests could come through adaptation of checking mechanisms at Cabinet level (within Commonwealth and State governments) to include environmental considerations. He comments that this would be an important advance in assisting Ecologically Sustainable Development, however, the trigger may not be as enforceable as Buckley would like.

Even without legislation to require SEA, Renton and Bailey (2000) report that from their late 1990s survey of a sample of government agencies there were clear indications that many of the agencies had policy processes that incorporated an assessment of environmental matters. Nonetheless, the authors concluded that policy development was idiosyncratic and there was still potential for improved processes associated with SEA.

The growing interest in SEA is a recognition of the need to consider the broad environmental implications of decisions. However, Horton and Memon (1997) caution that SEA as currently conceived does not necessarily provide ensured environmental protection. They point out that EIA "displaces uncertainty from the realm of science to that of politics ... (and) polarises environmental interpretation and assessment, as various groups advance their own interests" (1997: 166). Similar difficulties exist with SEA where the concept can lead to conflict between economy and the environment, together with the displacement of this conflict "SEA proposes, in effect, the intensified synchronous degradation and enhancement of the environment" (1997: 166). These effects would result from an assessment that proposes the zoning of activities, such as where some areas are given over to economic development while others are excluded from development to protect environmental values. However, Horton and Memon say that this does not condemn the use of SEA, but provides a warning of holding excessive expectations and applying it simplistically. They suggest that SEA has a role to play, but its use must confront the conflict of development and the environment, and avoid displacing this conflict with spatial solutions.

Despite the inclusion of SEA provisions in the EPBC Act, there has been limited application of SEA in Australia, although the use of SEA under the Act has increased over the past five years. A range of environmental assessments and studies which could be recognised as SEA across sectors and jurisdictions, is occurring. As mentioned in section 6.3.4, the Commonwealth and Western Australian governments conducted a strategic assessment under the *Environment Protection and Biodiversity Conservation Act 1999* to identify a hub site for liquefied natural gas to service the Browse Basin gas reserves in the Kimberley region. One of the first SEA's under the Act, a site selection process commenced in 2008 by the WA government in which more than 40 sites were considered. Throughout 2009 and 2010 the assessment process was undertaken for the site selected – James Price Point. The WA government formally approved the Browse Basin LNG precinct in November 2012. In April 2013 however, Woodside (the key proponent of the Browse Basin LNG development), determined that the development concept did not meet the company's commercial requirements for a positive investment decision and have put the project on hold (Woodside).

Marsden and Dovers (2002) identify a number of issues facing SEA – one of the biggest challenges is responsibility for implementation. They suggest that governments are unlikely to advance SEA unless pressure from the community is applied and it is also unlikely that other professions will implement SEA. However, if governments are serious about sustainability then SEA is certainly required and needed.

A critical word of warning is also sounded in respect of the implications of "getting the SEA wrong" for subsequent or "downstream project EIA". Reporting on the SEA for a proposed highway in Brazil Sánchez and Silva-Sánchez (2008) commented on the critical decisions associated with determining the significant strategic issues to be included in the SEA. Their conclusion was that:

> [P]articularly where no agreed-upon framework for SEA exists, when vertical tiering with downstream project EIA is sought, then a careful scoping of strategic issues is more than necessary. If an agreement on "what is strategic" is not reached and not recognized by influential stakeholders, then the unsettled conflicts will be transferred to project EIA. In such a context, SEA will have added another loop to the usually long road to project approval. (2008: 515).

Of course, similar comments can be made for the careful scoping (see 7.4.2) of significant issues in EIA to reduce the downstream effects on environmental management after a project has been commenced. In both cases the importance of the scoping stage cannot be underestimated.

As a final comment on SEA, we need to note that there are several terms used to cover the same concepts. So as well as the term Strategic Environment Assessment, you are likely to see: Policy, Plan and Program Environment Assessment (PPP EA); Policy Impact Assessment; Regional Environment Assessment; Sectorial Environment Assessment; Programmatic Environment Assessment; Environmental Overview; and possibly others (Arce and Gullón (2000)). In essence, however, they all refer to the same idea of making a "big-picture" assessment of the environmental effects of policies, plans and programs.

3.11 Integrated Impact Assessment (IIA)

IIA was conceived to draw together aspects of EIA, SIA and TA. Porter and Rossini suggest that EIA attracts the involvement of life scientists, SIA social scientists, and TA scientists and engineers. "[T]he emergent table of integrated impact assessment holds out promise to lessen these differences and increase interdisciplinary" (1983: 11).

All the assessment techniques considered previously have insight to offer. In many respects the only differences between them are in the scope of analysis and assessment contemplated, and the will and understanding of the assessors. The methodologies can be combined with one another, so the name used depends largely upon the definitions of terms (eg, as in "planning" and EIA). Carley and Bustfeld (1984) provide a review of relevant literature that emphasises a cross-disciplinary perspective which helps to illustrate the overlap between techniques.

The analysis of Becker (1991) suggests that the impetus for IIA in Europe has come from the need for program evaluation. He notes that the governments of the EU are having to make decisions on the basis of more than purely financial factors, so the role of IIA in project evaluation is likely to expand. Similarly, Bailey (1997) sees that integrated assessment lies at the heart of environmental policy, particularly at the global level. He comments on the use of "integrated environmental assessment" for the examination of global climate change, such as that related to greenhouse emissions, and how this assessment process feeds into policy-making. In this context he emphasises the importance of the users of the assessment, and proposes that the stakeholders (government, business, public and others) need to be directly involved in the process.

Despite some interest in the possibilities of this type of assessment, there are few examples of IIA to date. However, a closely related development is the exploration of Integrated Environmental Assessment. Vellinga (1998) has introduced the concept, pointing out that the research community is heavily involved, and that there is a strong emphasis on the analysis of policy. With this policy focus it appears that there would be overlap with SEA (see 3.10). To promote this form of integrated assessment, a European Forum has been established which involves scientists (Institute for Environmental Studies (2000)). The Forum's objectives are to: improve the scientific quality of integrated environmental assessment; and strengthen the interaction between environmental science and policy. A number of projects have begun to develop appropriate methodologies.

3.12 Sustainability Impact Assessment (SusIA)

SusIA is the assessment of the potential economic, social and environmental impacts of policies, plans and programs and differs from SEA in that the focus of SEA is to evaluate the potential environmental impacts of PPPs.

The European Union (EU) has developed trade SusIAs to integrate sustainable development (including environmental, social, and economic aspects) into its negotiating positions for trade agreements. First established in 1999, SusIAs were developed for the WTO Doha Development Agenda negotiations and it marked a turning point in international trade negotiations as questions of trade, environment and sustainable development were explicitly linked. The EU uses SusIA as a process for trade negotiations that seeks to inform negotiators of possible effects of a particular policy (United Nations Environment Programme (2008)).

3.13 Overview

A common theme pervades the assessment approaches discussed. Essentially they are designed to identify potential impacts of a development, action, program or policy, and to consider what may be needed to reduce adverse impacts. Particular approaches emphasise specific concerns (eg, health, social, risk identification). However, all have basically the same approach, although each may have its own individual language and detailed techniques.

All have derived from the earliest from of project impact assessment (EIA). However, we have seen an increase in the use of SEA, SusIA and CIA to assess impacts of policies, plans, programs and trade negotiations. If SEA is to be an effective tool for sustainability-based decision-making, then indeed the integration of biophysical, economic, social and political considerations needs to occur.

Chapter 4

THE PUBLIC FACE OF EIA

Previous chapters have emphasised the point that public participation is a major characteristic of EIA. The opening up of decision-making processes for a review of environmental consequences has been a catalyst for providing the general public with information about proposals, the ability to have input into these decisions and for seeking the public's comments.

The vast majority of EIA procedures incorporate some form of public participation. This chapter will provide background into what is meant by participation and how it can be encouraged. In particular, the role and experiences of public participation in EIA at the screening and scoping stages of EIA will be reviewed.

There is a substantial body of literature on the general topic of public participation, with each group of professionals tending to treat the topic from its own perspective. For instance, the International Association for Impact Assessment focuses on matters of particular relevance to EIA processes. The approach used here will be to provide a general introduction with suggestions of where more detailed information can be found.

4.1 Background

Like environmental attitudes, public participation (or public involvement) depends upon the interests and priorities of society. O'Riordan (1976) suggests that participation is an evolutionary process shaped by the political culture in which it develops, and that it is "essentially a means to social reform and political egalitarianism, a systems-transforming device that is regarded as evolutionary and subversive by many of the elite" (1976: 256).

The reasons participation has become an issue (since the 1960s) are varied. Cahn and Cahn (1971) list, among other points, complexity of life, increases in the functions of government and increased specialisation. In response to these factors, people have sought more control over their lives. O'Riordan (1976) proposes four suggestions as to why people would seek more control:

1. With conflicting political interests, and a largely fixed allocation of resources, adversely affected groups seek to look after their interests.
2. With greater access to information and education, people are more willing and able to participate.
3. When enjoying a sufficiently high standard of living, people become concerned about decreasing environmental quality.
4. The days of the paternalistic administrator, infallible expert and trusted politician have gone.

Another aspect is that of the right of citizens to participate under a democratic system. These points are given "real life" context by Goodman (1971) in his descriptions of urban planning and promotion of citizen involvement. A further perspective is provided by Wengert (1976), who considers the historical interpretation, political theory and social theory of public participation.

While there are a number of suggestions as to why people have sought participation, it is possibly due to a combination of a variety of desires coupled with an agreeable attitude within society. That public participation is an accepted "fact of life" can be indicated by the increasing opportunity provided by government for involvement; for example, through various governments' commitment to provide opportunities for participation in various policy processes, legislation, administrative arrangements and hearings/enquiries. In this context, as Sewell and O'Riordan (1976) suggest, the influence of the public can no longer be ignored.

It is sufficient to accept that public participation is now frequently undertaken. However, it is important to recognise that varying levels of public participation in the EIA process occurs, the advantages and disadvantages, and particularly what participation is expected to achieve.

4.2 Value of Public Participation

As public participation is an accepted part of many activities, it is worth attempting to identify the specific features of participation that have led to it being so widely accepted. Cahn and Cahn (1971) suggest that the values of participation fall into three broad categories. It provides:

1. A means of mobilising under-utilised resources (untapped labour or productivity).
2. A source of knowledge (both corrective and creative).
3. An end in itself (affirmation of democracy and elimination of alienation, hostility and lack of faith).

To these can be added a fourth and perhaps overriding advantage; that is, it provides:

4. Better decision-making.

Although this last point is somewhat nebulous, it is frequently considered to be the main value where objectives for public participation exercises have not been clearly thought out. "Better decision-making" would include aspects of the first three points, and particularly dissemination of information, identification of relevant issues (and perhaps values), and avoiding objections and delay at later stages (since the opportunity to participate has already been provided).

Sadar's (1994) perceptions of the value of public participation are somewhat more specific, in that he identifies the advantages of participation as being:

- the public is informed;
- different viewpoints are identified;
- concerns raised by the proposal are made clearer;
- potential areas of conflict are identified;
- trust and mutual respect are fostered; and
- the "comfort level" of decision-makers is raised.

Perhaps the value of participation is best summed up by Sewell and O'Riordan as "a consciousness-raising process through which people begin to understand their political roles and the need for legitimate conciliation and contribution" (1976: 17).

The International Association for Public Participation (known as IAP2), which is the preeminent international organisation advancing the practice of public participation, suggests that the core values of public participation are:

1. Public participation is based on the belief that those who are affected by a decision have a right to be involved in the decision-making process.
2. Public participation includes the promise that the public's contribution will influence the decision.
3. Public participation promotes sustainable decisions by recognising and communicating the needs and interests of all participants, including decision makers.
4. Public participation seeks out and facilitates the involvement of those potentially affected by or interested in a decision.
5. Public participation seeks input from participants in designing how they participate.
6. Public participation provides participants with the information they need to participate in a meaningful way.
7. Public participation communicates to participants how their input affected the decision. (IAP2: 3)

The IAP2 have developed a spectrum for public participation which moves from:

- informing the public;
- consult (obtain feedback from the public);

- involve (work directly with the public);
- collaborative (partner with the public); and
- empower (place final decision-making in the hands of the public). (IAP2 : 4)

A tangible example of the value of public participation is provided by the program to deal with "scheduled wastes" in Australia. These wastes are mainly polychlorinated biphenyls (PCBs), hexachlorobenzene (HCB) and organochlorine pesticides (OCPs) such as DDT and dieldrin. During the 1970s and 1980s the National Advisory Body on Scheduled Wastes (2000) reported that there were many attempts to construct a high temperature incinerator for the disposal of these wastes. However, there was considerable community opposition to the different sites selected by the technical panels set up by governments. In 1992 the Australian and New Zealand Environment and Conservation Council (ANZECC) announced the decision to abandon the idea of constructing an incinerator. Instead, to resolve the issue of waste disposal, a public involvement and consultation process was developed to achieve a broad consensus on the means of managing scheduled waste in Australia. This process took some three-and-a-half years, but at the end, waste management plans were agreed to by all the stakeholders; which included all three levels of government, environment groups, unions and business. The participation process was seen to be successful in achieving management plans, but also in:

- developing networks amongst the stakeholders;
- raising the credibility of the plans; and
- providing a model for bringing government, industry and community/ environment groups into working relationships.

For Sinclair et al (2008) this last point is especially important. They see that where laws or procedures require that public voices be part of the decision process, this has led to "the creation of fascinating, state-sanctioned, deliberative spaces for civic interactions" (2008: 415). Results from the study, based on 15 years of data, indicated that EIA "may be conducive to broad-based individual and social learning that could enable a transition to sustainability, but do not provide direct evidence of such a transition in terms of EA decision making" (2008: 425). In other words, participation helps people in the community to develop general participation skills and helps them to be better informed, but does not necessarily lead them to change their lifestyles.

So, as well as a participation program providing information to assist those running the EIA process and making decisions about a particular project, the program can provide educational and informational benefits to the general community.

4.3 Disadvantages of Public Participation

While participation has generally been embraced as a "good thing", there are some detracting aspects such as those discussed by Molesworth (1985):

- Only those with scientific or technical training are able to contribute to positive and constructive decision-making.
- It is more efficient to have a small number of people involved in making decisions (efficiency of time and decisiveness).
- Members of the public tend to be subjective whereas professionals (technical or bureaucratic) are thought to be objective.
- The existing political process works to take into account public opinion; public participation is almost interference.
- Third parties should not be allowed to interfere with another person's democratic right to do something.
- Public participation is not truly representative of public opinion.
- Public participation adds to the costs of projects or governing.
- Public participation encourages litigants to disrupt the proper processes of government/administration.
- The public cannot appreciate the importance of many affairs of State (which only government or its agencies can fully understand).

These points support the view that the public cannot contribute anything useful to decision-making processes and that the best decisions are made by experts. Each is raised from time to time as a reason for avoiding participation, although the accusations of delay and additional cost are most common (see 2.5.2 and 2.5.3).

Delay and additional cost are raised in relation to a variety of programs and processes (not just public participation). To use the analogy of the testing of new chemicals or medicines to ensure a degree of safety, public participation could be thought of as the testing of new ideas (or proposals) to ensure a degree of understanding and (perhaps) acceptance. If it is accepted that the community has the right to understanding, just as it has the right to safety, then public participation programs should not be decried on cost or time grounds.

Overall, it may be easy to say that the disadvantages noted above are insubstantial. However, while they may not carry much weight in dismissing the need for participation, they should be considered as aspects to be aware of in designing a participation program. If the design can reduce these disadvantages, it will be a better program than one which simply ignores potential problems.

4.4 Public Participation

The previous sections have provided general impressions from which a definition of public participation could be derived. However, the definition is likely to depend upon what the participation is for (the objectives), who would be involved and how. Hence, rather than attempting a precise definition, the aspects which would contribute to a definition will be discussed. The expectation is that anyone involved in participation will then think about the many facets of participation, instead of simply adopting an "off-the-shelf" program.

4.4.1 Objectives of participation

Following are a variety of suggestions as to how objectives of public participation could be expressed.

Wengert (1976) talks of the perceptions of participation (and what is expected from it) in terms of participation as:

1. Communication.
2. Conflict resolution.
3. Therapy (for those who participate).
4. Strategy (to accomplish other objectives).
5. Policy.

Lucas (1976) makes use of four reasons for adopting participation (stated by a United States federal task force on EIA) as:

1. Providing an opportunity for unrepresented persons to present their views.
2. Providing useful additional information to decision-makers.
3. Reinforcing accountability of political and administrative decision-makers.
4. Increasing public confidence in the reviewers and decision-makers.

Sewell and Phillips list a number of public participation case studies, with varying objectives depending upon who is considering the participation program. In respect of the perceived objectives of these participation programs, they comment:

> Agency personnel and representatives of citizen groups differed considerably in their perceptions of the purposes of public involvement. For most agency representatives, participation was seen as a means to develop programs which would have broad public acceptance, to enhance the efficient performance of agency responsibilities, and to improve the agency's image. In contrast, citizens perceived a much broader set of objectives. They viewed public participation as a means to reduce the power of planners and the bureaucracy and to ensure that people affected by government policies have influence over their design and implementation. Independent observers, while cognisant of both the philosophical and educational rationales of public participation programs, tended to emphasise the pragmatic, agency-oriented aspects. (1979: 352)

Bringing these many threads together, the authors suggest there are three basic objectives frequently desired in participation programs, namely:

1. A high degree of citizen involvement.
2. A high degree of equity among the public.
3. High cost efficiency for the agency conducting the program.

Clark (1981) looks at objectives from the point of view of a public servant and identifies a range of possible objectives under two categories:

1. Political objectives:
   (a) legitimacy of agency programs;
   (b) enhancement of agency image;
   (c) dissolution of organised opposition;
   (d) conflict resolution;
   (e) eradication of bias and combating of social injustice;
   (f) social therapy;
   (g) redistribution of political power; and
   (h) increase in individual and community satisfaction.
2. Functional objectives:
   (a) informing the public (developing informed "technical" inputs for decision-making);
   (b) seeking expression of value-preferences as input for decision-making; and
   (c) seeking greater accountability for decisions in decision-making.

Underwood, in describing public participation, identified a number of aspects that correspond to objectives:

> Effective community participation can be defined as a two-way investigatory and learning process of interaction and communication between the study team and the community. Basically, it is a process of collecting and analysing all the relevant facts, of presenting these data results and conclusions to the community, of obtaining the views and responses of all sections of the community, and of using all this information as input to the decision-making process. It seeks to ensure that the views of all individuals, interest groups, and other bodies are taken into account. Weight of numbers, investigatory capacity, or political influence are not meant to overshadow the determination of the potential effect on the individual who may not have the ability or the resources to make his [sic] views known. It should identify areas of agreement, of disagreement and of possible compromise. The process is not meant to be a political or consensus one – it simply provides input to decision-making. (1978: 1)

There is clearly an overlap between many of the above statements of objectives. Equally so, there is considerable variation, particularly when more precise explanation of objectives is attempted, or when the objectives are considered from different viewpoints. Overall, it may not be possible to establish a set of objectives which would satisfy everyone for a variety of reasons. For instance, Clark (1981) suggests that constraints will be evident that influence the choice of objectives. In the case of a public servant, these constraints could include lack of rationality by senior officers, human factors, budgetary factors and political factors.

Irrespective of whether a perfect set of objectives can be set, it is important to define the objectives. Only when the objectives are clearly expressed is it feasible to determine appropriate techniques to achieve the objectives and then to evaluate the effectiveness of the participation program. Clear objectives also help overall planning of the proposal (the reason for the participation program) and understanding among those involved as to "what's going on".

According to Gariepy (1991) public participation fulfils three functions:

1. Validation, or opening up the decision-making process to the public.
2. Internalisation, or enabling inputs from the external public.
3. Definition of the project environment, or determination of the factors to be considered in the investigation.

The effect of this participation may be difficult to judge. When considering hydro-electricity projects in Canada, Gariepy thought that the interest groups were not very effective with their participation through submissions and hearings. However, their potential to disrupt the implementation of the decision, by various means, meant that they were given a degree of attention.

Appleyard expresses a broad social objective for participation. He suggests that public participation, in environmental decisions, "is critically important, because this is the way in which people can become identified with a new environmental action, the way in which they can possess and feel responsible for it" (1979: 152).

4.4.2 Public participation and Social Impact Assessment

A particular objective of a public participation exercise may be to link with a Social Impact Assessment (SIA) (see 3.6). In this case the two would complement each other. Garcia and Daneke (1983) suggest the steps would include a rough SIA to identify the affected and interested groups, who would contribute issues and values to a public participation program. Alternative courses of action could be developed in response to this participation, and public feedback would be sought on the alternatives. Results from both the SIA and public involvement would be fed into the decision-making process.

If the SIA were to have a political perspective and adopt a consciousness-raising approach (see Gale (1983)), the SIA and public participation program would be very closely linked to the extent that they may be indecipherable (with one "feeding on" the other).

4.4.3 Levels of participation

Levels of participation are many and varied. The most concise groupings suggest three broad categories:

1. Informing the public of proposals.
2. Soliciting input to aid decision-making.
3. Public representation on decision-making bodies.

Subcategories can be identified such as the eight listed by Sewell and Phillips (1979) in their discussion of participation case studies. There is a strong correlation between these and the eight categories developed by Arnstein (1971) (see Figure 4.1 *below*), who also groups these into three broader groupings, using terms which are less positive but perhaps more realistic than those used in the three categories above. These categories below are similar to the more recent public participation spectrum of the IAP2 Association.

The point of recognising that participation can be directed at different levels is also to recognise that "levels" and objectives are closely linked. Specification of the level of participation will largely determine the objectives that can be expected to be met: an objective of "providing useful additional information to decision-makers" will not be met if the level of participation is restricted to "informing the public".

Figure 4.1 Levels of participation

Level	Descriptor	Effect
8	Community control	Empowerment of individuals
7	Delegated power	
6	Collaboration	Tokenism
5	Placation	
4	Consultation	
3	Information	Non-participation
2	Sense of security	
1	Manipulation	

Source: Based on Arnstein (1971)

The concept of communication is often confused. Wallace, Strong and Luckeneder make the point that "communicating information to the community is not sufficient in itself" (1990: 182). Rather, they see that the community must be involved in two-way communication. This requires that the organisation seeking the input must listen to and address issues of concern to the community, and provide information that can be understood by non-experts. Also, the community should be provided with access to sufficient information to participate in discussions, and should be able to participate in a constructive manner.

Whichever level(s) of participation is sought depends upon the many and varied factors involved in a particular proposal; for example, political will, community interest and budget. Associated with some levels is not just the power shared with the community, but also the responsibility for decision-making, and not everyone may want to share power, or accept responsibility for their actions.

4.4.4 Timing

The question of when public participation should be sought in relation to a particular proposal depends upon the objectives the proponent has for participation and the level of participation sought. These issues are evident in the following comments by Clark:

> Most "functional" and many "political" objectives will be enhanced by devising a participation program as early as possible in the planning phase. "Functional" objectives would be assisted by public input into the threshold decision whether an impact statement is required, and at subsequent stages. Confining public input to comment on a draft statement, which comes too late in the process to have a major effect on project design, does not enhance "functional" objectives. Too many participatory exercises also cease at the point of decision. They should continue through the implementation phase, especially if there is a move towards continuing environmental assessment. (1981: 320)

Underwood provides some practical comments based on experience with planning road proposals:

> In general, the extent of participation desirable will depend on the social and environmental factors involved, and on the possible degree of controversy of the proposals. Large-scale participatory studies should be commenced only after a careful consideration of all the relevant factors.
>
> There are pressures from some sections of the community for completely "open planning", ie for the whole planning process to be carried out in public. Again, though, this has the very serious disadvantage that the length of time of uncertainty, conflict and concern within the community, and the effects of planning blight, [Blight occurs if property prices are affected by studies investigating the possibility of some action in the vicinity] may be substantiated. The big dilemma is therefore to balance the demands for more open planning, and for greater community participation, against the need to minimise the period of uncertainty and of planning blight.
>
> Experience to date clearly indicates that, in order to have effective community participation, it is necessary to have specific alternatives as an initial basis of discussion, so that people can perceive these alternatives in terms of the specific impacts on their own interests. (1978: 3)

In the case of the EIA process, early participation so that key issues are identified before too much commitment is given to a particular proposal is important in the process. Aspects of timing include:

- allowing sufficient time to achieve the determined level(s) and objectives of participation;
- starting early enough in the planning/design process to enable the level(s) and objectives to be achieved;
- having ideas worked through sufficiently to give the public some information to respond to;
- having a sufficiently short participation phase to minimise uncertainty and blight; and

- fitting in with various requirements of governments (eg, meetings or elections).

Balancing these and other considerations is probably impossible, and perhaps should not be attempted. However, it is important to appreciate the effect timing can have on a public participation program. Brown (1997) provides an example of how early involvement can be achieved. Constructed for development project formulation, the United Nations Development Program guidelines that he discusses illustrate how a structured environmental overview can be used to guide the formulation of programs. Rather than adding EIA once a proposal has been formulated, he argues that the purpose of the environmental overview is to direct the development of the project. To achieve the successful application of an overview, four critical aspects were identified:

- The project/program must be in its earliest stage (draft formulation).
- There must be sequential completion of each question of the structured environmental review.
- The environmental review must be undertaken by a broad range of specialists and others to ensure that all pertinent environmental issues and perspectives are available to guide the questions of the overview.
- The process must include modification of the draft project/program if required, as an integral part of the environmental overview.

Brown comments that the process is very flexible, being suitable to development and other types of proposals. It can also be applied to proposals of a variety of scales (local to country level), and works with proposals related to policy ("soft" proposals) as well as specific projects and programs ("hard" proposals).

4.4.5 Who participates?

There are two aspects to who becomes involved in participation programs:

1. Those having an interest in the proposal and therefore expected to be involved in the participation program.
2. Those given the opportunity to participate.

In respect of the first point, the type of proposal and the stage to which the proposal has progressed will affect who is interested. Generally, policies, strategies and plans are abstract, whereas most people are more interested in how a proposal affects them (or their property). Although usually not directly affected, "public interest groups" (such as Australian Conservation Foundation and Wilderness Society) may also wish to be involved. It is reasonably common for "special interest" or "action" groups to form in response to a proposal and actively promote quite local issues, and to be heavily involved in any participation program. Burch (1976) suggests that people who express an interest and want to become involved are frequently middle class, have access to information and influence, and feel confident to participate. In addition to these individuals and groups, there are government agencies which would want to be informed and given an opportunity to participate.

O'Riordan (1976) provides a straightforward breakdown on the wider community that has relevance to how participation would be sought. These divisions are:

1. Body politic – the general public, sectional interests, pressure-group members.
2. Elites – politicians (national, regional, local), senior administrators, key technocrats, group leaders, opinion influentials.

Not all members of the general public will express an interest even though they may be affected by the proposal. It may be that they lack confidence to participate, are not aware of the proposal or that implementation is too far into the future for them to appreciate the effects (at which time a new group may be affected). Typically, the majority of participants are those directly adversely affected; those who benefit are frequently not heard from. Other sections of the community which may not express an interest include ethnic groups (if English is not commonly used) and disadvantaged groups (physically or socially), which may not find out about a proposal through the normal channels. In the same way, some of the "elites" (eg, politicians, senior administrators) may not express an interest through pressure of work or other priorities.

Sectional interests and pressure groups are frequently represented by community organisations. The variables most likely to produce organisations that are more effectively involved in participation exercises, and which influence public policy, are outlined by Gittle (1980). Overall, she concludes that the effectiveness of participation by these organisations is a function of both their internal structures and external forces, and she is generally not encouraging about their advocacy role (ie, the effect they can have on policy). While this work refers to North American experiences, it provides guidance for other Western countries and suggests that extra understanding may be needed if community organisations are to be involved to the point of making decisions about proposals.

In any of these cases, if the objectives of the participation program indicate the need to provide an opportunity to participate, the program will have to be designed to seek this participation; for example, by providing funds to assist groups. Providing the opportunity to participate also applies to those who have already expressed an interest; it is possible that the objectives and level of participation may need to be revised if an additional interest group is subsequently identified.

Again, it is difficult to identify the range of participants and allow for their particular requirements. However, recognition of their existence and sincere attempts to provide opportunities to participate are important.

4.5 Public Participation in EIA

4.5.1 General issues

With the passing of the US *National Environmental Policy Act* (NEPA) in 1970 (see 5.2), public participation was enshrined in legislation. Under NEPA, information on environmental effects of proposals is made publicly available and the public has the opportunity to comment on the draft EIS. Since then, the majority of EIA procedures have followed this lead and incorporate some public participation.

According to Clark, public participation in EIA procedures gives economy in decision-making:

> It is often argued that the impact statement process adds unnecessary costs and delay to the decision-making process and many proponents can adduce agonising testimony in support of that position. Yet once the policy to evaluate environmental consequences is adopted, to pursue public participation in the early stages of planning can reduce the possibility of subsequent delay through objection, lead to a qualitative enhancement of both the technical and non-technical information available to planners, and head-off the investigation of unacceptable alternatives. Indeed, in circumstances where public inputs reveal the need to accumulate further data, early public involvement can produce the possibility of an earlier go or no-go decision, instead of the possibility of attenuation of postponement. (1981: 310)

The IAIA also recognises that public participation is a core principle of EIA and suggests that EIA should include meaningful opportunities for public involvement, which should occur throughout the EIA process (ie, from scoping to the post-decision stage) and should be reflected in documentation and decision-making (IAIA & IEA (UK) (1999)).

EIA procedures around Australia (Chapter 6) indicate a variety of approaches. The minimal provision for public participation in procedures adopted by several States suggests a lack of commitment to the concept of public involvement in decision-making. In most jurisdictions in Australia, scoping does have an element of public participation although it may only be notification of draft scoping documents for comment

In 1982 Fowler commented on Victoria's EIA procedures and made the point that the *Environment Effects Act 1978* (EE Act) made very little provision for public participation. The only direct reference is in s 9(2) where "the Minister may at anytime invite and receive comments on the environmental effects of any works or proposed works from the public ...". The *Ministerial Guidelines for Assessment of Environmental Effects under the Environment Effects Act 1978* (June 2006), which are made under s 10 of the EE Act, make little reference to public participation, other than specifying periods for public comment (for scoping and the Environment Effects Statement (EES)) and requiring the proponent to prepare and implement a consultation plan (2006: 13).

Hence, "official" public participation occurs mainly around the midpoint of the EIA process. The public does not have the opportunity to contribute to the assessment (other than through comments on the EIS) or to comment on the assessment when it is published. However, a practice has developed where proponents have occasionally sought comment on the proposed contents of the EIS, providing the public with reasonably early information about proposals.

The Victorian EIA procedures are similar to others in that the levels of participation provided for are to inform the public about environmental effects of a proposal, and allow the opportunity to comment on the EIS produced, rather than involving the public in the development and design of the proposal at an early stage. Porter raises some related practical aspects of public participation in the EIA process:

> The purpose of the public review is often misunderstood, particularly by the proponent. It is not intended to be a type of mini-referendum, so the number of submissions opposing the project, or some aspect of it, is quite irrelevant. It is the rationality of the argument or the introduction of some unforeseen factor in the public submissions that makes them of value in the assessment. This is also often misunderstood by groups lobbying against a proposed development. In some cases, an active group will call on the public to make submissions opposing the project, sometimes supplying a coupon or a reproduced letter for the purpose. These actions may have some influence on those politicians or local authorities who ultimately may have to decide whether the development should proceed but they are of little value in the assessment process itself ...
>
> In the case of most major projects, the majority of public submissions will take the form of objections to all or part of the project. This is quite normal, although it has been known to distress some proponents who have prepared an EIS for the first time. The reason for a large proportion of the submissions containing objections is that citizens who support, or at least see no objections to, a project rarely take the trouble to respond to the EIS. Opposition groups sometimes seek to draw conclusions from statistics that may be in a ratio of five hundred in opposition to one in support.
>
> It is difficult to get a representative response to the publication of the EIS. A common attitude is that the individual is powerless to influence events, that public submissions will be ignored, if they are read at all, and that if the assessing agency has accepted the EIS for publication it has already agreed with the development. Overcoming these misapprehensions is an important task for environmental agencies. Some recognition of public submissions needs to be made in the assessment report, particularly where these have influenced the final recommendations. (1985: 171)

Public participation can also occur through public inquiries. For example, in Victoria inquiries can be set up under the *Environment Effects Act*, and they may be joint inquiries in association with other Acts (eg, *Planning and Environment Act 1987*), or may occur through the Environment and Natural Resources Committee (see 6.10.2). Unlike the procedures under NEPA, there is no recourse to the courts in Victoria. New South Wales is unusual in Australia in that some environmental matters can be taken to that State's Land

and Environment Court. Fowler (1982) notes that "overall, the (*Environment Effects*) *Act* provides a rather fragile basis for public participation in the assessment procedures" (1982: 42). However, the Victorian procedures are not unique in this respect.

Community input to EIA in Australia has been comprehensively documented and discussed by Martyn, Morris and Downing (1990). They have identified several issues that they see as crucial for public participation in the future:

- *Early input* – environmental values are in danger of being swamped by other considerations if they are not addressed as early as possible in the planning process (eg, see 4.4.4).
- *Scoping* – early consultation can reduce the time and cost to produce the EIS, while ensuring that it comprehensively covers the main issues in a compact rather than encyclopaedic manner.
- *Equitableness and efficiency* – it is important to ensure that participation is representative of public opinion, and that participation methods are implemented at the most effective time and are targeted.
- *Action on public input* – concerns and suggestions should be dealt with in the EIS, assessment report and the decision about the proposal.

Martyn, Morris and Downing (1990) identified a need for the public to have access to information about the level of assessment decided upon by the responsible authority (department or minister) so that citizens know what is happening with proposals, and have an opportunity to appeal the decision. Also, the position of public scoping should be formalised so that it actually occurs and the public, particularly at the local level, is kept informed; public meetings to do this are suggested. Further, Martyn, Morris and Downing propose that the assessment report should be made public as soon as it is completed. However, they single out monitoring as the area needing most attention so that the environmental management of developments can be reviewed by the public.

Some of these issues have been attended to in the consideration given to EIA by ANZECC. In particular, in the National Approach to EIA (ANZECC (1991a)) emphasises the need for the early involvement of the public in all EIAs. Subsequently the Guidelines for determining the need for EIA (ANZECC (1996)) identify public interest as one of the criteria for indicating the need for an EIA to be conducted, and the scale at which it is reported. However, there is little indication as to how the person using these guides should interpret the concepts of public involvement, and manifestations of public interest.

In summary, it is apparent that while provision of opportunities for public participation in EIA has been enshrined in legislation and administrative arrangements, often this is limited to the lower levels of participation; that is, largely information and comment opportunities.

Table 4.1 Summary of public participation in EIA processes in Australia (as at 1990) – access and adequacy

EIA process	Assessment type	Pre-EIS	During EIS	Post-EIS
Commonwealth	PER	✓✓	✓✓✓	✓✓
	EIS	✓✓	✓✓✓	✓✓
New South Wales	EIA	✓✓	✓✓✓	✓✓
Northern Territory	PER	x	✓	✓
	EIS	x	✓✓	✓
Queensland	EIS	x	✓✓	✓✓
South Australia	EIS	x	✓✓	✓✓
Tasmania	Level 1	x	✓	✓
	Level 2	✓✓✓	✓✓	✓✓
	Level 3	✓✓✓✓	✓✓✓✓	✓✓✓
Victoria	EES	✓✓✓	✓✓✓	✓✓✓
Western Australia	CER	✓✓	✓✓✓	✓✓✓
	PER	✓✓✓	✓✓✓	✓✓✓
	ERMP	✓✓✓✓	✓✓✓	✓✓✓

Notes

Pre-EIS – opportunity to raise all relevant environmental issues

During EIS – opportunity to influence study

Post-EIS – access to ensure implementation of major recommendations of assessment

Levels 1, 2 and 3 – documentation depending on type of proposal

Key

Nil = x

Minimal = ✓

Fair = ✓✓

Good = ✓✓✓

Very good = ✓✓✓✓

Source: Martyn, Morris and Downing (1990)

4.5.2 *Screening*

Early consultation is being increasingly emphasised by researchers such as Sammarco (1990). Consultation may occur even at the stage of screening proposals to see if they should be assessed by the EIA process (see 7.1).

The process of screening represents an early stage of EIA; it is the point at which a proposal is reviewed to see if it comes under the provisions of an EIA process, and, therefore, whether an EIA is to be performed. Tomlinson reviews five main methods available to assist screening:

1. *Project thresholds* – criteria have been established (these may be related to the proposal's size, location or type), and if the proposal fits the criteria, an EIA will be required.
2. *Sensitive area criteria* – as the term suggests, criteria are set regarding certain types of areas (these may relate to the "carrying capacity" of the area, or to the importance of some individual components of the area).
3. *Positive and negative lists* – positive lists usually prescribe the types of proposals which would be required to go through the EIA process, while a negative listing would be for proposals for which an EIA would not be expected (except under particular situations perhaps).
4. *Matrices* – a matrix which relates environmental factors with specific actions, such as the Leopold Matrix (see 8.3.2), may be used to provide an initial assessment of the environmental impact of the proposal, after which a better informed decision can be made about the need for an EIA.
5. *Initial environmental evaluation* – based on the use of some form of checklist of environmental factors, an assessment is made about the need for more information and analysis through a formal EIA.

However, these methods rarely involve the public. Rather, they are conducted by those developing the proposal, or those in the bureaucracy responsible for overseeing the EIA process. For example, the Australian and New Zealand Environment and Conservation Council (ANZECC (1996)) has outlined guidelines and criteria for determining the need for EIA. This approach is effectively a checklist which has been designed to be completed by professionals, and to remind them of the range of issues that could be important (these guidelines are discussed further in 7.1). However, the guidelines do require consideration of the degree of public interest in the proposal.

As a result, there has been dissatisfaction from environmental groups. An example of a response to this situation has been presented in Victoria. In this case a public register of decisions was established in 1988 to enable any member of the public to see which proposals had been required to undergo the EIA process, and which had been exempted. As with most States and Territories, the decision as to whether a proposal requires an EIS is placed on the relevant department's website for public information. In Victoria, the Minister for Planning's reasons for decisions are also published.

With the evolution of EIA there is also some expansion of the opportunities for public involvement in the earliest stages of the process. The most recent Commonwealth legislation dealing with EIA, the *Environment Protection and Biodiversity Conservation Act 1999* (EPBC Act) (see 6.3), requires the Minister for Environment to publicise (on the Internet) notification of proposals that have been referred for a decision about the need for an EIA. This notice also invites the public to comment on whether an EIA should be required for the proposal, however, comments have to be received within ten days. The Commonwealth also publish the decision as to whether EIA is required and the assessment approach to be followed on the website.

4.5.3 *Scoping*

Tomlinson proposes that scoping "is the term given to the process of developing and selecting alternatives to the proposed action and identifying the issues to be considered in an EIA" (1984: 186). He suggests that the aims of scoping are to:

- identify concerns and issues needing consideration;
- facilitate an efficient EIS preparation process;
- enable those responsible for EIA to properly brief the study team on the alternatives and the impacts to be considered, and the depth of the analysis;
- provide an opportunity for public involvement; and
- save time.

The details of scoping are discussed in 7.4.2, but most important is providing an opportunity for the community to have an input to considering alternatives and identifying the environmental factors to be examined as part of the study. By the community having an input at this stage, there is reasonable certainty that all the important points will be included in a serious way. If the community only has the chance to raise concerns after the study and EIS have been completed, it will be much more difficult to raise new concerns. Most EIA procedures in Australia allow for a public comment period on the draft scoping document and it is often the first time a person or community is made aware of a proposal. Once comments are received, the relevant government department finalises the scope and issues the scope to the proponent.

4.6 Techniques for Obtaining Public Participation

The methods by which participation has been sought are many and varied. For each participation program the technique(s) used to provide opportunities and seek participation is likely to vary. In particular, the technique(s) used will depend upon the objectives of the program. If an objective is to have the public involved in making decisions about a proposal, it is pointless using only information dissemination (eg, publishing a report). Considering environmental planning, Grima suggests that for the most effective participation:

> [Encouragement should be given to] the few who labour on behalf of the many. More specifically, (1) consultation should be an ongoing process; (2) the citizen participants should have the resources to commission a "peer review" of the experts' reports; (3) local problems should not be de-emphasised since they act as catalysts for public interests. (1983: 117)

To help the process, the expenses of "the few" could be reimbursed. This general technique could be applied to a wide variety of issues, either separately or in conjunction with other participation methods.

Specific techniques for obtaining participation are outlined in Table 4.2 *opposite*. These are the more usual techniques and each is provided with an assessment of when a particular method would be best used.

Holding a public inquiry is one technique that has generated considerable discussion, possibly because it can be time consuming and costly, and open up a range of issues. Woodward (1981) and Berger (1981) comment on enquiries in Australia and Canada, respectively, to help would-be users assess some advantages and disadvantages of the technique. Overall, while EIA procedures in Australia frequently make allowance for forms of public enquiries, they are infrequently used. Porter (1985) provides some practical comments about the operation of enquiries which indicate some reasons that few enquiries are held.

Environmental mediation is akin to a public inquiry. The relevant parties are brought together with the aim of exchanging views and information, and to reach some compromise. While an inquiry typically has a panel which hears "evidence" and makes a report, usually with recommendations, Mernitz (1980) comments that mediation involves a mediator who acts as a catalyst to encourage the parties to overcome their differences. Mediation sessions are likely to have a smaller number of participants than inquiries and to be closed to the public. Government departments occasionally initiate mediation (such as the Victorian Environment Protection Authority regarding levels of pollution control), but Talbot (1983) explains that the approach is becoming increasingly accepted in the United States where it has been used as an alternative to legal action.

Essentially, the benefits of successful mediation are the substantial reductions in the time and costs involved in agreeing to a course of action, and the avoidance of animosity. Gilpin (1995) refers to a number of situations in North America and Australia where mediation has been introduced.

While recourse to the courts occurs overseas, there are negligible opportunities for this form of participation in Australia. The *Environment Protection and Biodiversity Conservation Act 1999* (see 6.3) has slightly improved the opportunities for members of the public to challenge aspects of the Commonwealth's EIA process.

Indeed, over the past few years, there have been a number of highly publicised (and sometimes controversial) court actions taken by community groups or individuals against the Commonwealth Minister's decision(s).

4.6.1 Case Study: Port Phillip Bay channel deepening project

The Blue Wedge Coalition launched two legal challenges in the Federal Court of Australia concerning the Commonwealth Minister for Environment's approval of the controversial channel deepening project under the EPBC Act.

Table 4.2 Comparison of techniques for communication with the public

Communication characteristics			Public participation Communication techniques	Objectives					
Level of public contact achieved	Ability to handle specific interests	Level of two-way communication		Inform/ Educate	Identify problems / Values	Generate ideas/Solve problems	Feedback	Evaluate	Resolve conflict/ Consensus
2	1	1	Public hearings		x		x		
2	1	2	Public meetings	x	x		x		
1	2	3	Community group meetings	x	x	x	x	x	x
1	2	1	Operating field offices		x	x	x	x	
2	2	1	Information brochures & pamphlets	x					
1	3	3	Field trips and site visits	x	x				
3	1	2	Public displays	x		x	x		
2	1	2	Model demonstration projects	x			x	x	x
3	1	1	Material for mass media	x					
1	3	1	Letter requests for comments			x	x		
1	3	3	Advisory committees		x	x	x	x	
1	3	3	Task forces		x	x		x	
1	3	3	Ombudsman or representative		x	x	x	x	x
2	3	1	EIS review by public	x			x	x	

Source: Based on Munn   Key   1 = Low 2 = Medium   3 = High X = Capability

This project received Victorian assessment and approval in the later part of 2007.

The proponent, the Port of Melbourne Corporation, proposed to dredge 23 million cubic metres of sand, rock and toxic sediment from Port Phillip Bay's shipping channel to enable larger ships to dock in Melbourne. Initially, the proposal was to dredge an additional 2 metres from the channel. The Port of Melbourne Corporation and the Victorian government estimated the project to be worth $2 billion for the economy and was expected to create at least 2000 jobs. The project costs was estimated to be $969 million in 2008 (up from the original estimated cost of $200 million in 2001 when the project was first proposed (Pennicuik (2008: 1)).

In the first challenge, initiated before the Minister gave his approval, the Blue Wedge Coalition argued in the Federal Court that the Commonwealth Minister could not legally approve the project. This was because the channel deepening project as proposed at the end of the environmental assessment process had changed so dramatically since it was first referred to the Commonwealth in 2002 that is was no longer the same project. However, the judge dismissed the Blue Wedge Coalition's claim and ruled that, although there were differences in the project since first referred, the action of dredging and the general proposal – of increasing the depth of the channel for larger ships – remained the same.

A few weeks after the failed court injunction, and after the Minister had approved the project, the Blue Wedge Coalition launched a further court action claiming that the Commonwealth Minister for Environment had failed to consider key issues in his decision, including adverse social and health impacts of dredging the Port Phillip Bay, and that the Minister had failed to follow proper process by considering the environmental history of the proponent and informing other relevant Ministers before giving his approval. This court action delayed the commencement of dredging for a week and restricted the location of dredging activities; however, it did not stop the project with dredging commencing in February 2008 shortly after the judge dismissed the community group claims as not being evident from the facts or supported by the law.

The case study shows that under the Commonwealth legislation, third parties have standing, however, in this case the community actions to seek reconsideration of decisions about a project in the Federal Court were unsuccessful, and in the second case was required to pay the legal costs of the Commonwealth Minister, the State government and the proponent. Despite the community group's legal and financial losses, it was successful at ensuring a stringent environmental management and monitoring regime was established for the project both through public pressure and their participation in the EIA process, resulting in the strictest environmental controls that Port Phillip Bay has ever seen in over 100 years of dredging activity.

4.7 Evaluation of Public Participation

Clark (1981) suggests that one reason for defining clear objectives for a public participation program is to aid in the evaluation of the program, and its approach. Anyone involved in developing a participation program may wonder why evaluation is important. However, without some form of assessment there would be no guidance as to whether a particular technique worked well or to help develop future programs, nor would it be possible to produce the information contained in Table 4.2.

Participation programs consume resources (time and money), so it makes sense to try to use the most effective techniques, and measures of effectiveness only come from evaluation of previous programs. Evaluation can only take place after a program has been completed, so there is a tendency to forget about evaluation. Also a lack of guidance (or evaluation models) may have discouraged evaluation. To an extent, the study of 22 participation programs by Sewell and Phillips (1979) helps to provide this guidance (for an example of one approach, see Table 4.3 on *page 101*). However, these authors conclude that, in 1979 if not now, there was no universally applicable or generally accepted model. In the absence of such a model it is interesting that while specific criteria for assessment were seldom identified in the 22 programs examined, some general points emerged, namely:

- agency personnel tended to measure success in terms of the extent to which a program was accepted by those involved in it and the extent to which the image of the agency had been improved;
- citizen groups appraised programs by the success they had in preventing or modifying a proposed course of action, or the attainment of a broader recognition of the group, or public at large, in the decision-making process;
- independent observers looked for how well a program met its objectives, the degree of representation and the accuracy of information gathered; and
- cost, time, effort or resources were used as evaluation criteria by some agency personnel.

These observations suggest that the participation programs had fairly limited objectives.

To add to the difficulty of evaluation, Sewell and O'Riordan (1976) remind the reader that what is desirable and effective in one country may prove unsuccessful in another. To provide some help in evaluation participation, they propose a checklist of criteria for evaluating the responsiveness of the political and institutional culture to more broadly based participation. None of this provides the reader with much incentive to undertake an evaluation. However, even the simple checking of results against objectives is better than nothing. If nothing else, such a review will help identify lessons to be learnt and remembered.

A mid-1990s review of the Commonwealth's former EIA legislation, the *Environment Protection (Impact of Proposals) Act 1974* identified a number of barriers to effective public participation in the Commonwealth's former processes (Environment Protection Agency (1994a)). These were:

- participation not early enough, and often after substantial commitments had been made by proponents;
- concern for commercial confidentially;
- costs of keeping options open to move to more advanced stages of planning; and
- disparity of resources available to proponents and interest groups.

Unfortunately, these remain key barriers which could be observed in most EIA processes.

4.8 A Final Word

Public participation is an important component of EIA and needs to be considered early in the process to be effective. The material in this chapter demonstrates that participation provides benefits for EIA, and is an important aspect which needs to be given careful consideration; how important it is and how much consideration is required will depend upon the situation.

This chapter has avoided providing a prescription for public participation, however it outlines various approaches to public participation and the importance of public participation in the EIA process. As with EIA, participation programs need to be considered in the context of each proposal. The preceding sections have raised many points which should be taken into account in participation; at least, the designer of any participation program should be aware of them. How these aspects are handled rests with the designer. What is demonstrated in practice is that the earlier in the project planning a proponent begins public participation, the less likely critical issues will emerge later on in the EIA process.

Overall, public participation in the EIA process is an important mechanism for ensuring that the environment is protected and communities are considered in decision-making.

Table 4.3 Evaluation of a public participation

FRAMEWORK FOR EVALUATION						
Persuasion	**Education**	**Information feedback**	**Consultation**	**Joint planning**	**Delegated authority**	**Self-determination**
• Amount of resistance experienced towards program results • Number of participants reached in relation to the total population the program is attempting to affect • Number of people who have a positive attitude towards the program	• Amount of time spent participating in some educational activity, ie, reading lectures, interviews • The extent of behavioural change which occurs as a direct result of public education on environmental issues, eg, litter on roadsides • Increase in utilisation of community resources relative to the planning process, eg, libraries, community resources, Indigenous expertise	• Changes in the program caused by citizen feedback • Frequency of contact between the public and the authority as measured by meetings, letters, telephone calls • Participants' attitudes toward perceived influence over policy • Acceptance of final decisions (successful programs generate acceptance) • Extent to which ideas generated by the information feedback process contribute to management decisions	• Changes in the program caused by citizen feedback and consultation • Changes in the program caused by citizen feedback • Participants' attitudes toward perceived influence over policy • Output of the consultative process • The extent to which polarisation of public opinion was presented • Acceptance of final decision (successful programs generate acceptance)	• The degree to which pubic representatives actually influenced the group decision-making process • The degree to which the public perceives it has a voice via its representatives • Frequency of examples given which indicate changed self-perception of the committee representatives, eg, new leaders in the groups, previously inarticulate individuals speaking out	• Whether the program meets its objectives • Public reaction to products • The increase in the public's communication and problem-solving skills • Frequency of initiatives and acceptance of responsibility for problem resolution by the public	• Number of programs initiated by the public and number of people participating • Perceived control over environment by the public • Increased group formation and use of community resources for involvement purposes • Indications of constructive interaction between groups • Frequency of travel to outside areas by individuals to act as resource personnel in other communities on similar issues

Source: Based on Sewell & Phillips (1979)

Chapter 5

EIA WORLDWIDE

5.1 Introduction

> If we are to generate any change in our attitudes to the way we handle our environment and provide for future generations, we have to have an [environmental] ethic as a basis for action in the legal and economic field as we make the transition from a young to a mature society. For an environmental ethic to provide an acceptable basis for responsible conduct, it must lead to a legal framework. (O'Connor (1983: 87))

The key point is that to ensure responsible conduct (ie, conduct which is generally favoured by society) we usually resort to some form of legal process. The politicians of different countries have evolved a variety of arrangements to establish EIA, but these arrangements usually fall into the category of administrative directions or legislation. Overall, it seems that legislation is generally favoured; perhaps because it suggests a strong commitment by the government to environmental issues, and it may then be simpler to ensure that EIA is implemented.

This chapter provides an overview of EIA procedures employed by a variety of countries, to establish an appreciation of the extent to which EIA has been absorbed into decision-making structures, and to give insight into the variety of ways in which EIA can be undertaken. To complement this, the environmental assessment interests of the major international development funding bodies are also outlined. Readers should be aware that adjustments to EIA procedures are common so, while the following sections outline the situation, changes are likely to have been implemented since then. Generally, these changes will be related to issues of detail and administration.

The countries chosen for discussion are primarily those which have been instigators of EIA, or those with similar cultural and political backgrounds to Australia. Readers seeking additional information and detail on the international applications of EIA will need to delve into the references indicated and the other available documents, particularly the many journals which are now published dealing with EIA. Also, electronic sources of information are becoming increasingly important (see 5.9).

Broadly speaking, there are similarities in EIA procedures around the world. For example, Vlachos (1985) has identified commonalities running through assessment processes. In particular, there is a commitment to:

- account for the full range of impacts;
- delineate consequences in advance;
- control events (especially technology) and the environment for the common good; and
- interact with our destiny in such a way as to be part of it. (1985: 59)

However, the emphasis given to these aspects and the way the procedures are administrated show noticeable differences, as illustrated by Wood (1995).

5.2 United States of America National Environmental Policy Act 1970

During the 1960s, the United States federal agencies were subjected to increasing criticism about the environmental effects of their projects and the apparent lack of concern about those effects by the agencies. As Canter (1997) comments, agencies such as the Atomic Energy Commission and the Army Corps of Engineers were considered to pursue single-purpose economic goals with little or no regard for other considerations. In addition, the public was not involved in any decisions with regard to these federal government departments, and had little access to information about the projects. Overall, there was a lack of a broad approach to environmental problems and works; projects and policies tended to be looked at segmentally.

During the 1950s there had been increasing concern over the effects of clearing of forests, soil erosion and the threat of extinction of native animals. In 1959, the Resources and Conservation Bill was proposed to cope with some of these problems, but was dropped after opposition from the President. The activity of federal government agencies in pursuing their single goals during the 1960s brought increased political pressure to ensure that environmental considerations were taken into account when projects were designed and undertaken. To fulfil this requirement, in 1969 Senator HM Jackson introduced the *National Environmental Policy Act* (NEPA), which was passed in 1970. The United States Environment Protection Agency provides details of the Act.

Black (1981), Canter (1977) and Porter (1985) all discuss the background and characteristics of NEPA. Briefly, however, the features of NEPA include:

- operation at the federal level of procedures which do not overlap with State or local land-use controls;
- being related only to federal government projects; and
- being aimed at major actions significantly affecting the environment.

An important aspect of the Act is that it enables recourse to the courts if it is considered that the proponent of a project has not fulfilled the obligations of the Act, including the adequacy of an EIS. Also, the Act required that each federal government department review its procedures, regulations and policies to bring these in line with the Act; that is, it was required that these departments should consider environmental effects as part of their overall objectives. NEPA does not require the federal agencies to select the environmentally preferable alternative or prohibit adverse environmental effects. However, NEPA does require that decision-makers be informed of the environmental consequences of their decisions (Council on Environmental Quality (2007)).

In a study of the effectiveness of NEPA in its first 25 years, the Council on Environmental Quality (which is the central agency responsible for its implementation) found that overall, NEPA is a success and has made federal agencies take a better look at the potential environmental consequences of their actions, as well as bringing the public into the decision-making process. Although, as with EIA processes elsewhere, it was found that the environmental assessment process under NEPA frequently "takes too long and costs too much, agencies make decisions before hearing from the public, documents are too long and technical for many people to use, and training for agency officials at times is inadequate" (Council on Environmental Quality (1997: ix)).

More recently, in September 2013, during her testimony to the US Senate Committee on Environment and Public Works, Nancy Sutley, Chair of the Council on Environmental Quality stated:

> It wasn't that long ago that the public had little voice in the Federal decision making process regarding all aspects of the human environment, which includes the social and economic aspects of Federal decisions, for projects that affected them. Prior to NEPA and Federal agencies beginning to embrace environmental stewardship, dams displaced Tribes or water projects harmed important ecosystems. Our country is still addressing such impacts decades later for some of these projects. Before NEPA, there were limited opportunities for preventing the Federal Government from ignoring the environmental concerns of affected communities. (Council on Environmental Quality (2013))

The Act applies only to individual federal government proposals, but had a follow-on to State legislation and activities, so after a few years the majority of States had some form of EIA legislation or procedures. In California, this desire to take account of environmental effects and embody this in fairly specific legislation, which also enabled recourse to the courts, led to some problems; for instance, a court ruling that "project" included all private activities permitted by the government resulted in the production of something like 6000 EISs per year. While this was obviously good for the consideration of the environment, it did require considerable resources to be committed to the EIA process.

A substantial amount of information is available about the operation of NEPA, and to assist the development of EIAs. As Jessee (1998) notes, since the early 1990s, these data have been made available through the Internet. In particular, the Council on Environmental Quality website (<www.whitehouse.

gov/administration/eop/ceq>) and the Department of Energy website (<www.energy.gov/nepa/office-nepa-policy-act-compliance>).

During the first five-and-a-half years of the operation of NEPA, some 3 per cent of all proposals instigated by federal government departments were subject to preparation of an EIS. At the beginning of NEPA, there were some 5800 EISs between 1970-1972, however over the years this has been reduced to an average of around 500 EISs per year. Jain, Urban and Stacey (1981) provide an indication of the trend in the number of EISs filed over the early years of NEPA's operation. Their analysis indicates that after a backlog of proposals were taken under the EIA process the numbers of EIA settled down to a more consistent annual workload.

Since NEPA's introduction, many States have introduced EIA provisions and States that have not enacted EIA law have procedures which are very similar to NEPA and require EIAs for activity taking place at the State level. Other States do not have specific EIA regulations, but have requirements that are in addition to NEPA.

5.3 Canada

As in the United States, public concern with regard to activities of federal government agencies, and recognition of the need for environmental protection, resulted in environment-related legislation in the 1960s and 1970s. The passing of NEPA in the United States provided a prod for Canada and the Act was examined by a task force within the Canadian Federal Department of Environment.

Eventually in 1973 the Federal Cabinet adopted the Environmental Assessment Review Process (EARP). Sewell (1981) discusses how, in this situation, EIA was undertaken via a Cabinet Directive with the objectives of:

- having management of the environmental effects and their review being the responsibility of the proponent of the proposal (to avoid delays);
- providing a system of review, advice and expertise; and
- informing and involving the public in decision-making.

While the Cabinet issued this Directive, the implication of not embodying EIA processes in legislation was that the government perhaps did not have a full commitment to the process. This was also suggested by the activities of the federal government which indicated EIA to be more of a provincial matter. Consequently, it appears that overall EIA was undertaken through the exercise of political and/or administrative discretion. Sewell has described the essential features of the Canadian federal process at that stage of its evolution. Importantly, under the Canadian Directive, there was no appeal to either the courts or the administrative procedures.

In 1992, Canada joined the vast majority of countries which operate EIA through legislation. The *Canadian Environmental Assessment Act* has brought with it a variety of administrative changes for EIA to the federal level. However, the majority of features of the previous system are maintained, while the Act provides a more secure legal base for EIA. Couch (1991) has identified several important elements of the Act, including:

- The establishment of the Canadian Environmental Assessment Agency which oversees EIA.
- Proponents can undertake "class assessments" for frequently recurring projects.
- Assessments have to consider need, alternatives, cumulative effects and resource sustainability.
- The Minister is able to initiate the EIA process.

The Act applies to projects where the Government of Canada has decision-making authority – whether as a proponent, land manager, source of funding or regulator.

In the implementation of the Act some interesting aspects have been developed. One aspect is that assessments are overseen by a review panel, appointed by the Minister of the Environment, when the environmental effects of a proposed project are uncertain or likely to be significant or when warranted by public concerns. Review panels offer individuals and groups, with different points of view, a chance to present information and express concerns. Another aspect is where the Minister of the Environment appoints an impartial mediator to assess a project and help interested parties resolve issues. This approach may be used when interested parties agree, are few in number, and consensus appears possible. Alternatively, projects with known effects that can be easily mitigated may be assessed through a "class screening" (Canadian Environmental Assessment Agency (2004a)).

The *Canadian Environmental Assessment Act 2012* is an updated Act that provides a more modern approach to EIA that responds to Canada's economic and environmental context. The revised Act "implements central elements of the government's plan for responsible resource development to modernise the regulatory system and allow for natural resources to be developed in a responsible and timely way", CEAA (2013) (<http://www.ceaa-acee.gc.ca/default.asp?lang=en&n=16254939-1>). EIAs under the revised Act focus on potential adverse environmental effects that are within Canadian federal jurisdiction, including:

- fish and fish habitat;
- aquatic species;
- migratory birds;
- federal lands;
- effects that cross provincial or international boundaries; and
- effects that impact on Aboriginal peoples.

In parallel with the federal process, the Canadian Provinces have introduced their own EIA procedures, most of which involve special EIA legislation. Ontario has specific EIA legislation, while the others have embodied their requirements in pollution and resource legislation. The EIA processes adopted within Canada, at federal and provincial levels, have been outlined by Jeffery (1989), while the Canadian Environmental Assessment Agency (1997, 2004a) provides a source of current information. Cooperative agreements are in place to cover situations where both federal and provincial processes are involved.

In Canada, there is also an opportunity for the public to apply for participant funding to aid individuals and not-for-profit organisations interested in participating in the EIA process. An example is the provision of funds to support public participation in the Joint Review Panel Process for the proposed Mackenzie Gas Project. For more information on the Canadian EIA process, go to <http://www.ceaa-acee.gc.ca/default.asp?lang=en&n=16254939-1>.

5.4 United Kingdom

The United Kingdom has a long history of planning legislation covering private development. Government projects were not covered by these Acts, but the government departments generally cooperated with the planning department to ensure that government proposals were in line with the legislation. The scope of these planning Acts, according to Clark, Bissett and Watham (1981), allowed authorities to require documents equivalent to an EIS, and so theoretically EIA was built into existing legislation. This meant that before joining the European Union (EU), the United Kingdom did not have separate EIA procedures and/or legislation. Public information and input to the EIA were limited to public hearings and special committees. Public hearings were allowed for under the planning legislation, and special committees were set up to deal with the specific proposals; for example, the Roskill Committee, which investigated the alternatives for a third London airport. To assist planners assess the environmental effects of proposals, a manual was produced. This manual helped to formalise the investigation of environmental effects and perhaps ensure that environmental effects were actually considered in a proposal's investigation, but the manual did not change the decision-making system or ensure that environmental effects were fully considered in any decision-making process.

Investigations into the need for separate EIA processes were undertaken in the mid-1970s, but there was reluctance to produce such procedures. Catlow and Thirwall (1976) see the reasons for this reluctance as being related to the belief that the procedures would result in additional project costs, the possibility of delays occurring and the contention that such EIA procedures would first require a national and regional priority to be assumed for natural resources (so that EIA would not then be required to set these priorities).

Since joining the EU, the United Kingdom has been obliged to conform to the EU's Directives. In particular, a Directive passed in 1985 requires the United Kingdom to adopt the European Union's broad principles for EIA, and to develop mechanisms for applying EIA in the United Kingdom. (The EU Directive is briefly outlined in 5.6.)

The *Town and Country Planning (EIA) Regulations 2011* (Statutory Instrument 2011 No 1824), known as the "EIA Regulations", form part of the development control system in the UK and relate to certain types of development. Schedule 1 to the EIA Regulations list different types of development that would always require EIA irrespective of their location. The UK EIA Regulations relate to the European Union Directive (Directive 85/337/EEC).

5.5 New Zealand

In 1973 the Commission for the Environment devised procedures for EIA to relate to major government works having environmental significance. These procedures were known as the Environmental Protection and Enhancement Procedures. These procedures were embodied in legislation passed in 1979, being the *National Development Act*. Piddington (1981) notes that this Act covered private and public projects. The Act is interesting in that it provided the Commissioner for the Environment with a vote to act independently of the Minister in auditing the environmental impact report (EIR, the New Zealand equivalent to the EIS). Also, this EIA process was not a separate or an independent process, but was part of the land-planning process. A further interesting feature was that the EIR could not consider alternative sites. The Planning Tribunal (set up to consider the proposal) was required to consider only one site, and the "do nothing" option was not evaluated.

Current legislation passed to guide EIA places a somewhat different emphasis on the roles of planning and EIA. New Zealand passed its *Resource Management Act 1991* (RM Act) to achieve sustainable management of natural and physical resources (see New Zealand Ministry for the Environment (2001)). The concept of sustainable management in the RM Act provides for a balance between environmental protection and development and the focus is on ecological rather than social or economic considerations (Ministry for Environment (2003)). Wells (1991) points out that the Act supersedes many previous Acts dealing with resources, and takes over some aspects of planning legislation. It does not cover EIA explicitly, but includes EIA and social impact assessment (SIA) (see 3.6) in a generic way. The Act provides for both project and policy based environmental assessment (Dixon (2002)). Impact assessment is integrated through s 32 of the RM Act into the plan-making and consent processes by requiring decision-makers to assess impacts of policies, and requires proponents to assess impacts of their proposals. Section 88 of the Act sets out the EIA process for a resource consent application. The other feature of the RM Act (also under s 32) is the specific requirement to consider alternatives in the policy making process (Dixon (2002)). A schedule outlines matters to be included in an assessment and the scope of the assessment, but detail depends on the scale and significance of potential effects.

Explicit coverage of the assessment of plans, policies and rules, along with project assessment, makes the Act noteworthy and provides clear opportunities for Strategic Environment Assessment (SEA) (see 3.10). Through the Act, EIA is not a separate exercise to be activated by political discretion; hence, as Wells comments, "EIA is less vulnerable than it has been in the past" (1991: 20).

Fookes and Schijf consider that changes through the Act have been so fundamental that policies, plans and decisions about resource use are described as "effects-based", meaning that:

- regional policy statements, and regional district plans, are expected to specify the effects (beneficial and adverse) which are being sought through objectives, policies and rules; and
- decisions on whether to grant a resource consent or not include an assessment of the actual or potential effects of the proposed activity. (1997: 58)

This broad scale approach means that the effects of both public and private development are identified.

As suggested above, although the term strategic environmental assessment (SEA) does not appear in the Act, provisions exist for SEA through the assessment of broad government planning proposals such as national and regional policy statements, regional and district plans, national environmental standard regulations and coastal policy statements. At the project level of proposals, EIA is embedded in the local authority procedures for making decisions about applications for land use and subdivision, and for permits to make discharges to water bodies. Where a resource consent is required (eg, coastal permit, discharge permit or land-use consent), then an "assessment of environmental effects" needs to be prepared. Those proposals that might have a "more than minor" effect on the environment include some form of public notice, however in reality the public is not notified about most resource consent applications (Ministry for the Environment (NZ) (2001)). Where there are appeals over decisions made by these authorities, the Environment Court (formerly the Planning Tribunal) sits in judgment. Fookes and Schijf (1997) also note that another aspect of the EIA process in New Zealand is the Parliamentary Commissioner for the Environment, who provides independent advice to parliament on the agencies involved in environmental planning and management.

A number of reviews of the Act has occurred since 1991, the latest being 2013. The 2013 Amendments help create a resource management system that delivers communities' planning needs, enables growth and provides strong environmental outcomes in a timely and cost effective way.

5.6 Europe

5.6.1 Western Europe

The context for the introduction of EIA was typified by the patchiness of control in the early years of the EU. There was no overall policy for planning at the community level; rather, planning was all undertaken by the member States. Technical rationality, or the technical complexity and sophistication of planning, within those member States was another factor, as it led to a variety of approaches to planning. Also, the nation states of Europe have a long history, so the national context within Europe, essentially the diversity of approaches within the member States, mitigated against the speedy adoption of uniform procedures.

Wandesforde-Smith (1980) suggests that in comparison with the United States approach, via NEPA, the EU placed greater emphasis on internal links between government departments and environmental groups to ensure that environmental considerations were taken into account. This was thought to alleviate the need for formal EIA procedures. In addition, there was a general view within the European countries that if EIA procedures similar to NEPA were introduced, this could result in recourse to the courts, delays in projects and increases in overall costs. Apparently the European member countries considered that all these aspects were to be avoided.

Proposals to formalise existing arrangements and relationships to incorporate EIA at the policy and program level decisions within the EU had been discussed for some time. In the late 1970s the European Community Commission considered a proposal of the Environment and Consumer Protection Section Service to use EIA to control pollution and environmental degradation by reshaping the pattern of development.

Wandesforde-Smith describes how the proposal was to introduce EIA into particular a stage, that is, to apply to certain major proposals but not to apply to policies. EIA was also suggested to supplement, but not replace, existing procedures, and it was considered important to leave decisions about implementation of EIA to the member States.

Jones (1984), Monbaillin (1984), and Von Moltke (1984) have all reviewed the status of EIA in European countries before 1985, and each commented on the hesitancy of governments to adopt EIA. For example, the existence of regulations in most European countries for the consideration of land-use and environmental issues, particularly zoning arrangements, lessened the need for specific EIA legislation. Von Moltke also suggested that constitutional factors made European countries more cautious about specific legislation.

Lee and Wood (1991) report that around the late 1980s France had EIA legislation. Also, since the EEC Directive (see below) regions of Belgium, Denmark, Germany, Greece, Ireland, Italy, the Netherlands, Portugal, Spain and the United Kingdom have EIA covered in their legislation.

The situation for other European countries is variable. Switzerland, Maystre (1991) notes, has EIA legislation which does not make allowance for public participation, nor does its assessment of environmental issues cover social or economic factors. Norway and Sweden were early adopters of EIA legislation, Lind (1991) comments, but Finland and Iceland have been slower in providing support through legislation; rather, there is some coverage from informal procedures. In Norway, regulations in the *Planning and Building Act* are complemented by those laid down by Royal Decree pursuant to the *Public Administration Act* to cover the proposals requiring EIA and the process (Ministry of the Environment (2004)). Poland has been more traditional in linking its EIA to land-use planning via several Acts and executive orders (see Rzeszot (1991)).

For countries formerly in the Soviet Bloc (now the Russian Federation) it remains to be seen how the changed political structures take account of environmental concerns. Some have joined the EU (see below). The status of others is discussed in the following section on Eastern Europe.

In the mid-1980s, the earlier ambivalence to EIA changed. Following a decade of drafts and discussions, Wathern (1988b) explains, the EEC resolved to formalise EIA within the member countries and the EIA Directive on Environmental Impact Assessment of the effects of projects on the environment (EU Directive 85/337) was passed in 1985 (see Europa (2001) for the Directive). This Directive covered the specification of projects requiring EIA, the scope of assessment, consultation requirements during the assessment and the role of the European Community Commission. It also required member States to adopt mechanisms for applying EIA to both public and private activities, but the details of implementation were the responsibility of each State. Subsequently Directive 97/11/EC amended the original EIA Directive 85/337/EEC, and included the key provisions:

- Broad definition of the effects to be considered.
- Mandatory application for specified projects.
- Requirement to submit an EIA report.
- Types of information to be provided by developer.
- Outline of alternatives studied and reasons.
- Submission to be made available for public comment.
- Results of consultations and information must be taken into consideration in decision-making.
- Content and reasons for decisions made public and detailed arrangements for public consultation to be drawn up by member states.

The EIA Directive was again amended in 2003 following the signature of the EU to the Aarhus Convention on public participation in environmental matters. The Aarhus Convention lays out a number of requirements for public participation in the EIA process, primarily in articles 6, 7 and 8 (Bektashi and Cherp (2002)). There is also the EU Directive for strategic environmental assessment – SEA Directive (2001).

The EIA Directive was again amended in 2009 to amend Annexes I and II of the EIA Directive by adding projects related to the transport, capture and storage of carbon dioxide (CO_2). The initial Directive of 1985 and its three amendments have been codified by Directive 2011/92/EU in December 2011.

With the benefit of the experiences of EIA in other countries, the EU has been able to introduce measures which avoid some of the implementation problems of other countries. For example, Glasson, Thérivel and Chadwick (1994) note that, whereas NEPA requires an EIA for "major federal actions", the Directive lists types of projects for which an EIA is automatically required. Two classes of projects are listed: those that must undergo an EIA in any country of the EU (mandatory projects); and those which are listed at the discretion of the member State (discretionary projects). Projects on the mandatory list (Annex 1) include oil refineries, power stations, motorways and ports, while the types of projects appearing on the discretionary list (Annex 2) cover agriculture, extractive industries, metal processing and the like. For the Annex 2 proposals the screening process involves two layers of consideration. First they have received the general scrutiny at the level of the European Union, by being listed. Secondly, the proposals go through the application of criteria or thresholds of the relevant member state, to determine if, in that state, the proposal should be subject to EIA (Clement (2000); Wood (1995)).

5.6.2 Eastern Europe

With the collapse of the USSR and the Soviet Block in the early 1990s there was a substantial realignment of EIA procedures in the region. Cherp (2001a) reports that during the 1980s socialist countries had been developing EIA procedures, but whereas the procedures of most other countries were formalised and transparent, those in the old soviet countries were mainly internal government processes that were not transparent and which emphasised compliance with processes rather than seeking for solutions. These were known as the State environmental review (SER) system of the USSR. According to Bektashi and Cherp (2002: 2), the SER is a "mechanism by which state authorities verify conformity of virtually all proposed economic activities to environmental requirements, taking root in socialist environmental appraisals used in centrally planned economies". While there are many levels of difference across the procedures that are evolving in the countries of Eastern Europe, Cherp (2001a) has summarised the key patterns as at the beginning of the 21st century (see Table 5.1 *opposite*).

The Newly Independent States of the former USSR, Central and East European countries, and the Baltic States were categorised as "countries in transition" by Cherp (2001b). He notes that most of these countries started reforming their EIA systems in the late 1980s or in the early 1990s, with the result that more than one hundred new legal acts on Environmental Assessment have been adopted in all of the 27 countries (an attachment to his paper provides the details if you are interested, also the website for Environmental Assessment in Countries in Transition provides up-to-date information). As an example, Russia has its Law on Environmental Protection (1991, 1993), supported by several sets of regulations and rules to direct specific aspects of assessment, including Environmental Expert Review. Cherp (2004: 1) makes an important statement in a further paper where he suggests that, "understanding the societal context of countries in transition, or 'emerging democracies', is critical in appraising the progress of environmental assessment systems in these countries". Some of the issues these countries face include the lack of transparency and accountability of decision-makers, as well as the social and economic problems that exist within these countries that are often a higher priority for the government (and its people) than EIA systems.

Table 5.1 The pattern of EIA procedures in Eastern Europe

Country Grouping	Socio-economic Context	Evolution of EIA
Central and Eastern Europe – European Accession countries: **Bulgaria**, Croatia, **Czech Republic**, **Hungary**, **Poland**, **Romania**, **Slovakia**, **Slovenia**	These countries had been seeking admission to EU.* They have strong ties to Western countries and have market economies and a democratic structure.	EIA procedures have been reformed to fit with the EEC's directive and Western models. Generally legal provisions are followed.
The Baltics: **Estonia**, **Latvia**, **Lithuania**	These countries also had been seeking admission to the EU, but economic and political dislocation is more severe than the CEE countries.	Reform is similar to the above, but current practice follows provisions similar to those of the USSR.
Newly Independent States: Belarus, Moldova, Russia, Turkmenistan, Ukraine	Only weak economic and political ties to Western countries are evident and market economies and consolidated democracies have not been established.	EIA practices have evolved from the USSR model, but are evolving and being guided by the procedures of more economically advanced countries and regions.
Albania, Armenia, Azerbaijan, Bosnia and Herzegovina, Georgia, Kyrgyzstan, FYR Macedonia, Tajikistan, Uzbekistan, Yugoslavia	Generally there is social and economic dislocation with political instability and declining economies	EIA procedures are behind those of other Eastern European countries. As their political and economic situations improve more substantial EIA processes can be expected.

**The countries in bold are now members of the EU.*

Source: Cherp (2001a)

5.7 Other Countries

5.7.1 Developing countries

In terms of EIA, developing countries have particular problems. Most importantly there are:

- the pressures which result from overpopulation and poverty on land in general, and particularly those pressures in subsistence areas;
- the typically limited database for existing environmental conditions and for environmental standards. In recognition of this, and that in these situations assessments are likely to be a purely technocratic process, Yap (1989) states that community needs have to be included in the data, and that discussion of potentially negative impacts must be ensured; and
- the use of foreign technology by the commercial sector (eg, the resource use which is associated with high-technology processes, and the pollution associated with uncontrolled manufacturing processes).

Able and Stoking (1981) suggest that the First Development Decade launched by the United Nations in 1961 to stimulate growth and remedy socio-economic and environmental problems has failed. It appears that the Second Development Decade has suffered a similar fate. The implications then are that greater pressure is likely to be applied to the local environments of developing countries to attempt to maintain the increased standards of living which have been encouraged by development and, consequently, there is an even greater need to predict and assess environmental effects. As Wood states, "in an era of rapid industrialisation and population growth in some of these regions [developing countries], and increasing recognition of the regional and global environmental impacts of certain development projects, the need to apply EIA in these countries effectively is apparent" (2003: 3).

The education of all involved with EIA in developing countries is also an important factor. Following a workshop considering how EIA could be made to work more effectively, Brown and McDonald (1989) reported that participants specifically identified the need to train personnel (both technical staff and EIA reviewers). There is also the need to educate decision-makers in the objectives and processes of EIA, develop relationships between environmental and sectoral development agencies, improve public participation and encourage the active participation of non-government organisations.

In relation to EIA requirements for developing countries, Able and Stoking note that the Western nations frequently have required EIA procedures to be undertaken with development projects. For example, projects funded by the United States Agency for International Development have required a mandatory assessment of environmental impact.

In the very early days of EIA, Gour-Tanguay (1977) noted that one-third of developing countries (surveyed in 1975, including India and Israel) reported having some formalised land-use planning. Only one country noted the need to implement a formalised system of EIA (that was Israel). Since then a number of countries have introduced formalised processes, either as administrative arrangements or as legislation (see below).

It appears that countries which have adopted some form of EIA procedures have tended to use the NEPA and other Western procedures as models for their own internal EIA processes. EIA is said to be universal, but situations change from country to country, and from culture to culture. The conditions under which projects are undertaken in industrialised and developing countries differ sharply, and some modifications to EIA are needed to adapt the process to developing countries. Curi (1983), and Pearson and Pryor (1978) discuss some of the special conditions that may require adaptation of EIA for use in developing countries:

- Market prices in developing countries frequently fail to reflect economic scarcity.
- Development projects in developing countries are more apt to be large relative to the total economy.
- Developing countries often lack the national pollution standards and policies within which projects can be evaluated, and administration and enforcement are frequently less effective.
- In private-sector-dominated industrial countries, negotiation between polluters and damaged parties takes place over environmental externalities, but this approach would have little merit in developing countries where the government is often the principal sponsor of development projects with spill-over environmental problems.
- A highly-developed network of environmental groups that can be relied upon to monitor development projects does not exist in most developing countries.

The points made above in the late 1970s and early 1980s are still relevant to developing countries today.

As discussed above, EIA in developing countries can be very different from EIA in developed countries. Wood (2003) notes that this difference can be contributed to the fact that the first EIAs in developing countries were usually demanded by development assistance agencies on a project-by-project basis, rather than for the purpose of determining environmental impacts and thereby protecting the environment.

Possibly as a consequence, progress in adopting EIA is slow in developing countries, with Ebisemiju (1993) recording only nine of some 120 countries having established frameworks for its implementation by the mid-1990s. He attributed this poor performance to flaws in their legislative, administrative, institutional and procedural arrangements. Two of the suggestions he made for improving the adoption of EIA were to use a legislative rather than an administrative option to establish authority for EIA, and to establish an independent environmental agency with considerable political influence over sectoral agencies.

As an overview, in the context of developing countries, Pearson and Pryor (1978) conclude that EIA "is a first, but not final, step", a comment that is just as applicable to EIA in industrialised countries. Even within a particular country, EIA procedures must be flexible enough to ensure adaptability to a variety of proposals and social and cultural conditions.

5.7.2 Asian-Pacific countries

Within the Asian-Pacific area EIA has a reasonably wide acceptance, and most of the larger countries have some procedures in place. Htun (1988) and Chuen (1989) note that EIA legislation exists in Indonesia, Republic of Korea, Malaysia, Pakistan, Papua New Guinea, the Philippines, Sri Lanka, Bangladesh and Thailand. In particular Indonesia was a reasonably early adopter of the EIA process, in 1982 through Act No 4/1982, which refers to Basic Provisions for Environmental Management. This was supported by the 1986 Government Regulation No 29/1986 regarding Environmental Impact Analysis, and guidelines a year later (Purnama (2003)). Likewise, through an amendment to the *Malaysian Environmental Quality Act*, that is Environmental Impact Assessment, the late 1990s saw the commencement of the formal Malaysian procedures, in which types of proposals that require EIA are specified (Department of the Environment (2003)).

Although without going as far as legislation, EIA provisions exist in statutory requirements in India, Nepal and Singapore. For example, in India, the "Notification on EIA of Developmental Projects" issued in 1994 under the *Environment (Protection) Act 1986* makes EIA mandatory for 29 categories of developmental projects (one more item was added to the list in 2000) and in Nepal, EIA is formalised through a number of laws including the *Environmental Protection Act 1996*, the *Environmental Protection Regulation 1997* and the National Environmental Impact Assessment Guidelines 1993. The Nepal example is a common occurrence of where guidelines or a policy was introduced to guide EIA prior to any legislation, both for developed and developing countries. Singapore relies heavily on its planning and control system to examine potential environmental impacts arising from development projects (Briffett et al (2003)).

After numerous activities relating to EIA, Briffett et al report that formalisation of procedures was seen in Vietnam in 1994, through enactment of the Law on Environmental Protection. Around the same time in Taiwan, the EIA Law was passed by the Legislative Yuan (the Parliament) and came into effect on 30 December 30 1994, thus establishing the legal basis for EIA implementation, compliance monitoring, and enforcement (Leu et al (1996)). After many years of operating with statutory regulations, Japan has promulgated the Environmental Impact Assessment law in June 1997, to govern its procedures (Anon (1997)). Around the same time Hong

Kong introduced its Environmental Impact Assessment Ordinance – the Hong Kong Environmental Protection Department website provides details of how the Ordinance operates, and examples of assessments. More recently, China, which has had a long association with EIA processes as Wang et al (2003) explain, enacted its 2003 Law of the People's Republic of China on Environmental Impact Assessment.

5.7.3 Latin American countries

Countries within the Latin American region show a range of commitment to EIA. Legislation which embodies EIA has been in operation in Colombia since 1974; Venezuela, 1976; Mexico, 1980 (although this early attempt was considered weak and was strengthened by the 1996 General Law of Ecological Equilibrium and Environmental Protection); and Brazil, 1981 (see Moreira (1988)). Also, while Uruguay and Peru have no formal procedures, EIAs have been undertaken, usually in response to the requirements of aid agencies.

5.7.4 African and Mediterranean countries

South Africa introduced EIA Regulations in 1998 to implement sections of the *Environment Conservation Act*. These regulations are overseen by the Department of Environmental Affairs and Tourism and have the objectives to:

- ensure that the environmental effects of activities are taken into consideration before decisions in this regard are taken;
- promote sustainable development, thereby achieving and maintaining an environment which is not harmful to people's health or well-being;
- ensure that identified activities which are undertaken do not have a substantial detrimental effect on the environment, and prohibit those activities that will;
- ensure public involvement in the undertaking of identified activities; and
- regulate the process and reports required to enable the Minister or his or her designated competent authority to make informed decisions on activities.

Otherwise, information about the status of EIA in central African countries indicates that formal processes have generally not been adopted. However, some assessments have been conducted as a consequence of the requirements of international aid and development organisations.

For North African countries and those in the region of the Mediterranean, CITET (Centre International des Technologies de l'environment de Tunis (2004)) has outlined the status of EIA laws (see Table 5.2 *over page*).

In addition, EISs have been used in Israel from the mid-1970s. To formalise this, regulations governing the requirements of EISs were promulgated under the Planning and Building Law in 1982 (Israel Ministry of Foreign Affairs (1994)).

Table 5.2 The status of EIA laws in North African and Mediterranean countries

Country	Status of EIA laws
Algeria	Environmental Law 83-03 of 1983, supported by specific Decrees in 1987 and 1990
Cyprus	Council of Ministers approved a comprehensive system of Environmental Impact Assessment in 1991, and a new law on EIA (No 57(I)/2001) was enacted in 2001
Egypt	Law No 4 on Environmental Protection 1994, with Executive Regulations 1995 (Prime Minister's Decree 338)
Jordan	Law on Environmental Protection No 12 (1995), with Law on Environmental Protection No 12 (1995)
Lebanon	Law on Protection of the Environment No 444, dated 2002
Morocco	Law on Protection of the Environment passed in 2003, with EIA Operational Directive
Palestinian Authority	Environmental Law 2000, with Palestinian Environmental Assessment Policy (intended to have same status as by-law) 2000
Syria	Environmental Protection Law No 50 of 2002, with specific process details covered in a Draft EIA Decree
Tunisia	Law No 91 of 1988 provides the context for the details of EIA provided in Decree No 362 of 1991, and Law 2001-14 of 2001
Turkey	Environment Act (No 2872) of 1983 with detailed processed covered in EIA Regulation 23/6/97

Source: CITET (2004)

5.8 International Funding Bodies

International organisations can exert considerable pressure on the environmental assessment practices of countries. In particular, aid and development funding has often been contingent on there being an EIA process in place in the country receiving the assistance. Yet, throughout the 1970s and 1980s major development activity funded by the international funding bodies, especially the World Bank, was often criticised for focusing on economic development, while showing little consideration for environmental or social impacts. Large development projects, such as dams, have created monumental environmental and social disruption, but these impacts were poorly assessed (World Commission on Dams (2000)). The case study below on the Bujagali Falls hydropower project highlights some of the issues which remain current.

During the 1990s, the international funding bodies put considerable effort into improving their procedures for identifying environmental impacts and mitigation measures. For the World Bank, a suite of environmental assessment processes is used to examine the environmental risks and benefits associated with Bank lending operations. Lending operations are broadly defined to include credits, sector loans, rehabilitation loans through financial intermediaries and investment components of hybrid loans. Broadly, the World Bank notes that environment assessment should be thought of as a process rather than a specific product. Environment Assessment is used by the World Bank to examine the potential environmental risks and benefits associated with Bank lending operations (Worldbank (2009)).

The Bank's Operating Procedure OP 4.01 Environment Assessment (World Bank (1999)) notes that the Bank:

- advises the borrower on the Bank's EA requirements;
- reviews the findings and recommendations of the EA to determine whether they provide an adequate basis for processing the project for Bank financing; and
- reviews the EA to ensure its consistency with the procedures when the borrower has completed or partially completed EA work, and before the Bank's involvement in a project.

In addition, the EA process has been developed to be undertaken alongside other activities related to identifying potential projects, their design, and their implementation. Further, close association with other types of project analysis ensures that environmental considerations are given appropriate weight in project selection, siting and design.

The Bank screens proposed projects to determine the appropriate extent and type of EA. These proposals are categorised into one of four groups, depending on the type, location, sensitivity, and scale of the project and the nature and magnitude of its potential environmental impacts:

- Category A – the proposal is likely to have significant adverse environmental impacts that are sensitive, diverse, or unprecedented.
- Category B – where there may be potential adverse environmental impacts on human populations or environmentally important areas – including wetlands, forests, grasslands, and other natural habitats. Impacts are less adverse than those of Category A proposals.
- Category C – where the proposal is likely to have minimal or no adverse environmental impacts.
- Category FI – where the proposal involves investment of bank funds through a financial intermediary, in subprojects that may result in adverse environmental impacts.

For all Category A and B proposals, the borrower is required to consult affected groups and local NGOs and to take their views into account. Early consultations are sought. With Category A proposals, the borrower consults these groups at least twice: just after environmental screening and before the terms of reference for the assessment are finalised; and once a draft EA report is prepared. Additional consultations may be needed to work on EA-related issues as they arise.

The Bank's procedures have been subject to a number of formal internal reviews, in 1989, 1996, 2004, 2007 and 2011. The 2011 update clarifies the use of framework instruments and to add strategic environmental and social assessment to the list of available instruments. Rees (1999) reports on the recognition of the progress made in integrating assessment practices into the Bank's operations. He also comments that "much remains to be achieved both 'upstream' (through the development of sectoral and regional EAs) and 'downstream' (through the preparation and more effective use of environmental management plans)" (1999: 333).

In 1993, the World Bank formed the Independent Inspection Panel (IIP) to "increase the transparency and accountability in Bank operations" (World Bank (2003: vii)). The IIP also resulted from pressure from the international community regarding some controversial projects, such as the Sardar Sarovar Dam in India's Narmada Valley in 1993. The IIP enables communities to raise concerns about a project where they believe that their interests have been, or could be, directly affected by the proposal which is financed by the World Bank.

With a focus on the region of Africa, the African Development Bank (2000) also has procedures for the identification and management of environmental impacts. This Bank has established guidelines for proposals that relate to Forestry and Watershed, Industrial Sector, Irrigation, and Coastal and Marine Resources. As an example, the guidelines for Coastal and Marine Resources Management have six parts. The first four provide an introduction to the Bank's overall environmental policy, and to key coastal issues: integrated coastal zone management; demographic and infrastructure pressures; global sea rise (key factors threatening sustainable development in coastal areas in Africa); marine habitats and resources; institutional constraints (such as processes for coordinated planning and implementation of projects. The fifth part is designated Environmental Assessment in Coastal and Marine Resources Management – Technical Guidelines, and provides direction of particular relevance for coastal and marine resources organized by sector. The final part presents Operational Guidelines, outlining the project cycle adopted by the African Development Bank and the overall planning context required for a proposal to be in accordance with sustainable development planning in coastal areas.

With a focus on investments in the Asian region, the Asian Development Bank (2002) noted that promotion of sustainable development and environmental protection is important. The Asian Development Bank's Environment Policy addresses the need for "more upstream environmental assessment at the level of country programming, the need for more structured consultation in the conduct of environmental assessments, the need for greater emphasis on monitoring and compliance with environmental requirements during project implementation, and finally the need to view environmental assessment as an ongoing process rather than a one-time event" (2009: web page).

As a consequence, a range of assessments have been undertaken spanning most DMCs (Asian Development Bank (2003)).

The focus of EIA in the Bank is the Regional and Sustainable Development Department which:

- reviews the environmental impacts of its projects, programs and policies;
- encourages DMC governments and executing agencies to incorporate environmental protection measures in their project design and implementation procedures and provides technical assistance for this purpose;
- promotes projects and programs that will protect, rehabilitate, and enhance the environment and the quality of life; and
- trains ADB and DMC staff in, and provides documentation on, environmental aspects of economic development.

The Department also provides an overview of the Bank's Environment Policy. This, according to the Asian Development Bank (2002) is grounded in ADB's Poverty Reduction Strategy that recognises that environmental sustainability is a prerequisite for pro-poor economic growth and efforts to reduce poverty. Recognising that Asia is rapidly transforming, the ADB commissioned a number of studies for addressing the increase in development and investment, which resulted in a new focus for the ADB in early 2007. The "New ADB" envisages that by 2020, the Asian region will have "conquered widespread poverty in most countries, with more than 90 per cent of its people living in 'middle-income' countries" (Asian Development Bank (2008: 1)). The "New ADB" thus acknowledges that the ADB will need to move from economic growth as its focus to environmentally sustainable growth.

The following case study reflects the numerous issues that developing countries face when developing significant projects, participating in EIA and relying on international funding bodies to support these projects. The Bujagali Falls hydropower project in Uganda has a complex history; however the information below is a simple summary of the development proposal demonstrating the many social and political influences of EIA.

5.8.1 *Case Study: Bujagali Falls hydropower project, Uganda*

Less than 5 per cent of Uganda's population has access to electricity, and power shortages have been stated as an impediment to economic growth and decline of Uganda's GDP (World Bank (2003)). The $530 million Bujagali Falls hydropower project began in the late 1990s when the US-based AES Corporation won a contract with the Ugandan government to build the hydropower project at Bujagali Falls on the Victoria Nile in eastern Uganda.

AES Nile Power prepared SEA documentation that was approved by various agencies between 1999 and 2001. However, AES discontinued with the project in 2003, and in 2005 Bujagali Energy Ltd entered into a power purchase and implementation agreement with the Government of Uganda (Burnside International Ltd (2006)).

The project involved the construction of up to 250 megawatt run-of-the-river power plant (no capacity to store water), dam works and a new 100 kilometre, 132 kilovolt transmission line to bring the power to the national grid. The dam will impound a reservoir that extends upstream inundating Bujagali Falls (Burnside International Ltd (2006)). The project requires 125 hectare permanent land and 113 hectare of temporary land for the projects ancillary facilities. The project will relieve power shortages in Uganda and substantially reduce the need for more expensive thermal power (International Finance Corporation (2007)). The project began construction in 2007 and concluded in 2012.

The issues

The project is upstream of nearby natural protected areas, including the Jinja Wildlife Sanctuary and Bujagali Falls. Through the social and environmental impact assessment process for the project, it is proposed that these areas will be offset through the designation of new areas to replace those lost due to the development of Bujagali.

Other impacts identified were the cultural significance of the Bujagali Falls, its value for ecotourism and the resettlement of people living on the shore of the Falls. Also of concern were Uganda's legal obligations under the power purchase agreement with AES which was a source of resentment towards the project from various groups within Uganda. These legal issues included lack of competitive bidding for the project, lack of financial information and potential for corruption in the development, construction and implementation phases of the project (World Bank (2003)).

The role of the World Bank

Although there had been controversy surrounding the Bujagali Falls hydropower project including whether or not the World Bank should be involved, the World Bank decided to approve the equivalent of US$115 million to support the hydropower project in December 2001. The World Bank has continued its funding support for the project throughout its duration of social and environmental impact studies.

Five months before the World Bank provided funding support, a request from the National Association of Professional Environmentalists of Kampala, Uganda Save Bujagali Crusade and other local institutions and individuals was submitted to the Inspection Panel (an independent arm of the World Bank) to investigate a number of issues regarding the project. The World Bank agreed to this investigation being conducted (World Bank (2003)).

Specifically, the requesters claimed that the proposed Bujagali project (as well as another hydropower plant nearby) was likely to result in social, economic, and environmental harm to the local population including negative effects on tourism activities, adverse impacts on fisheries, and increased electricity tariffs. There was also potential for cumulative impacts due to the number of power plants already running and proposed within the area.

The EIA process

EIA was required under the Ugandan *National Environment Act 1995* and an assessment of potential social and environmental effects is also a key requirement for all the international financial institutions involved in the project. The Social and Environmental Assessment (S&EA) was completed in 2006 in line with the requirements of the Ugandan Government and the various international financial institutions that are expected to finance the project. The authors of the S&EA state the purpose of the document is to "ensure that the project is designed and developed in a manner that minimises negative social and environmental effects while maximising project benefits" (Burnside International Ltd (2006: 13)). Cumulative impacts were addressed in this second round of environment assessments that looked at such issues as possible change to flow regimes; likelihood of sediment, erosion and degradation of water quality; and loss of natural habitats and resources.

The S&EA report was made available for public review. The public consultation program involved numerous public meetings and discussions with a wide range of stakeholders including local communities, government agencies, NGOs, businesses (including tourism operators) and cultural groups including the Kingdoms of Buganda and Busoga (Bujagali Hydropower Project (2008)).

The outcome

Noting all of the above, after numerous environmental and social impact assessment studies, consultations, an investigation by the World Bank's independent panel, the Bujagali Falls hydropower project was approved by the World Bank (for funding of US$360 million in loans and guarantees) in April 2007. A detailed environmental mitigation and monitoring program has been prepared for the project, incorporating water quality monitoring during and after construction, native vegetation planted on islands and riverbanks in the new Bujagali reservoir, monitoring of fish stocks, and provision of new tourism facilities in the Bujagali area (Bujagali Hydropower Project (2008)).

5.9 EIA on the Internet

For those with access to electronic communications, further information about EIA procedures worldwide can be obtained from the Internet. If you use the search parameter "Environmental Impact Assessment" you will find thousands of "hits".

The International Association for Impact Assessment (at <www.iaia.org>) provides information on international organisations involved in EIA. There is also information related to electronic networks and EIA contacts to facilitate the development of networks which support EIA training and processes. This site lists journals, leaflets, books and other publications where additional information can be obtained, and provides an overview of various specialty areas. In particular, it provides links to professional interest sites and environmental learning exchanges that outline international experience and practice. The Commonwealth Department of the Environment site <www.environment.gov.au> also provides links to these sites, and to key government sites both internationally and in Australia.

5.10 Final Comments

Over a period of some 40 years, the governments of the vast majority of industrialised, and industrialising, nations have adopted EIA procedures. There is considerable variety in the way in which these procedures are applied, but they all have the same broad intent; to assess potential effects before proceeding with actions. As would be expected, the procedures used in Australia also show variety. These Australian procedures are presented in Chapter 6.

Chapter 6

EIA PROCEDURES IN AUSTRALIA

6.1 Introduction

In some instances, EIA procedures at State level predated those of the Commonwealth, and in most cases the processes adopted were considerably different from the Commonwealth's approach. With experience in the process of EIA and increasing cooperation between governments, the processes are evolving to show considerable similarities. During the 1990s, there was a move to embed EIA concepts and procedures in integrated land-use planning and resource management structures. Over the past decade there has been a focus on reviewing and updating EIA legislation and procedures across Australia, with some jurisdictions, such as Western Australia and the Commonwealth, incorporating strategic environmental assessment (SEA) and risk based approaches into their legislation and procedures.

A general background to the development of EIA in the States and Territories is provided by Fowler (1981, 1982), Gilpin (1995) and Porter (1985). The principles embodied in the national approach adopted by the Australian and New Zealand Environment and Conservation Council (ANZECC (1991a)) established the direction during the 1990s in which governments moved their EIA procedures.

The following sections outline the approaches to EIA that have been developed by the Commonwealth, State and Territory governments throughout Australia. Each jurisdiction is considered separately through a brief history of the changes in its procedures (although a more detailed history can be found in previous editions of this text). An outline of the current application for each approach is presented in broad terms to provide a general understanding of the particular process. Readers wanting more information about these procedures should contact the relevant government departments responsible for administrating the EIA processes.

Much of the material in the following sections has been derived from the various Acts and guidelines which have been made available to direct EIA processes, and comments from officers associated with the EIA processes in the State, Territory and the Commonwealth. This information demonstrates that, while EIA is an established process of all governments in Australia, many changes in the administration of EIA has occurred over the past decade and additional modifications continue to be proposed and/or recently introduced. Hence, anyone involved in undertaking or reviewing EIA should obtain the most recent information. The availability of information on internet sites will assist, although information on these sites is often limited, and may not represent all aspects of the most recent practice. The Commonwealth Department of Environment provides information about Commonwealth procedures and links to State and Territory activities (see specifically <www.environment.gov.au/epbc/index.html>). This site also has links to sources of information about EIA practice and professionals. However, when you are involved in an EIA, to be clear on what is required, you need to contact the responsible government agency to ensure that your understanding of the processes is accurate. If you are the proponent, or representing the proponent, ideally you should contact the relevant State/Territory agency during the project planning stage rather than waiting until a final development proposal has been decided upon.

The following section outlines the national approach to environmental assessment in Australia that was adopted in the mid-1990s, and which provides an appreciation of the goals for EIA. This establishes a basis for considering, in subsequent sections, the current (as at mid-2013) Commonwealth, State and Territory processes.

6.2 Background to a National Approach to EIA

The variety of assessment procedures existing across Australia has been of some concern in the past, although improved procedures and better coordination of EIA processes between government jurisdictions has seen some shift away from the duplication of assessment procedures. This had especially been the case when a proposal came under the jurisdiction of two or more separate processes; such as where a proposal required approval from a State government and the Commonwealth government.

For some time, issues of duplication and consistency in EIA procedures have been concerns for regulators, proponents and the community. Discussions of these issues culminated in the adoption of "A National Approach to Environmental Impact Assessment in Australia" (ANZECC (1991a)). The purposes of this national approach are to:

- reach a common understanding and agreement on principles and, where appropriate, the practice of EIA in Australia;
- improve the EIA process, including increasing the efficiency of the contribution made by the process to environmental decision-making;
- reduce uncertainty about the application, procedures and function of the process;
- promote public understanding, and provide and facilitate consistent opportunities for public involvement;

- improve consistency of approach between jurisdictions in Australia responsibility for EIA, and, where proposals may have environmental impacts across jurisdictions, to apply consistent environmental protection measures;
- avoid duplication where multiple jurisdictions apply; and
- identify and apportion responsibility for participants in the EIA process.

The benefits identified in having a national approach included the integrated management of environmental assets plus advantages for the community, proponents and government, such as coordination of processes (ANZECC (1991b)). Also recognised was the enhanced credibility of the EIA process in general. The components associated with these benefits are presented in Figure 6.1 *over page*.

To achieve these benefits, several aspects of EIA procedures were singled out by ANZECC (1991a) for explicit description. Some relate more to administrative concerns, but three have implications for the way in which EIA is generally undertaken, and how existing procedures could evolve.

In particular, environmental significance was identified as a key factor in determining the need for EIA. To provide guidance on "significance", the national approach notes that it is a judgment on the degree of importance and consequence of anticipated change. This judgment is based on:

- the character of the receiving environment and the use and value which society has assigned to it;
- the magnitude, spatial extent and duration of anticipated change;
- the resilience of the environment to cope with change;
- confidence about the predicted change;
- the existence of policies, programs, plans and procedures against which the need to apply the EIA process can be assessed;
- the existence of environmental standards against which the proposal can be assessed; and
- the degree of controversy on environmental issues likely to be associated with the proposal.

Significance is seen to have an influence on the level of assessment undertaken. In particular, there should be different levels to take account of the scale of the proposal, the significance of the environmental context in which it is proposed and the associated degree of public interest. Also, while there are links between the type of assessment and the outcomes of an assessment, the national approach has proposed aspects of outcomes that are applicable to most levels. These stress public access to information and the EIA process, and accountability of decision-makers (they are outlined in Figure 6.1).

It was these principles, and others included within the national approach, that influenced the revisions of the EIA procedures that mostly occurred in the past decade around Australia. There are already many examples of where these principles are part of existing procedures, but the national approach continues to lead to greater precision, and consistency, in the way in which EIA is implemented.

Figure 6.1 Benefits and outcomes of the national approach to EIA

BENEFITS

Factors relating to the community benefits included:

- accessible information about development proposals;
- opportunities to participate, offer opinions and advice, and to influence decision-making;
- accountability by proponents and decision-makers for the environmental consequences of proposals and decisions;
- a proven mechanism to help achieve community objectives for ecologically sustainable development; and
- a visible means of protecting the environment.

Proponents were thought to gain from:

- clarification and analysis of proposal objectives and alternatives for meeting goals;
- integrated project planning (technical, financial and environmental);
- potential savings of input costs arising from site selection analysis, and through effective controls/reductions of waste and emissions;
- orderly process for hearing and responding to public concerns; and
- opportunities to gain credibility and support on environmental aspects of proposals.

Advantages to governments were identified as:

- environmentally responsible development;
- a tool for achieving government policy;
- opportunities for community participation in proposals before decisions have been taken;
- opportunities for reducing conflict, and for resolving problems on environmental issues;
- a tool for achieving community-wide environmental objectives;
- orderly process for the provision of information and advice before decision-making; and
- better (but not necessarily earlier) decision-making.

OUTCOMES

The outcomes which are anticipated for proposals undergoing an EIA based on the National Approach are that:

- all outcomes should be public;
- the assessing authority responsible for the EIA process should have the power to recommend environmental conditions;
- authorities responsible for decision-making should take environmental advice into account, and make the reasons behind their decisions public;
- enforceable, auditable conditions to protect the environment should be set by decision-makers and made public;
- any disagreement about environmental acceptability should be resolved between the ministers, or by Cabinet;
- trade-offs for any unavoidable differences (benefits and costs) should be part of the government's decision-making after the EIA has been completed;
- environmental conditions should ensure that environmental management can change as a consequence of monitoring information;
- should a proposal not proceed within a reasonable time frame, re-assessment of the proposal may be required.

Source: ANZECC 1991b

Development of consistency is a key element of the Schedule (3), dealing with EIA, in the InterGovernmental Agreement on the Environment (IGAE) (National Strategy for Ecologically Sustainable Development (1992)). This Schedule sets out a common set of principles, which will achieve greater certainty about the application of EIA throughout Australia, and avoid duplication and delays in the process. The IGAE sets out 12 broad points for EIA, including the role of public consultation, ecologically sustainable development (ESD) guidelines, and environmental monitoring and management programs. ANZECC (1991a) notes that this is intended to provide general consistency throughout Australia and to reduce duplication, as well as meeting the principles of the national approach.

The IGAE was signed in 1992 by the Heads of Australian governments, excluding Western Australia, and commits the signatories to basing their EIA practices on a number of principles, which have built on the ideas and direction given by the national approach (see Appendix B). This agreement has not resulted in substantial revision of procedures across Australia, since the Agreement has essentially maintained the independence of the States' and Territories' activities. However, as EIA procedures are amended the principles of the IGAE and the national approach provide direction for change towards greater consistency.

Overviews of the EIA procedures operating at Commonwealth and State levels have been given by Fowler (1982), and Martyn, Morris and Downing (1990). These procedures are constantly evolving, and will no doubt change in response to the national and international approaches to EIA. Hence, the following sections outline the basic procedures within Australia as they were operating in mid-2013. Since some jurisdictions were still undertaking reform of their current EIA and/or planning procedures at the time of printing, it is worth checking the relevant government websites for further detailed information.

6.3 Commonwealth of Australia

6.3.1 Introduction

Parallels can be drawn between the situation, since the 1940s, in the United States and in Australia with regard to the actions of the government departments and the effects on the environment.

More recently, the 1960s gave us environmental controversies of oil drilling and limestone mining on the Great Barrier Reef, and inundation of Lake Pedder in Tasmania. As was the case in the United States, environmental impacts from humans' actions were becoming more apparent and with such publications as Rachel Carson's *Silent Spring* in 1962, the community and governments were beginning to question the way the environment was managed (more detail on the US *National Environment Policy Act 1970* can be found in section 5.2).

At the same time in Australia there was an increasing interest in conservation as evidenced by the formation of the Australian Conservation Foundation (the membership of this organisation grew from 1562 to 9110 between 1968 and 1975). Other indications of conservation interest were the increase in the number of environmental groups (in New South Wales the number quadrupled from the early 1970s through to the 1990s), and the green bans that were imposed by the Builders' Labourers Federation.

Pastoral and industrial development in the post-war era led to the growth of population and production, which put increasing pressure on the environment. The passing of the *National Environmental Policy Act 1970* (NEPA) in the United States provided an impetus for the Commonwealth government, and the electorate, to seek ways of both conserving the environment and natural resources as well as improving procedures for protecting the environment. By 1973, the Australian Environment Council had established a working party to investigate the implementation of EIA procedures at the federal level. The recommendations of this working party resulted in the passing of the *Environment Protection (Impact of Proposals) Act* (EP(IP) Act) in 1974. (After serving for 25 years, the Act was replaced by the *Environment Protection and Biodiversity Conservation Act 1999* (EPBC Act), discussed below in 6.3.4).

6.3.2 *Environment Protection (Impact of Proposals) Act 1974*

The objects of the Act were to ensure, to the greatest extent practicable, that matters affecting the environment to a significant extent would be fully examined and taken into account before decisions were made as to whether a development captured under the Act could proceed. Another key aspect was the broad definition used for environment under the EP(IP) Act. This was defined as all aspects of the surrounds of humans whether affecting them as individuals or in their social groupings. Other aspects of the Act, and the process by which Environmental Impact Statements (EIS) were produced, are explained by Fowler (1982) for those who have an interest in the history of EIA at the Commonwealth level.

Like many other pieces of legislation relating to EIA, the Commonwealth's Act was discretionary in nature. While this left the decision about some matters to the Minister for the Environment, and sometimes the department, rather than precise direction being given by the Act, Formby (1987) points out that this gave the government flexibility in implementing the Act.

The effect of the Act's operation between 1975 and 1978 has been reviewed by Formby (1981), who notes that 49 EISs had been required during that period. More specifically, between 1974 and June 1991 a total of 19,271 proposals were referred for advice. Of these, 2533 were considered environmentally significant: 129 required the preparation of an EIS; 21 required a Public Environment Report (PER); and three, a public inquiry. The overall result then was that some 0.5 per cent of projects examined resulted in the preparation of an EIS under the EP(IP) Act during this timeframe.

6.3.3 Reviewing the EP(IP) Act

In 1987, the concept of the Public Environment Report (PER) was introduced to provide an intermediate level of assessment below that of the EIS, and to provide for public comment on the environmental consequences of proposals. From 1987, and particularly throughout the mid-1990s the Act was subject to almost continuous review. In 1993, this involved public comment on all aspects of the Act and its administration, consultant research (into matters such as cumulative and strategic assessment, and public involvement) and the release of discussion documents. Proposals in the discussion documents related mainly to tightening up the administration of the Act, and clarifying its application relative to the EIA procedures of the States and Territories. In practice, the proposals were not radical and would result in little noticeable change for those outside the administration of the Act.

At the beginning of these reviews the emphasis was mainly on administration and procedures, but with a change of government in 1996 a broader political interest was introduced. At the heart of this has been the tensions between the Commonwealth and the States over responsibility for the environment, which had been apparent for many years. Doyle and Kellow (1995) point out that the Commonwealth's involvement in environmental issues comes from the Australian Constitution, especially s 51. The Commonwealth has also become involved through the use of its external affairs powers. Based on this section, there are arguments that the Commonwealth should be more directly involved, and while the Commonwealth has had the constitutional power to intervene in matters of great environmental significance, Toyne (1994) observes it has often declined to intervene, leaving environmental management the responsibility of the States and Territories. This no doubt arises from the fact that our Constitution does not mention the environment. In explanation, Toyne presumes that no one drafting the Constitution in the late 1800s anticipated the impending emergence of severe, widespread and chronic environmental problems.

6.3.4 Background to the Environment Protection and Biodiversity Conservation Act 1999

Arguments to include the environment in Australia's Constitution have raged for years as discussed briefly above. Ramsay and Rowe (1995) present a range of views on the matter and provide detail from the Final Report on the Constitutional Convention 1988 that canvassed the inclusion of the environment. After considering the main arguments related to the matter, the authors note that the Convention recommended the Constitution should not be altered "by adding an express provision to empower the Federal Parliament to make laws with respect to the environment" (1995: 287).

A decade later, the Senate Environment, Communications, Information Technology and the Arts References Committee saw the situation somewhat differently. Through its inquiry into the powers of the Commonwealth related to environmental management, the majority report of the Committee made recommendations that would have strengthened the role of the Commonwealth; although not to the point of changing the Constitution. The principal recommendations that touch on EIA included:

> The Commonwealth should exercise a leadership role in the protection and improvement of the Australian environment. This role should be supported by the unsparing use of all Constitutional power available to the Commonwealth to act in the field of the environment.
>
> The use of the concept of "national environmental significance" should be abandoned as a means of delineating the appropriate role of the Commonwealth in the regulation of environmental matters.
>
> The Commonwealth should be responsible for environmental impact assessment process whenever it is involved in making a decision about an activity or matter (its own or that of a third party) that may have a significant effect on the environment. (1999: xiii, xvi)

While these were the recommendations of the Committee's Chairperson, government (Liberal/National Party) Senators submitted a dissenting report which supported submissions from business and government that a "less centralised approach based on various criteria of appropriateness, such as efficiency, transparency, simplicity and the value of local ownership ..." (1999: 133) was preferable. Senators from the Labor Party also provided a minority report in which they expressed support for many of the recommendations, but saw the need for a greater degree of the involvement of State/Territory governments.

In practice, the recommendations of the inquiry have been ignored. The inquiry process had been run in parallel with another Senate inquiry into the Environment Protection and Biodiversity Conservation Bill (Senate Environment, Communications, Information Technology and the Arts Legislation Committee, 1999). This inquiry was presented with many submissions covering the spectrum of political positions, with strong representations from conservation and environment groups for tight control of EIA by the Commonwealth. In parallel, discussions between political parties, including some conservation groups, were held to determine an agreeable outcome.

Subsequently, and after many amendments negotiated between the government and opposition parties, and with the involvement of some parts of the environment movement (see Doyle (2000)) the *Environment Protection and Biodiversity Conservation Act* (EPBC Act) was passed in 1999. The EPBC Act establishes a national approach to a wide range of environmental protection and biodiversity conservation matters. Generally it has been described as the biggest rewrite of environmental law for over two decades. Commentators have identified a number of positive aspects of the Act, particularly the integration of previously separate management of biodiversity issues. However, others consider fundamental concerns are considered to remain with the procedures for EIA. Macintosh (2009) suggests that the EPBC Act is deficient in a number of respects, particularly in relation to the structure of the EIA regime and listing processes concerning threatened biodiversity and heritage areas.

As required under s 522A of the EPBC Act, an independent review of the Act is currently being undertaken to consider the operation of the Act and the extent to which the objects of the Act have been achieved (Department of the Environment, Water, Heritage and the Arts (2008)).

6.3.5 Environment Protection and Biodiversity Conservation Act 1999

Information sources for the Act

A considerable amount of material is available about the Act through the Internet, especially the websites of the Department of the Environment and those of some States and Territories. However, the key sites at the time of publication are:

- Department of the Environment <http://www.environment.gov.au/topics/about-us/legislation>; and
- Details on the EPBC Act <http://www.environment.gov.au/topics/about-us/legislation/environment-protection-and-biodiversity-conservation-act-1999>.

Introduction

The Commonwealth's *Environment Protection and Biodiversity Conservation Act 1999* (EPBC Act) came into operation on the 16 July 2000. The EPBC Act regulates proponents directly and civil and criminal penalties (fines and imprisonment) apply for breaches of the Act. Specifically, it establishes an environmental assessment and approval system that is separate and distinct from the EIA systems of the States and Territories (although there is provision for the possible integration with these systems either by a bilateral agreements (discussed further down) or accreditation of a State's EIA process on a case-by-case basis). This Commonwealth Act does not affect the validity or conduct of the States and Territories environmental and development assessments and approvals; rather the EPBC Act runs as a parallel system to the State/Territory systems. Nonetheless, State/Territory agencies will have to be aware of the operations of the Act, and will help stakeholders (such as developers) to become aware of their obligations under the EPBC Act. It is important to note that if a development proposal requires EIA under the Commonwealth as well as within a State or Territory, the development will need to obtain two approvals and one is not dependent on, nor overrides, the other.

The EPBC Act applies directly to proponents

The EPBC Act requires proponents of actions, to which the EPBC Act may apply, to seek a determination from the Commonwealth Environment Minister regarding whether or not their proposed action is a "controlled action". Proponents must then, if the Act applies, seek approval for the action from the Commonwealth Minister directly.

A State or Territory cannot provide advice to a proponent on whether a proposal falls within the definition of a controlled action or on whether or not any particular proposal requires referral to the Commonwealth government. This advice can only come from the Commonwealth Environment Minister, although under s 74(2)(b) the Commonwealth Minister must invite the relevant State/Territory to make any comments on a referral if applicable to that State/Territory.

What is an action?

The EPBC Act defines the term "action" as including a project, a development, an undertaking, and an activity or series of activities. An action does not mean a decision. Importantly, an action outside the boundary of a matter of national environmental significance (see *page 136*), for example outside the boundaries of a world heritage area, might still trigger the EPBC Act if it is likely to have a significant impact on a matter of national environmental significance.

What is an impact?

Impact is defined in section 527E of the EPBC Act. The Department has published Significant Impact Guidelines to assist proponents and others to determine whether a proposed action is likely to have a *significant impact*.

In considering the significance of the potential impacts of an action, the Minister must consider all *direct* impacts as well as *indirect* impacts which are:

- facilitated, to a major extent, by the action; and
- within the contemplation of the person taking the action; and
- reasonably foreseeable.

Downstream and/or indirect impacts to matters of national environmental significance are also meant to be considered by the Commonwealth in its decision-making about a proposal.

In all situations the Minister will decide whether the action will, or is likely to, have a significant impact on a matter of national environmental significance or on Commonwealth land. If the Minister decides that this is the case, then the action requires approval under the EPBC Act. In order to decide whether an action is likely to have a significant impact, it is necessary to take into account the nature and magnitude of potential impacts. In determining the nature and magnitude of an action's impact, it is important to consider matters such as:

- all on-site and off-site impacts;
- all direct and indirect impacts;
- the frequency and duration of the action;
- the total impact which can be attributed to that action over the entire geographic area affected, and over time, the sensitivity of the receiving environment; and
- the degree of confidence with which the impacts of the action are known and understood.

The Commonwealth has been challenged in the Federal Court over decisions where the community has considered that indirect impacts were not taken into consideration (eg, *Minister for Environment and Heritage v Qld Conservation Council Inc* (2004), known as the Nathan Dam case. See 6.3.6).

When may the EPBC Act be triggered? What is a "controlled" action?

The EPBC Act provides that a person must not take an action that has, will have or is likely to have a significant impact on a matter of national environmental significance (NES) except where certain processes have been followed and/or approvals obtained (see *What are the exemptions?* below). Chapter 2, Division 2 also provides that matters affecting Commonwealth land may also trigger the Act. An action will require approval from the (Commonwealth) Environment Minister if:

1. The action has, will have, or is likely to have a significant impact on a matter of national environmental significance (and the action is not subject to one of the exceptions identified below);
2. The action will have or is likely to have a significant impact on the environment associated with Commonwealth land, ie, the action will take place:
    - if the action is taken on Commonwealth land, the environment; or
    - if the action is taken outside Commonwealth land, the environment on that Commonwealth land; or
    - if the action is taken by the Commonwealth on Commonwealth land anywhere in the world.

An action falling under either of these two categories requires assessment and approval under the EPBC Act and is known as a "controlled action". However, where particular processes have been followed and/or relevant approvals obtained the action may be exempted from the Act.

The Act provides that the Minister must, in deciding whether an action is likely to have a significant impact on a matter of national environmental significance, take account of the precautionary principle. The precautionary principle is explained in the EPBC Act as follows:

> Lack of full scientific certainty should not be used as a reason for postponing a measure to prevent degradation of the environment where there are threats of serious or irreversible environmental damage. (EPBC Act s 391)

Accordingly, the fact that there is a lack of scientific certainty about the potential impacts of an action will not itself justify a decision that the action is not likely to have a significant impact on a matter of national environmental significance.

Also, the Minister may agree with a person responsible for the adoption or implementation of a policy, plan or program that an assessment be made of the relevant impacts of actions under the policy, plan or program that are controlled actions. This would be a strategic assessment under the Act and, for example, is used for Commonwealth Fishery Management Plans, Defence activities and many others as well.

What are matters of national environmental significance?

The nine matters of national environmental significance under the EPBC Act are:

- world heritage properties;
- national heritage places;
- wetlands of international importance (listed under the RAMSAR Convention);
- listed threatened species and ecological communities;
- migratory species protected under international agreements;
- Commonwealth marine areas;
- the Great Barrier Reef Marine Park;
- nuclear actions (including uranium mines);
- a water resource, in relation to coal seam gas development and large coal mining development.

The Commonwealth amended the EPBC Act in January 2004 to include a national heritage trigger under the matters of national environmental significance. A national heritage list has been established which includes places that are owned or managed by the Commonwealth government. These amendments meant repealing the *Australian Heritage Commission Act*, which duplicated State responsibilities and failed to provide substantial protection for heritage places of national significance (Department of Environment and Heritage (February 2004)).

Other matters of national environmental significance may be prescribed by regulation. The Commonwealth government prepared a consultation paper on the possible application of a greenhouse trigger in December 1999 (Department of Environment and Heritage (June 2004)) and a draft regulation released in November 2000. Under the draft regulation, the Act would be triggered by major new developments likely to result in greenhouse gas emissions of greater than 0.5 million tonnes of carbon dioxide equivalent in any 12 month period, for example, a new coal-fired power plant. The greenhouse gas trigger re-emerged on the agenda with the lead up to the Federal election in November 2007 and was an election promise of the Labor government, as was the introduction of an emissions trading scheme. Discussion on such a trigger has now ceased.

The Great Barrier Reef Marine Park environment was added in 2009 as a matter of national environmental significance, providing greater protection of the park under the Act.

What is a significant impact?

The Commonwealth government has published criteria in Administrative Guidelines that will indicate when an impact on a matter of national environmental significance will be considered a "significant impact" (see 6.3.8 *below*). The criteria are broad in nature, hence it would frequently be unknown whether an action will be caught by the EPBC Act until at least preliminary investigations have been undertaken.

The Department of the Environment provides examples of how significance can be tested. For instance, regarding "Wetlands of International Importance" an action will require approval from the Environment Minister if the action has, will have, or is likely to have a significant impact on the ecological character of a declared RAMSAR wetland (a wetland protected under the International Ramsar Convention, to which Australia is a party). Specifically, a significant impact on the ecological character would be identified if the action is likely to result in:

- areas of the wetland being destroyed or substantially modified; or
- a substantial and measurable change in the hydrological regime of the wetland; or
- a substantial change to the volume, timing, duration and frequency of ground and surface water flows to and within the wetland, or the habitat or life cycle of native species dependent upon the wetland being seriously affected, or a substantial and measurable change in the physico-chemical status of the wetland; or
- a substantial change in the level of salinity, pollutants, or nutrients in the wetland, or water temperature which may adversely impact on biodiversity, ecological integrity, social amenity or human health; or
- an invasive species that is harmful to the ecological character of the wetland being established in the wetland.

What are the exemptions?

An action does not require approval from the Environment Minister under the Act if:

- the action is approved under, and taken in accordance with, a State management plan that is accredited by the Commonwealth for the purposes of a bilateral agreement (eg, Regional Forestry Agreements); or

- the action is approved under, and taken in accordance with, a Commonwealth management plan that is accredited by the Environment Minister for the purposes of a Ministerial declaration;
- the action is a forestry operation taken in a Regional Forest Agreement region; or
- the action is taken in the Great Barrier Reef Marine Park and is authorised by the *Great Barrier Marine Park Act* 1975; or
- the action has been authorised by a government decision on which the Minister's advice has been sought. (Environment Australia (2013))

What about existing actions?

The EPBC Act provides that an action does not require approval under the Act if it is a lawful continuation of a use of land that was occurring before 16 July 2000.

Importantly, an enlargement, expansion or intensification of use is defined not to be a continuation of a use and therefore may be caught by the EPBC Act if it has, will have, or is likely to have a significant impact on a matter of national environmental significance.

6.3.6 Case study: Nathan Dam

In 2002, Sudaw Development's Ltd referred a proposal under the EPBC Act to construct and operate a dam, known as the Nathan Dam, near Taroom on the Dawson River in central Queensland. The purpose of the dam was to provide irrigation water to 30,000 hectare of farmland, mostly to grow cotton. Through connections with other rivers, the water flows east to the coast and eventually meets the Great Barrier Reef World Heritage Area (GBRWHA) near Rockhampton.

Issues with the proposed dam were raised by a number of Queensland groups including the Queensland Conservation Council (QCC), WWF for Nature (Australia), and the Great Barrier Reef Marine Park Authority. A particular concern was the potential use of the insecticide endosulfan, which is commonly used in cotton growing, being able to flow (and therefore potentially polluting) the GBRWHA (McGrath (nd)). Other indirect impacts may include increased sediment and nutrient pollution to the GBRWHA.

The QCC and the WWF (Australia) sought a review in the Federal Court of the Commonwealth Environment Minister's decision to approve the Nathan Dam proposal and argued that the Commonwealth Minister did not consider the downstream impacts of the action. The case was mostly concerned with whether the impacts on the Great Barrier Reef from the agricultural activities should be considered an impact of the Dam itself (Environmental Defenders Office (Qld)). Indeed, the judge, Kiefel J, found that the Commonwealth

Minister erred by refusing to consider the impacts from the irrigators (in other words, the associated agricultural development) when assessing the impacts of the dam. The Commonwealth Minister appealed this decision, but it was later upheld by the Full Court.

What this meant is that the decision by the Commonwealth Minister on the controlling provisions (ie, the relevant matters of national environmental significance to be addressed in the assessment) as well as the level of assessment, were required to be remade by the Commonwealth Minister (Environmental Defenders Office (Qld)).

As the Nathan Dam case illustrates, the consideration of both direct and indirect impacts is an important principle to emerge under the EPBC Act and has certainly been cited in other case examples (eg, Hazelwood coal mine expansion in Victoria). The definition of impact was inserted into the EPBC Act in 2006 to include both direct and indirect impacts.

6.3.7 Referrals

A person proposing to take an action that the person thinks may be, or is likely to be, a controlled action must refer it to the Commonwealth Environment Minister in accordance with the requirements of the EPBC Act. The information to be included in a referral can be found on the Departments website and within a referral from. A manual that details this information can be found at <http://www.environment.gov.au/system/files/resources/0b0cfb1e-6e28-4b23-9a97-fdadda0f111c/files/environment-assessment-manual.pdf>.

The Commonwealth Minister must decide whether or not the proposed action requires assessment and approval under the EPBC Act. The Minister can determine the following outcomes:

- not a controlled action;
- not a controlled action, particular manner;
- controlled action; or
- project is clearly unacceptable.

In the 2011-2012 reporting year, the Commonwealth Department of the Environment received 412 new referrals and 132 of these were determined to be 'controlled actions' requiring further assessment and approval under the Act.

6.3.8 Assessments

When the Commonwealth Environment Minister determines that the proposed action is a controlled action, an environment assessment will be required.

If a bilateral agreement is in place, that accredits a State assessment process (see below). The relevant State will assess the action under the terms of that agreement. If a relevant bilateral agreement or Ministerial declaration

Figure 6.2. Referral Process under the *Environment Protection and Biodiversity Conservation Act 1999*

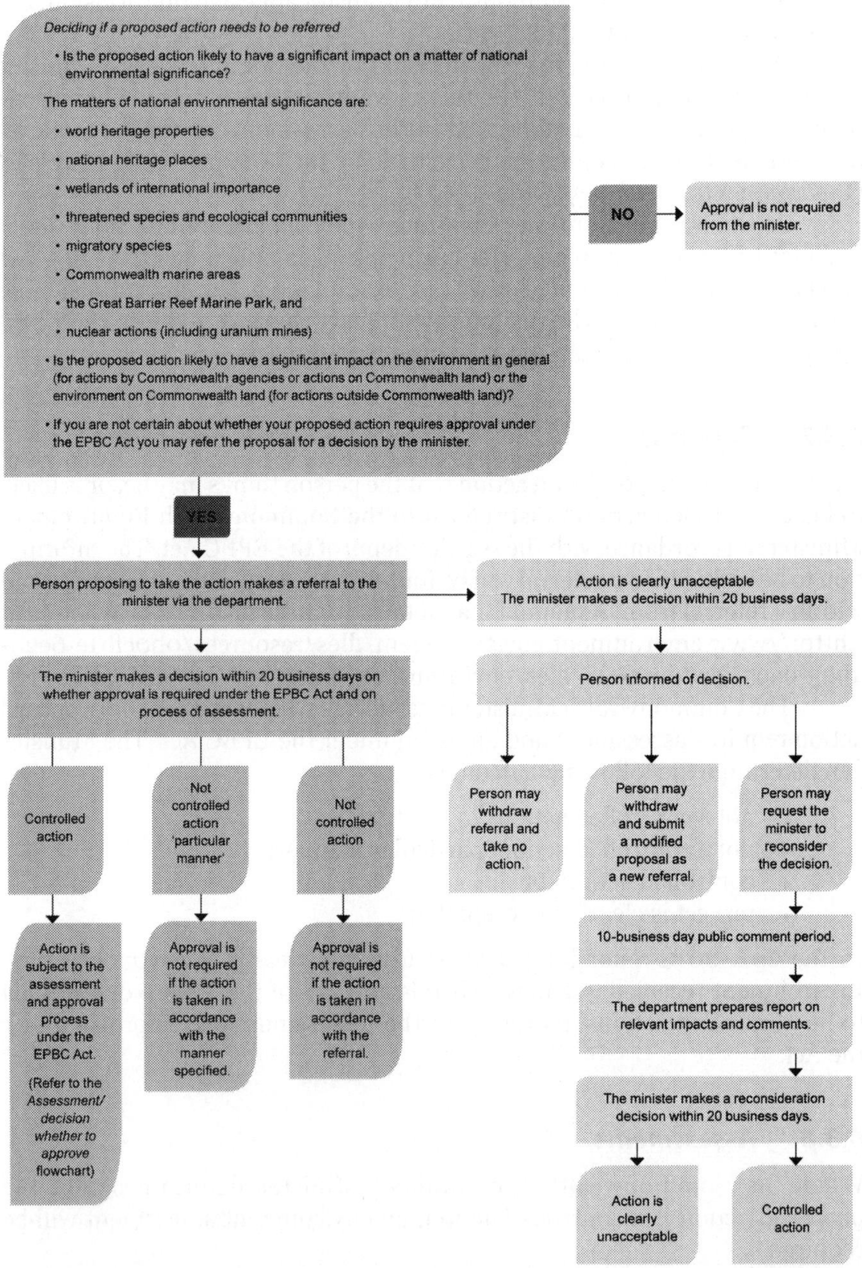

Figure 6.3 Assessment Approaches under the *Environment Protection and Biodiversity Conservation Act 1999*

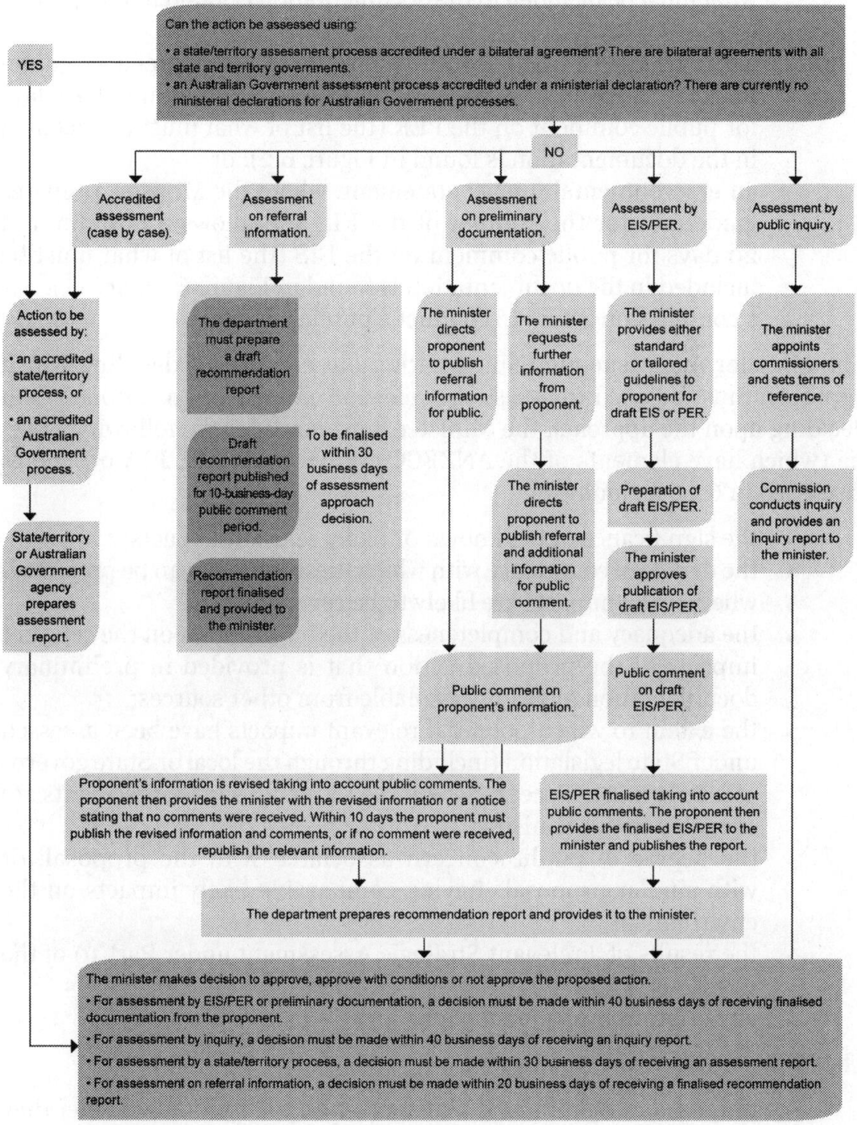

is not in place, the Commonwealth Environment Minister will decide on an assessment approach. This approach may be either:

- an accredited assessment process (ie, accreditation on a case-by-case basis); or

- preliminary documentation, where the proponent must publish a notice inviting public submissions on the proposed action (the list of what must be included in the documentation is found in Figure 6.2); or
- a public environment report, where the Minister prepares guidelines for the content of the PER and allows a minimum of 20 days for public comment on the PER (the list of what must be included in the documentation is found in Figure 6.2); or
- an environmental impact statement, where the Minister prepares guidelines for the content of the EIS and allows a minimum of 20 days for public comment on the EIS (the list of what must be included in the documentation is found in Figure 6.2); or
- a commission set up to conduct a public inquiry.

The Minister may issue guidelines setting out criteria for deciding which approach must be used to assess the relevant impacts of an action. When deciding upon the approach the Minister must consider the following criteria (which have elements of the ANZECC criteria and IGAE EIA principles discussed in 6.2), including:

- the significance of the known or likely relevant impacts;
- the degree of confidence with which these impacts can be predicted;
- whether the impacts are likely to be reversible;
- the adequacy and completeness of the information on the relevant impacts of the proposed action that is provided in preliminary documentation or readily available from other sources;
- the extent to which potential relevant impacts have been assessed under State legislation (including through the local or State government planning process), where the results of such assessments are available to the Minister;
- the degree of public concern associated with the proposal, or with similar proposals having comparable likely impacts on the environment;
- the results of a relevant Strategic Assessment under Part 10 of the EPBC Act; and
- any relevant bioregional plan.

The Minister may also consider:

- proposed mitigation and management measures and whether they are of proven effectiveness in circumstances similar to those in which the action is proposed;
- the costs and benefits, including environmental costs and benefits, to the community and the proponent of further data collection and analysis relating to the relevant impacts of the proposal;
- whether one assessment approach is more likely than others to provide information of a quality and extent required by the Minister to make his or her decision on approval of the proposal; and

- the extent to which public concerns and views about the proposal, including the concerns and views of any Indigenous people affected by the proposal, have been documented and are available to the Minister.

Assessment on preliminary documentation, or by PER or EIS, involves the preparation and publication of draft assessment documentation, a period for public comment, and finalisation of assessment documentation taking public comments into account. All of these steps are taken by the person proposing to take the action, following regulations and guidelines prepared by the Commonwealth Environment Minister.

6.3.9 Decisions

After the chosen assessment process for an action has been completed (which may include the preparation and consideration of a public environment report or environmental impact statement), the Commonwealth Environment Minister must decide whether or not to grant unconditional or conditional approval of the action. There are a number of matters that the Minister may consider in making his or her decision which are listed in Part 9 of the Act. Such matters include, but not limited to, social and economic matters, principles of ESD, the PER or EIS (where prepared) and the relevant assessment report.

Strategic assessment

Under Part 10 s 146 of the Act, strategic assessments can be undertaken of a policy, plan or program. This happens if this recommendation is put to the Minister by anyone responsible for the adoption or implementation of the policy/plan/program, and the Minister agrees (Raff (2000)). In this case, the assessment is of the relevant impacts of controlled actions that come from the policy/plan/program. The assessment may deal with other impacts (with some limitations) if the Minister is requested, by the relevant State or Territory, to assess these impacts. Strategic assessments are landscape scale assessments and unlike project-by-project assessments, which look at individual actions (such as mining, construction or operation of a pipeline or wind farm), they can consider a much broader set of actions (DSEWPC (May 2013)). In its submission to the Productivity Commission in 2013, the Commonwealth Department suggested that the benefits of strategic assessments include:

- clear requirements for protection of matters of national environmental significance being set up-front, at the planning stage; greater certainty to local communities and developers over future development;

- reduced administrative burden for strategic assessment partners and government through:
  - a substantial reduction in the number of environmental assessments required for an area; and the avoidance of potentially duplicative and separate environmental assessments by different types of government (such as Australian, state, territory or local governments)
  - capacity to achieve better environmental outcomes and address cumulative impacts at the landscape level;
  - coordinated establishment and management of offsets; and
  - flexible timeframes to better meet planning processes. (DSEWPC (May 2013:13))

A strategic assessment has to cover the publication and public comment on terms of reference for the assessment, publication of a draft assessment report, public comment on the report, finalisation of the report, and preparation of recommendations. Through a specific provision, the Act outlines the conduct of strategic assessments in relation to fisheries management. Strategic assessments have been undertaken for major military exercises, offshore petroleum exploration, fisheries management, gas projects and urban expansion projects and many more. The Commonwealth government is increasing the use of strategic assessments to improve certainty for proponents by reducing the need for project-by-project assessments.

Bilateral agreements

The EPBC Act provides for "accreditation" of certain State or Territory processes under a bilateral agreement (BA) between the Commonwealth and State or Territory governments.

Under a so-called "assessment bilateral", particular State or Territory environmental impact assessment processes may be accredited. The effect of accreditation of a State or Territory assessment process is that a proponent would need to complete only the accredited State or Territory assessment and receive a decision on the proposal from the State or Territory.

The Commonwealth Environment Minister would then use the information from that accredited assessment to form the basis for making his or her own decision on whether to grant or refuse an approval (conditional or unconditional) under the EPBC Act, thus eliminating duplication of EIA processes.

However, there are a significant number of additional requirements that apply to "approvals" bilaterals. Most significant is the need for the State or Territory assessment and decision to be carried out under an accredited management plan that is in force under State law and which complies with Commonwealth regulations. The accredited management plan may be disallowed by either House of the Commonwealth Parliament. It is also notable that the Commonwealth Senate has passed a resolution to disallow any instrument that the Commonwealth Environment Minister may enter into that accredits a State or Territory's decision for the purposes of the EPBC Act.

At the time of publication, assessment BAs are in place with all States and Territories. If no assessment bilateral is in place and where EIA is required at a State/Territory level and at the Commonwealth level, the Commonwealth accredits the State/Territory EIA process on a case-by-case basis to assess the Commonwealth matters.

It is important to note that, even if an assessment BA is in place and therefore EIA is undertaken by the State or Territory, a separate approval decision by both the State/Territory and the Commonwealth still need to be made, that is, two "green lights" are required before a proposal can proceed.

Public involvement

In addition to the opportunities available to the public to contribute to a PER, EIS or an inquiry, there are requirements under the Act for the public to have input to:

- the drafting stages of the BAs – the Minister must take account of public comment, give reasons why the agreement is being entered into, and report on the comments received;
- proposals referred to the Minister – the public must be notified, via the Internet, of referred actions, and the Minister must consider public comments and provide reasons for the final decision on the action;
- assessment approach – while the public has no input to the decision about the type of approach to be used for an EIA, the Minister must publish a notice related to the decision;
- assessment on preliminary documentation – the proponent must invite the public to comment on the proposed action and submit these comments to the Minister;
- assessment guidelines for a PER or EIS – the Minister may, but is not required to, seek public input to the these guidelines; and
- draft approval conditions – comment may be sought (eg, Bell Bay pulp mill in Tasmania).

Further to these points of community access, Anton (1999) also identifies several opportunities related to conservation of biodiversity (the other part of the Act). Overall, a wide range of decisions about the assessment and approval process must be published weekly on the Internet.

Third party rights

Unlike the previous Commonwealth EP(IP) Act, the EPBC Act provides extensive appeal rights (for both individuals and organisations) in relation to decisions (albeit for whether the correct process has been followed, not on the merits if the decision) made by the Commonwealth Minister, including a decision on whether a proposal should be a controlled action; the level of environmental assessment (for example, in the case of the *Wilderness Society v Minister for Environment (Cth)* (EDO, Qld) regarding the Gunns Bell Bay pulp mill) and the right to appeal the final approval decision (for instance, the *Blue Wedges Coalition v Minister for Environment, Heritage and the Arts (Cth)* (2008) 165 FCR 211). Interested persons can apply to the Federal Court for an injunction where a person commits an offence against the EPBC Act. Standing is given if the individual's interests have been affected or the individual (or organisation) has been engaged in the protection or conservation of the environment at any time in the two years before the conduct or decision (s 475(6)).

6.3.10 Case study: Melbourne's Urban Growth Boundary

The Victorian and Commonwealth governments agreed to undertake a strategic assessment under the EPBC Act in March 2009 for the expansion of Melbourne's urban growth boundary. A number of reports were prepared, public comment followed and in February 2010, the Commonwealth Minister for Environment endorsed the program document *Delivering Melbourne's Newest Sustainable Communities: Program Report* (December 2009). Classes of actions can be approved under the endorsed program. To date, this has included the regional rail project; Melbourne's 28 precincts; and Melbourne's Northern, North Western and Western Growth Corridors.

The Melbourne Strategic Assessment allows 41,000 ha of developable land which does not require further referral under the EPBC Act, and for approvals of the Regional Rail Link and Outer Metropolitan Ring/E6 Transport Corridor, rather than the need to refer various individual project proposals/precinct plans under the EPBC Act (because of the potential impact to listed species and communities under this Act) for the four growth corridors.

The endorsed program under the EPBC Act includes actions associated with urban development in four growth corridors in the 2010 Melbourne Urban Growth Boundary. The endorsed program included a commitment by the Victorian government to prepare a Biodiversity Conservation Strategy for the four corridors as well as regional strategies for the Growling Grass Frog, Golden Sun Moth and the Southern Brown Bandicoot.

Native vegetation time-stamping

The time-stamping project captured and "time-stamped" native vegetation information to establish a native vegetation dataset and maps showing the type, extent and condition of all native vegetation in Melbourne's growth corridors. This was a program that the Victorian government had committed to doing under the Melbourne Strategic Assessment.

As the Victorian Department of Environment and Primary Industries suggests:

> this information was used together with information on threatened species habitat mapped for the Biodiversity Conservation Strategy and sub-regional species strategies to calculate habitat compensation requirements for the removal of native vegetation and threatened species habitat associated with urban development in the growth corridors. (DEPI (2012: 2))

These growth areas are currently in the process of precinct structure planning where the planning processes need to consider the obligations under the Melbourne Strategic Assessment.

Access Economics (2011) estimated that the Melbourne strategic assessment will deliver cost savings in net present value in excess of $3.2 billion for the private sector over its lifetime to 2039, by reducing delay and bringing forward project approvals. It is estimated that the Melbourne Strategic Assessment has removed the need for referral of approximately 252 projects over the life of the program, and has also saved the substantial associated monitoring and compliance costs for those individual projects.

The benefits for the environment will result in establishment of a new 15,000 hectares grassland reserve to the west of Melbourne and a new 1200 hectares grassy woodland reserve north-east of Melbourne. The 15,000 hectares reserve includes 10,000 hectares of critically endangered grasslands (DSEWPC (May 2013: 14)).

6.3.11 Emerging issues

An independent review of the EPBC Act was conducted in 2009, as required under the Act. The "Independent Review of the Operation of the EPBC Act" report was released in October 2009 and the Commonwealth's response to this review released in August 2011. The review emphasised the importance of reducing regulatory burdens and duplications within and across jurisdictions. The review made over 70 recommendations to the operation of the EPBC Act including redrafting of a new Act to replace the EPBC Act (which was not supported by the Cth government at the time).

The Commonwealth government's response (particularly regarding EIA) was that:

> improving the efficiency of the environmental impact assessment regime is a top priority for the Australian Government. In particular, the government is looking to increase its strategic and early engagement with proponents, and in development planning, including by working with states and territories through strategic assessment and regional environment planning. These strategic approaches will better protect matters of national environmental significance, while supporting sustainable development. (DSEWPC (2011: 10))

The COAG agreed in 2012 to fast-track the completion of EIA and Approval Bilaterals under the EPBC Act with the States and Territories by March 2013. However, no Approval Bilaterals had been published by late 2013.

As part of reaching the goal of having Approval Bilaterals in place with each State and Territory, the Commonwealth released a *Draft Framework of*

Standards for Accreditation of Environmental Approvals under the EPBC Act 1999 in July 2012. In October 2013, the Commonwealth government approved the Framework. The approved Framework includes a three-stage process with each of the willing jurisdictions, comprising:

- signing a Memorandum of Understanding;
- agreement on bilateral assessments and updating any existing agreement with the state; and
- negotiation of approval bilateral agreements within 12 months.

The current Commonwealth government has proceeded to negotiate the Bilateral Agreements with the State and Territories as the Commonwealth's position is that the Approval Bilateral Agreements will remove duplication in regulation between the Commonwealth and states/territories. There remains some issues associated with a 'one stop shop' and whether this will be achievable.

6.4 Australian Capital Territory

6.4.1 *Background*

Before the Territory government assumed responsibility for its own administration (in the mid-1980s), EIA was undertaken within the Commonwealth's procedures. With independence, the Territory developed EIA within Part IV of the *Land (Planning and Environment) Act 1991* (Land Act).

This Act provided for the environmental assessment of both development applications and variations to land use under the Territory Plan. The fundamental purpose of the Territory Plan is to manage land use change and development within the ACT in a manner consistent with strategic directions set from time to time by the ACT government and community (Planning and Land Management (2002)).

In December 2004, the ACT government announced major reforms to the EIA process in the ACT as part of its wider planning system reforms. The *Planning and Development Act 2007*, together with the Territory Plan, provides the basis for the system of environmental assessment that came into effect on 31 March 2008.

6.4.2 *Development proposals*

The EIA process is similar to other States and Territories where after a determination that EIA is needed, scoping requirements are prepared, the proponent prepares the EIS and there is an opportunity for the public to comment. The Minister then decides whether to call for a public inquiry, present the EIS to the Legislative Assembly, or take no action. The format and content of an EIS is specified in the Regulations at <http://www.legislation.act.gov.au/sl/2008-2/default.asp>.

A proponent needs to request a scoping document from the ACT Planning and Land Authority and following consultation with relevant agencies the proponent is provided with a written notice of matters to be addressed by the EIS. The proponent then prepares a draft EIS.

In the ACT, environmental impact assessment can be required under the following legislation, or mechanisms:

- The development is listed in Schedule 4 to the *Planning and Development Act 2007*.
- The development is listed in the relevant Territory Plan development table for the zone as impact assessment.
- The development isn't mentioned in a development table of the Territory Plan.
- The Minister for Planning has asked for an EIS.
- The proposed development is prohibited by the relevant development table in the Territory Plan, but an existing lease over the land already allows for that development.
- The Minister responsible for the *Public Health Act 1997* has declared the impact track applicable. (ACT Planning and Land Authority (2013))

A number of changes to the EIS process were introduced in the ACT via the *Planning and Development (Environmental Impact Statements) Amendment Act 2010*, which took effect in February 2011. One of the major changes was the introduction of the "Environmental Significance Opinion (ESO)" where a proponent can apply for if the proposed development is "unlikely to have a significant adverse environmental impact" and is only applicable to Schedule 4 items under the *Planning and Development Act 2007*.

Once an EIS is completed, a development application can be lodged in the impact tract and considered against the Territory Plan. When the development is publicly notified, representations about the application may relate to how the development meets, or does not meet, any finding or recommendation of the EIS for the development. It may not relate to the adequacy of the EIS for the development.

An inquiry panel for an EIS may be established to consider any or all aspects of an EIS. If a panel is to be established, the Minister must prepare terms of reference for the inquiry and give the proponent notice of the inquiry. The inquiry panel must report in writing on the result of the inquiry no later than 60 days after the panel is established, however the period for reporting may be extended by the Minister on application by the panel.

The Environment Protection Authority may use the EIS provisions for assessing an application for an environmental authorisation under the *Environment Protection Act 1997*. The EPA may only do so where the activity for which an environmental authorisation is sought is not the subject of a development application.

6.4.3 Environmental assessment of plans (Strategic Environmental Assessment)

Environmental assessment of major plans occurs through the use of Strategic Environmental Assessment (SEA) and Planning Reports under the *Planning and Development Act*. SEAs are primarily used to assess the environmental impacts of major plans and planning policies and specific, discrete policy initiatives, or major changes to the Territory Plan rather than individual development proposals.

SEA would typically be of major government initiatives, such as a significant urban land release program. SEA is undertaken at the earliest stages of planning and will ensure that any further environmental assessment is targeted and focused on matters that, for example, require a site-based assessment.

Where there are significant planning issues evident but these do not have major strategic planning implications then the Authority would prepare a planning report, or direct a proponent/developer to prepare a planning report. Planning reports have four primary functions:

- to identify and assess the impacts of granting a proposed Crown lease and inform the decision about whether it is to be offered;
- to assist in the crafting of the terms of a draft Crown lease to be made available to prospective purchasers;
- to inform the Authority's consideration of whether a Variation to the Territory Plan is appropriate; and
- to inform a Plan of Management for public land.

Information on the EIA process in the ACT can be found at <http://www.actpla.act.gov.au/topics/design_build/da_assessment/environmental_assessment>.

6.5 New South Wales

6.5.1 Background

Declaration of an environmental impact policy based on NEPA was made in 1972 and acted as a forerunner of EIA in the State. The procedures were given a more formal backing in 1974 when a Cabinet Directive was released which related to government projects. This Directive was administered by the State Pollution Control Commission. Guidance for these procedures was given through the booklet *Principles and Procedures for Environmental Impact Assessment in NSW*, published by the Commission.

Fowler (1982) notes that the next stage for EIA in New South Wales was the introduction to parliament of the Environmental Planning Bill in 1976. However, this Bill, which would have established a form of EIA within a revised land-use planning system, lapsed due to the proroguing of parliament. Nevertheless, the ideal of EIA embodied in legislation was pursued, and in 1979 the *Environmental Planning and Assessment Act* (EP&A Act) was passed. The purpose of the EP&A Act is to facilitate environmental planning and to protect the environment (social, economic and biodiversity).

The criterion specified to give guidance as to whether environmental effects require detailed environmental assessment is that the activity/proposal would be likely to significantly affect the environment (activity includes formulation of proposals, incurring expenditure or carrying out works).

The EP&A Act and inherent processes have much in common with other town and country planning schemes, in that "environmental planning instruments" form the basis of State environmental planning policies (for issues of State significance), regional environmental plans (prepared by the Department of Planning) and local environmental plans (prepared by local government). Occasionally, environmental studies undertaken as part of the development of these plans, provide input to these three types of plans, and could in some cases theoretically offset the need for the preparation of EISs, as the potential effects have already been assessed.

6.5.2 Current EIA process

Changes in the late 1990s included several categories of development and types of proposals considered under different parts of the Act, depending on the type of proposal and magnitude of perceived impacts. Proposals are considered under either Part 4 or 5 of the Act. The categories determine how environmental factors are taken into account with particular developments and determine what level of environmental assessment is to be undertaken.

In 2011, the NSW government introduced a new assessment system for projects of state significance, which includes state significant development or state significant infrastructure. The State significant development assessment system has been established to guide planning decisions on large scale industrial, resource and other proposals or developments in precincts identified as important for the State by the NSW government. Development classes are listed in Schedule 1 of the State and Regional Development State Environment Planning Policy (SEPP).

6.5.3 State significant development assessment system

Part 4

Most development proposals in NSW are assessed under Part 4 of the Act. There are a number of different development types under Part 4.

- *Designated* – These proposals are listed either in planning instruments or in Schedule 3 to the EP&A Regulation 2000 and have the potential to have adverse impacts because of the scale or nature of the development proposal, or because of the area in which the development is proposed (eg, coastal or adjacent to wetlands).

Applications for such developments must be accompanied by an EIS, and there is recourse to the Land and Environment Court.

- *Local development* – development which requires consent, but is not "State significant development", and usually local councils will be the decision-makers. The council will assess the impacts of the proposed development taking into consideration any comments received.
- *Regional development* – notified and assessed by a local council and then determined by the relevant joint regional planning panel. Regional development includes:
  - development with a capital investment value (CIV) over $20 million;
  - development with a CIV over $5 million which is council related, lodged by or on behalf of the Crown (State of NSW), private infrastructure and community facilities or eco-tourist facilities;
  - extractive industries, waste facilities and marinas that are designated development;
  - certain coastal subdivisions;
  - development with a CIV between $10 million and $20 million which are referred to the regional panel by the applicant after 120 days; and
  - Crown DAs (with a CIV under $5 million) referred to the regional panel by the applicant or local council after 70 days from lodgement as undetermined, including where recommended conditions are in dispute. (Department of Planning and Infrastructure NSW (2013(1)))

Part 5

For proposals which do not fall under Part 4. These are often public projects including infrastructure proposals that are approved by local councils or State agencies, which are undertaking them.

For information on the above categories of development, refer to the Department of Planning website at <http://www.planning.nsw.gov.au/en-us/developmentproposals/developmentassessmentsystems.aspx>.

The Minister for Planning can declare certain projects as State significant. This Minister is the consent authority for State significant development, and other development identified in particular State environmental planning policies and regional environmental plans.

Under the NSW procedures, both private and public proposals are involved. For all categories of development, public participation is relatively extensive in that there are opportunities to be involved in the exhibition of environmental plans, in objections to developments via the Land and Environment Court, by appearing at commissions of inquiry and through submissions to EISs.

Other proposals, essentially those of government agencies and State-owned corporations but not the Minister for Planning, are considered under Part 5. Where there is potential for environmental impacts, these impacts will still need to be assessed via an EIS. Changes to the EP&A Act in October 2011 mean that some activities under the Part 5 assessment system will be determined by the Minister for Planning and Infrastructure, following an assessment by the Department, (Department of Planning and Infrastructure (2013(2))).

In any situation where an EIS is required, the process begins with the relevant approving authority considering the need for an EIS by ascertaining whether the proposal is designated by referring to the Regulation. If development consent is not required, that is, it falls under Part 5 of the Act, then the proponent and any Minister or public authority approving the activity decide if an EIS is required on the basis of whether the activity would significantly affect the environment. The proponent consults with the Department of Planning and Infrastructure about the form and content of the EIS, and the Director General's requirements for the EIS are provided to the proponent.

Once the EIS is completed, it and the associated DA are exhibited for public comment. The approving authority organises the public review and receives the submissions, passes copies of the submission onto the Department of Planning and Infrastructure and, finally, determines the overall DA. For this determination, advice may be provided by the Director of Planning to the approving authority. The Minister for Planning may call for a formal commission of inquiry to be held.

In June 2008, the New South Wales government established the independent Planning Assessment Commission (PAC) to determine approximately 80 per cent of major development project applications dealt with by the Minister (Department of Planning (March 2006)). The Planning Assessment Commission is independent of the NSW government, the Minister for Planning and Infrastructure and the Department of Planning and Infrastructure. The role of the PAC is set out under section 23D of the EP&A Act and includes:

- determining applications for major developments under delegation from the Minister;
- reviewing any major development including conducting of public hearings; and
- providing independent expert advice on planning and development matters. (PAC (1) (2013))

Public hearings may occur for any planning and development matters that the PAC are considering.

Information about the New South Wales EIA processes can be found on the Department of Planning and Infrastructure's website <http://www.planning.nsw.gov.au/en-us/developmentproposals/developmentassessmentsystems.aspx>.

6.6 Northern Territory

Before 1979, EIA procedures in the Northern Territory were undertaken through the Commonwealth government's *Environment Protection (Impact of Proposals) Act*. With the transition from Commonwealth government jurisdiction to a State government-style administration within the Territory, EIA procedures moved from Commonwealth government administration to administrative arrangements made by the Northern Territory's Chief Administrator. This system continued until 1982 when the *Environmental Assessment Act* was passed by the Northern Territory government.

Procedures were outlined in the Act, in summary form, and these have subsequently been detailed in the Administrative Procedures. The Act formally commenced with the gazettal of the procedures in 1984. This Act was updated in 1994 to bring the EIA process into line with the requirements of the 1992 InterGovernmental Agreement on the Environment. Based on the previous decade of experience, the revised Act brought about a number of refinements in some aspects of the assessment process. The key objective of the Act is to ensure that matters affecting the environment to a significant extent are fully examined and taken into account in decisions by the Northern Territory government. A *Guide to the EIA Process in the Northern Territory* was released in January 2013 which details the procedures for EIA in the Northern Territory under the Environment Assessment Act 1982 and can be found at <http://www.ntepa.nt.gov.au/environmental-assessments/eiaguide>.

While the *Environmental Assessment Act* is the prime location for EIA, there is provision for further examination of proposals (that is after the EIS has been assessed) in the *Inquiries Act 1985*. This takes place when it is considered necessary by the Minister.

The *Environmental Assessment Act* applies to both public works and private projects; that is, private development proposals such as construction and mining projects, as well as public policy and decisions made by government departments (eg, water, road or rail infrastructure). However, the procedures are essentially concerned with land-use and development decisions. The procedures apply where, in the opinion of the Minister, a matter could have a significant effect on the environment. Currently, the NT Environment Protection Authority (EPA) is responsible for administering the environmental impact assessment process in the Northern Territory.

There are two levels of environmental assessment within the Northern Territory; the Public Environmental Report (PER) and an Environmental Impact Statement (EIS). The requirement for either a PER or EIS will depend on the sensitivity of the local environment, the scale of the proposal and its potential impacts. A decision on whether formal environmental assessment is required is based on a judgment about the environmental significance of a proposal and the scale and complexity of the development, rather than on a list or schedule of types of projects. The Minister determines which proposals should be subject to assessment under the Act and decides the level of assessment. The responsibility for beginning the process lies with the Minister responsible for authorising the proposed action who notifies the Minister responsible for EIA as soon as practicable after being informed of the formulation of a proposed action. Also, the Minister may initiate the process if a notification has not been given and he or she are of the opinion that the matter may be a proposed action. In practice, it is the authorising government department that provides the notification or causes the proponent to provide notification – this occurs in the form of a Notice of Intent. A notice of intent is examined on the basis of several criteria, and if the proposal does not involve a significant effect on the environment, the NT EPA provides comment on it to the authorising government department. Otherwise a PER or EIS will be required; the PER is called for to assist in assessing environmental impacts which are considered significant but limited in extent.

Figure 6.4 (*see over page*) illustrates the EIA process in the Northern Territory.

Submissions from the public and government bodies received during the public review period are forwarded to the proponent where a proposal is being assessed as either an EIS or under the BA with the Commonwealth as a PER. The Northern Territory government signed an amended BA relating to EIA with the Commonwealth in May 2007. The BA accredits the PER and EIS processes under the *Environmental Assessment Act 1994*.

The proponent then revises the draft EIS/ bilateral PER in the light of these comments and submits a supplement to the EIS/bilateral PER to the Minister.

NT EPA then provides an assessment report based on the EIS/ bilateral PER and other relevant information (eg, public comment). If a proposal is being assessed as a PER and the BA does not apply, the end of the public exhibition period triggers a 14 day statutory period in which the EPA is to prepare an assessment report. This assessment report, together with the final EIS/PER, is then used by the Minister to formulate advice to the Minister responsible for making the decision about the proposal. This advice will be in the form of an environmental assessment report. The Minister can make recommendations on the acceptability or otherwise of the proposal, on environmental grounds, and on any conditions, safeguards or standards which should apply. Also, the Minister must make public the recommendations relating to environmental safeguards. As with most other States and the ACT, the Northern Territory environmental impact assessment process is not an approval process; it plays an advisory role.

To provide the public with information about the concluding step in the assessment process a public register has been established to outline information about the assessment outcomes and environmental recommendations of those proposals that have been formally assessed (see <www.nt.gov.au/nreta/environment/assessment/register/assessed.html>).

The following example of the EIS process for a proposed gold mine in the Northern Territory demonstrates the importance of the scoping phase of

Figure 6.4: Northern Territory Environmental Impact Assessment process

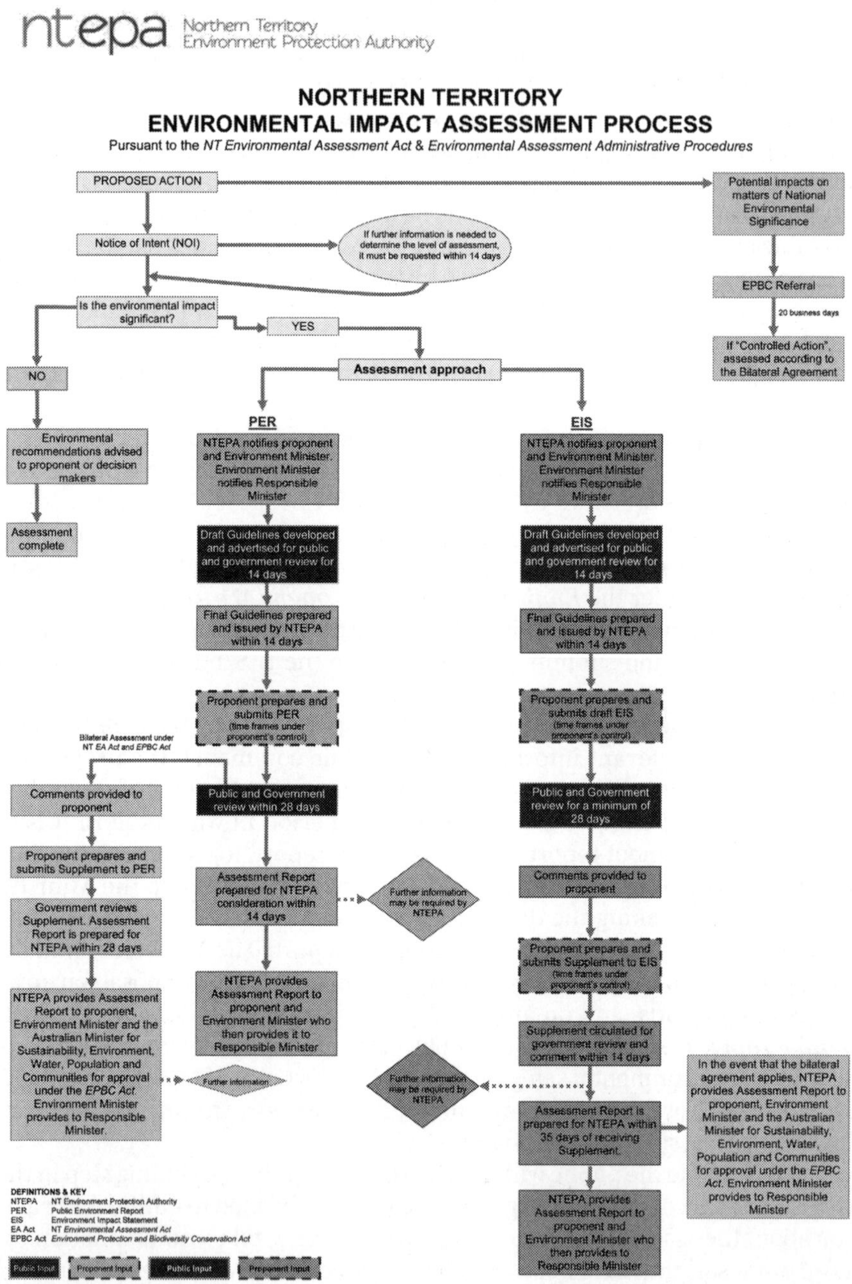

Source: NT EPA (2013)

EIA and highlights the positive outcomes of proposals being changed early in the process due to potential environmental impacts.

6.6.1 Case study: Maud Creek gold mine

The proponent, Terra Gold Mining Ltd, proposed an open cut gold mine along Maud Creek in an already mined location, just downstream from the Katherine River which supplies the township of Katherine with 75 per cent of its drinking water. The gold mine is also adjacent to Nitmiluk National Park. Terra Gold proposed to continue to use open-cut mining to extract the ore, necessitating a small creek diversion as well as the need to have a waste rock dump (and potential acid drainage problems). The proposal also included processing of ore requiring a tailings storage facility on site.

As part of the scoping phase, an extensive community and government consultation program was undertaken. During this phase, the impacts of a large open cut mine and processing facility located in the Katherine River catchment was deemed to be a potential significant risk to the Katherine water supply. Consequently, Terra Gold revised its original proposal and changed it to an underground mine with processing off-site (URS (2008: ES7)). The EIS Guidelines prepared to support the assessment of the proposal reflected the findings of this scoping exercise.

The benefit of applying an early risk based approach during the scoping stage of the EIA process is that the proponent was alerted upfront to the key issues of concern to government (and community).

6.7 Queensland

6.7.1 Background

The State of Queensland was one of the first governments in Australia to introduce EIA procedures. This historic occasion occurred in 1971 when amendments to the *State Development and Public Works Organisation Act 1971* (SDPWOA) were passed to establish the Environmental Control Council.

A further amendment in 1974 obliged public authorities to have regard to environmental consequences of their decisions. The usual procedure was to require impact assessment studies (IAS, or EIS – these terms were used interchangeably). Procedures for such studies were outlined in a procedural manual, where a developer, or proposer of some substantial action, was required to submit details of the proposal. If the proposal was considered to have significant environmental effects, the review of these effects was undertaken. Advisory bodies, such as government departments or private organisations with particular interests in the proposal, submitted comments regarding the form of the EIS to be produced.

Because the *SDPWOA* was not commonly used by the numerous public authorities who issued development approvals, specific provisions regarding the environment were also incorporated into several other Acts, including the *Local Government Act* 1936-1985. Under the procedural manual, and the *Local Government Act*, the EIA procedures related to government departments, and their works, but not to private works. Subsequent abolition of the Environmental Control Council in 1974 was in line with decentralisation policies, and put responsibility for EIA procedures back to individual government departments and local government authorities.

This was followed by further changes in 1979 when responsibility for administration of EIA procedures was given to the Coordinator-General's Department.

In 1988, the former Department of Environment and Conservation was created and assigned the functions of environmental management and coordination. To assist coordination, responsibility for several Acts was transferred to the department. Further changes ensued in 1991 when amendments to the *Local Government Act* established the *Local Government (Planning and Environment) Act 1990* (PEA) and the *Local Government Act 1990* (which is concerned with the powers and function of local government). Specific powers covering impact assessment were incorporated in the PEA. This amendment was to ensure that local government would have proper regard for the environment and identify situations when an impact assessment study was required.

6.7.2 Current practice

As indicated, the historical context of EIA in Queensland has taken place under many different pieces of legislation especially the PEA, SDPWO and *Environmental Protection Act 1994* (EP Act). A review in the mid-1990s of development approval processes found that there were over 400 separate EIA processes that related to development in 60 different pieces of legislation and regulation (State of Queensland (1999a)). This system resulted in duplication of impact assessment and approval processes and increased costs and delays for proponents.

Essentially, three pieces of State legislation govern EIA in Queensland:

- *Sustainable Planning Act 2009* for development other than mining and petroleum activities;
- *Environmental Protection Act 1994* for mining and petroleum activities; and
- *State Development and Public Works Organisation Act 1971* for "significant projects".

The processes under these Acts are described below.

The *Sustainable Planning Act 2009* replaces the *Integrated Planning Act 1997* and is, administered by the Department of State Development, Infrastructure and Planning and seeks to achieve sustainable planning outcomes through:

- managing the process by which development takes place;
- managing the effects of development on the environment; and
- continuing the coordination and integration of local, regional and State planning.

The *Sustainable Planning Act 2009* forms the foundation of Queensland's planning and development legislation. The *Sustainable Planning Act 2009* has established the State Assessment and Referral Agency (SARA) that will provide a whole of government approach to state assessment by creating a central point for development applications (DSDIP (1) (2013)).

The impact assessment provisions contained in Part 4 (s 29) of the SDPWO are activated if a proposal is declared to be a "significant development" by the Coordinator General (COG). In essence, a project is considered to be "significant" when it requires the approval of a State government department, corporation, statutory authority or local body and is likely to have significant economic and environmental impacts (eg, major infrastructure developments). To assist the COG in making this decision the proponent prepares an initial advice statement describing the project and the existing environmental values and heritage values (DSDIP (2) (2013)).

Once a project has been declared significant, it is removed from the IDAS and planning system and an EIS is prepared. The COG or his delegate manages the EIA process to the point of government approval. This is taken as fulfilling the requirement under the information and referral, and notification stages of the IDAS. Input is sought from government agencies that might otherwise have been referral agencies in the assessment of such projects. If a proposal is determined a "significant project" under SDPWO Act, SARA timeframes are suspended until the COG provides a report to the Assessment Manager that evaluates the environmental impacts. The report can specify conditions to be attached to the approval or can state that the application must be refused. This report passes onto the Assessment Manager who makes the final decision. The COG can also become the Assessment Manager in some circumstances.

Once it has been determined that an EIS is required, draft terms of reference with or without public notification will be prepared, to identify the key issues and impacts that need to be addressed in the EIS. Usually, there must be sufficient grounds provided by the chief executive not to publicly notify (15 business days) regarding the terms of reference. Once the public consultation period has been completed, the chief executive must finalise the terms of reference and give these to the proponent within 10 business days. The proponent then prepares the EIS in accordance with the Terms of Reference. A draft EIS is provided for public comment (for a period of 30 business days), submissions are received by the chief executive who considers these submissions before deciding to accept the draft EIS as final or asking the proponent to amend the draft EIS. Once the EIS is accepted, the chief executive must prepare an EIS assessment report (within 30 business days), which may include recommendations for conditions, such as monitoring and enforcement.

Of interest, the Queensland EIA process has a focus at the end on providing future mitigation, management and monitoring strategies for the proposals. Specifically, the EIS should include an appropriate follow-up process with strategies for minimisation and management of impacts, monitoring and evaluation that can be effectively audited (Department of Infrastructure and Planning (2007)).

6.8 South Australia

6.8.1 Background

Establishment of the Department of Environment in 1972 led to the formation of the Environment Protection Council, which advised the State government in 1973 to adopt EIA legislation. Cabinet circulated proposals for this legislation, but, Fowler (1982) notes, nothing was adopted. The proposed legislation had some interesting features. There were to be similar administrative arrangements to those of the Australian government legislation with a government department responsible for administering the legislation, and using a notice of intent (NOI) for the provision of information, so that the Minister could decide if an EIS was required. However, more interestingly, the proposed legislation included a provision whereby failure to supply a NOI would entitle the Minister to stop any works that were proceeding and if necessary dismantle the works. Approval, rejection or modification of a proposal would have been the responsibility of the government, as distinct from the more usual delegation of responsibility to a particular Minister.

Nevertheless, without legislation, from 1973 until 1982 EIA was carried out on numerous proposals within South Australia under administrative arrangements, or through the provisions of the Commonwealth *Environment Protection (Impact of Proposals) Act*. Cabinet agreements provided the basis for State government support for the EIA process during this period. These processes were only applicable to proposals in which there was either Commonwealth or State government involvement. In 1978, the Department of Environment produced an EIA handbook with regard to internal procedures to be followed by government departments for their own proposals or projects. This handbook outlined procedures that were similar to those of the Commonwealth government.

6.8.2 Current practice

EIA is undertaken in South Australia under the Development Act 1993 which includes provisions for EIA to be undertaken for major development proposals only. Specifically, where a person proposes to undertake a development which in the opinion of the Minister for Urban Development and Planning may be of major social, economic or environmental importance, the Minister makes a declaration that the development is to be assessed within the scope of the Major Developments or Projects division of the *Development Act* (ss 46-48). This section is also applicable to developments proposed by the Crown, when the Minister considers a government proposal to be a Major Development (Department of Housing and Urban Development). A proponent generally will refer the proposal to the Minister, however the Minister may also be asked by members of the community to consider making such a declaration, alternatively the Minister may become aware of a development proposal that may require referral to him or her. Importantly, the Minister can declare a project with an already approved development proposal a Major Development if works have not already commenced on the site (Planning SA).

The beginning of the assessment process is marked by the Minister deciding that a proposal is of major importance, and the publication of this declaration in the South Australian Government Gazette. The Development Assessment Commission then decides on the appropriate level of assessment and is responsible for formalising the contents of the guidelines to be used by the proponent. The Development Assessment Commission (DAC) considers the development application and identifies key social, environmental and economic issues relevant to the assessment of the proposal and may consult with other State agencies as part of this process. The DAC will also seek public submissions about the level of assessment and the scope of matters to be dealt with in the assessment. It will then determine which level of further environmental assessment is required and issue formal assessment guidelines to the proponent. Once the DAC issues guidelines and determines the level of assessment, its role is complete and the development assessment process goes back to the Minister (assisted by Planning SA).

Three levels of assessment are possible under the Act; an environmental impact statement (EIS), a public environmental report (PER) or a Development Report (DR). An EIS is relevant to Major Developments or proposals with many unknown issues and includes detailed description and analysis of a wide range of issues relevant to a development, incorporates information to assist assessment of the environmental, social and economic effects and how these might be managed. Public exhibition for EIS is at least 30 business days.

The next level of assessment is the Public Environmental Report (PER), sometimes referred to as a "targeted EIS", which includes detailed description and analysis of a limited number of issues and examines those relevant to the specific development. The PER may also cover a wide range of issues where an extensive amount of information is already available. Public exhibition for PER is the same as that for an EIS. For both PER and EIS, a public meeting is mandatory in an area close to the site of the proposed development when the document is on exhibition.

Finally, the Development Report (DR) is the level of assessment for major developments, primarily related to planning, and described as development under the *Development Act*. A DR includes description and analysis of general issues relevant to a development and how these can be addressed. Public exhibition for the DR is at least 15 business days and a public meeting is not a statutory requirement.

PERs and DRs are intended to be less lengthy, time consuming and expensive to prepare than and EIS. These two alternative assessments have been developed in recognition that the all-embracing notion of an EIS is not the only appropriate way to assess a major development and that assessment can be tailored in accordance with the type and scale of the proposed development.

The process for all three levels is essentially the same and differs mainly in the scope of investigation required of the proponent. The process involves:

- the preparation of guidelines identifying the issues to be discussed in the documentation;
- the preparation of an EIS, PER or DR, using the guidelines recommended by the DAC, which addresses the issues, is made publicly available and invites submissions;
- the holding of a public meeting (EIS and PER only);
- the proponent receiving copies of the submissions and comments from relevant bodies/authorities;
- the proponent responding to the issues raised in submissions and their preparation of a Response Document that is released for public information;
- the production of an assessment report by the Minister which is made available to council(s) and public;
- the EIS, PER or DR, any responses and the assessment report are being made available for public inspection and are provided to the relevant council(s); and
- the Governor making a decision, on the basis of the information collected, on the development including further requirements, limitations and time frames.

All decisions on Major Development applications are made by the Governor who cannot grant an authorisation (full or part) unless an EIS, PER, or DR and assessment report have been prepared. There is no right of appeal against decisions by the Governor, but the Governor can delegate decision-making powers to the Development Assessment Commission which can in turn, sub-delegate those powers.

This broad process also pertains to any proposal decided to be a Major Project by the Minister. Instead of a development application a project proposal is required. The Minister makes a declaration in the same manner as for a Major Development and the DAC performs the same role except that only two levels of assessment (EIS or PER) apply.

At the completion of either of the three levels of assessment, an assessment report is prepared by the Minister. This assessment report is intended to be an independent assessment of the proponent's proposed development. It includes the Minister's assessment of the development or project and any comments on the proponent's document and the submissions and responses to submissions, comments from the Environment Protection Authority, council and other bodies, comments by the Native Vegetation Council where required and any other matters or comments relevant to the development or project. The Minister may require a monitoring program to be established, and reporting directives may be stipulated to convey the results of the program. As a result of monitoring, conditions that had been attached to a development approval may be amended by the Governor.

Under the Act, the Minister may amend or require the amendment of an EIS, PER or DR to correct errors or to take into account new or more complete information. Also through the Act, official recognition of an EIS does not constitute a decision on the proposal; rather, it is a recognition that the EIS process has been satisfactorily completed.

Some mining proposals are addressed under the *Development Act*, and specific details can be sought from Planning SA. Generally, for these proposals, either the Minister for Mines or the Minister for Urban Development and Planning may require an EIS or PER to be conducted as defined in the *Development Act*. Either Minister may exercise the power of the DAC in determining the level of assessment and the overseeing of EIS/PER processes. The DAC consults with the Minister for Mines about whether an application should be granted, and the conditions of any approval. Any dispute regarding the exercise of powers under the *Development Act* is resolved by the Governor.

Once a decision has been made by the Governor on a development proposal, there is no right of appeal by any party against that decision. There are also no appeal rights to challenge any aspect of the assessment or decision-making process in court, even in relation to a failure on the part of the government to follow its own laws (see *Development Act* s 48E).

Sequential amendments to the *Development Act* have brought the operation of EIA processes in South Australia closer to the national approach (see 6.2). In particular, early public involvement is allowed for, and there are provisions for giving specific guidance to proposals that would be unacceptable for environmental reasons. The Minister for Urban Development and Planning can ask the Governor at any stage to consider a development proposal to be given a "no" decision if it is clear to the Minister that the proposal would have unacceptable impacts.

6.9 Tasmania

In Tasmania, there is no legislation specifically referring to environmental assessment. Before the mid-1990s the *Environmental Protection Act 1973* informally embodied provisions for EIA.

Introduction of the Resource Management and Planning System in the mid-1990s provided a more integrated approach to the assessment of land and water based activities in the State. With the aim of moving towards sustainable development, the System relies on a suite of legislation; that is:

- *Environmental Management and Pollution Control Act 1994* (EMPCA);
- *Land Use Planning and Approvals Act 1993* (LUPAA);
- *State Policies and Projects Act 1993* (SPPA); and
- *Resource Management and Planning Appeals Tribunal Act 1993*.

Mostly, environmental assessment is undertaken for activities which may cause environmental harm under the EMPCA. The EMPCA defines environmental harm as “any adverse impact on the environment (of whatever degree or duration) and includes an environmental nuisance ...” (Division of Environmental Management (1996: 13)). The Act expands on this definition in detail and discusses what is meant by “serious” and “material” environmental harm. While human health and property are specifically mentioned, the term “environment” is left as a broad concept.

EMPCA also deals with three classifications of activities that may cause environmental harm, and how the potential for harm should be handled:

- Level 1 activities are those that require a permit under the LUPAA and which may cause environmental harm – some examples of activities and screening questions are provided by the Department of Environment, Parks, Heritage and the Arts. The point is also made that some activities may cause harm by virtue of their location, size, or type and that it is the responsibility of the Planning Authority to decide if an activity could cause environmental harm.
- Level 2 activities are specified in the EMPCA (Schedule 2) and usually require a land use permit under the LUPAA. Once a permit is submitted for a Level 2 activity, the planning authority must refer the application to the EPA Board for environmental assessment. A Notice of Intent is prepared by the proponent and provided to the EMPC Board who will determine whether or not the project requires assessment by the Board, and the likely level of assessment. A project can be assessed as a Level 2A, 2B or 2C; most projects will be assessed as Level 2B activities (EPA Tasmania (2008)). Level 2A activities are generally small projects with minimal environmental impacts, whereas Level 2C projects are much more complex and may have significant environmental effects. An Environment Effects Report (EER) is usually prepared for Level 2A proposals, while for Levels 2B and 2C, a Development Proposal and Environmental Management Plan (DPEMP) is prepared. Both the EER and the DPEMP are made available for public comment; the proponent provides a response to submissions and then an assessment report is prepared by the Department. The EMPC Board considers the report and makes a determination about whether the project should proceed or not, and under what conditions.

- Level 3 activities are those declared to be of State significance according to the SPPA (eg, the 2007 Gunns Bell Bay pulp mill proposal).
- In addition, certain other activities that are not included in one of the levels, and which may be environmentally harmful, are recognised in the EMPCA (s 27) – these may be referred by the Director of Environmental Management to the Board of Environmental Management and Control (established in 1996 under EMPCA) for review of the need for environmental assessment.

More detail can be found in the *EPA Guide to EIA* (April 2013) <http://epa.tas.gov.au/regulation/document?docid=1222>.

The objectives of the Resource Management and Planning System place obligations for furthering sustainable development. In practice, this means that municipal Planning Authorities are responsible for undertaking some form of environment assessment at the initial stages for all development proposals. Where an assessment is carried out according to the Environmental Impact Assessment Principles (specified in the EMPCA, ss 73 and 74), it is called an "environmental impact assessment" (ie, an EIS). If it is undertaken by a planning authority for a Level 1 activity it is called simply an "environmental assessment".

The department point out that environment should be considered in its broadest sense, to include social, public health, economics, resource planning, biological and physical aspects of the environment. To assist with the assessment of Level 1 activities, a pro-forma "environment effects report" has been developed. To assist preparation of an EIS detailed contents are provided, including the incorporation of a section dealing with a management plan to identify measures to avoid or mitigate environmental impacts. It is recommended that the public be made aware of the environmental effects report or EIS and that submissions, to provide information for these documents be encouraged. All Level 2 activities are advertised and the public has the opportunity to make representations to the Planning Authority.

With Level 2 activities, the EPA Board (members include representatives from State and local government, industry and the community) is responsible for the EIA; but the proponent prepares the EIS. However, Level 3 activities are handled by an integrated assessment process carried out by the Resource Planning and Development Commission (RPDC) on a case-by-case basis. The RPDC is a statutory body established by the *Resource Planning and Development Commission Act 1997*.

As in other States where there is a close relationship between EIA and land and resource planning, proponents have the opportunity to appeal decisions made about their applications. In Tasmania, the Resource Management and Planning Appeal Tribunal hears appeals.

6.10 Victoria

6.10.1 Background

EIA procedures were enshrined in legislation when the *Environment Effects Act* (EE Act) was passed in 1978. Few amendments have occurred since the Act was proclaimed. This short Act, currently twelve pages, provides the statutory framework for EIA procedures; the detailed procedures are set out in the publication *Ministerial Guidelines for assessment of environmental effects under the Environment Effects Act 1978*, the most recent, eighth edition, being released in 2006. These guidelines are prepared under s 10 of the EE Act and are currently administered by the Department of Transport, Planning and Local Infrastructure (DTPLI), which is responsible for the EIA process. These guidelines are available on the department's website at <www.dtpli.vic.gov.au> (under Planning, then Environment). The status of current referrals under the EE Act, projects subject to EIA and the relevant assessments is also available at this website.

6.10.2 Environment Effects Act 1978

Once the Minister has determined that an EES is required, draft scoping requirements are exhibited for public comment for a minimum of 15 business days. The Minister considers any public submissions prior to finalising the scoping requirements. Once the EES is considered satisfactory by the Minister, he or she will seek public comment for a period of between 20 and 30 business days.

Amendments to the EE Act were made in 2006 after a review of Victoria's environmental assessment process began in late 2000. Although the amendments were not substantial, the most significant is that the Minister can make a conditional "no EES" decision. There are now a number of examples, one of the earlier ones being the Sugarloaf Interconnector pipeline project which the Minister determined was not an EES, subject to a number of conditions. The proponent, Melbourne Water has constructed a 70 kilometre pipeline to transfer water from the Goulburn River north of Yea to the Sugarloaf Reservoir in the Yarra Ranges. Here the ministerial conditions included preparation of a project impact assessment report for public comment and provision of an advisory committee to be appointed to hear submissions on the report.

Other amendments to the Act in 2006 include: the Minister can direct that a project be referred; the Minister can specify the form of EES required, and; the Minister can require a supplement EES for all projects, not just public works. Proponents can also now directly refer their proposal to the Minister for a determination as to whether an EES is required, whereas prior to the amendments only decision-makers or other relevant Ministers could refer proposals.

The Ministerial Guidelines (2006) were a major component of the review and certainly detail screening (or referral) criteria as well as expanding upon the decision criteria used by the Minister in determining whether an EES is required. In line with a national approach to assessment of coal seam gas projects, an amendment to these guidelines occurred in September 2012 that specified that any project proposal to extract coal seam gas, develop a new coal mine or expand an existing approved area for coal mining must be referred under the EE Act.

The Act applies to works, which could reasonably be expected to have or be capable of having a significant effect on the environment. Section 8(1) enables any person or body required to make a decision (under any law or Act of parliament in Victoria) in relation to works that could have a significant effect on the environment to seek the advice of the Minister for Planning as to whether an Environment Effects Statement (EES) is required. In addition, as specified in s 8(3) of the EE Act, a proponent may seek the advice of the Minister regarding the need for an EES.

The Act allows for EIA of both public and private works and establishes the basis of procedures for referral, triggering EIA, public comment, inquiry and the final assessment of environmental effects. The process of deciding whether an EES is required begins with the proponent, a relevant decision-maker or proponent Minister for public works, considering whether a potential exists for significant environmental effects to result from proposed works, and, if so, seeking the advice of the Minister for Planning by referring the proposal under the Act. If there is any doubt about whether the potential effects warrant referral of a proposal, early consultation with the Department is encouraged.

Referral criteria set out in the Ministerial Guidelines (see Figure 6.5 *over page*) help proponents and decision makers in making this determination.

Referrals are placed on the Department's website for information, along with the decision of the Minister. There are three different decisions that the Minister can make on a referral:

- an EES is required;
- an EES is not required if conditions specified by the Minister are met; or
- an EES is not required.

If the Minister determines that the proposed works could have significant environmental effects, he or she may require an EES to be prepared. Once prepared to the Minister's satisfaction, the EES is advertised to seek public comment; the EES is normally available for public comment for a period of between 20 and 30 business days. There is provision in the EE Act for the Minister for Planning to appoint one or more persons to conduct an independent inquiry to consider and advise on the environmental effects of proposals, subject to the Act. An inquiry reviews the EES and public submissions, rather than conducting its own investigation. The forms of inquiry available to the Minister, as specified in the Ministerial Guidelines, include:

Figure 6.5. Referral criteria for EIA in Victoria

Referral criteria: individual potential environmental effects	Referral criteria: a combination of potential environmental effects

Individual types of potential effects on the environment that might be of regional or State significance, and therefore warrant referral of a project, are:
• potential clearing of 10 ha or more of native vegetation from an area that:
  • is of an Ecological Vegetation Class identified as endangered by the Department of Sustainability and Environment (in accordance with Appendix 2 of Victoria's Native Vegetation Management Framework); or
  • is, or is likely to be, of very high conservation significance (as defined in accordance with Appendix 3 of Victoria's Native Vegetation Management Framework); and
  • is not authorised under an approved Forest Management Plan or Fire Protection Plan
• potential long-term loss of a significant proportion (eg, 1 to 5 per cent depending on the conservation status of the species) of known remaining habitat or population of a threatened species within Victoria
• potential long-term change to the ecological character of a wetland listed under the Ramsar Convention or in "A Directory of Important Wetlands in Australia"
• potential extensive or major effects on the health or biodiversity of aquatic, estuarine or marine ecosystems, over the long term
• potential extensive or major effects on the health, safety or well-being of a human community, due to emissions to air or water or chemical hazards or displacement of residences
• potential greenhouse gas emissions exceeding 200,000 tonnes of carbon dioxide equivalent per annum, directly attributable to the operation of the facility.
	A combination of two or more of the following types of potential effects on the environment that might be of regional or State significance, and therefore warrant referral of a project, are:
	• potential clearing of 10 ha or more of native vegetation, unless authorised under an approved Forest Management Plan or Fire Protection Plan
	• matters listed under the *Flora and Fauna Guarantee Act 1988*:
	  • potential loss of a significant area of a listed ecological community; or
	  • potential loss of a genetically important population of an endangered or threatened species (listed or nominated for listing), including as a result of loss or fragmentation of habitats; or
	  • potential loss of critical habitat; or
	  • potential significant effects on habitat values of a wetland supporting migratory bird species
	• potential extensive or major effects on landscape values of regional importance, especially where recognised by a planning scheme overlay or within or adjoining land reserved under the *National Parks Act 1975*
	• potential extensive or major effects on land stability, acid sulphate soils or highly erodible soils over the short or long term
	• potential extensive or major effects on beneficial uses of waterbodies over the long term due to changes in water quality, streamflows or regional groundwater levels
	• potential extensive or major effects on social or economic well-being due to direct or indirect displacement of non-residential land use activities
	• potential for extensive displacement of residences or severance of residential access to community resources due to infrastructure development
	• potential significant effects on the amenity of a substantial number of residents, due to extensive or major, long-term changes in visual, noise and traffic conditions
	• potential exposure of a human community to severe or chronic health or safety hazards over the short or long term, due to emissions to air or water or noise or chemical hazards or associated transport
	• potential extensive or major effects on Aboriginal cultural heritage
	• potential extensive or major effects on cultural heritage places listed on the Heritage Register or the Archaeological Inventory under the *Heritage Act*.

Source: DSE (2006: 7)

- an inquiry based on the EES written submissions; or
- an inquiry also involving submitter conference; or
- an inquiry with a formal hearing.

An inquiry by written submissions means that the Minister may appoint one or more persons to conduct an inquiry on the basis of the written submissions to an EES, without presentation from submitters. An inquiry by submitter conference is where the Minister may invite submitters to attend a "submitter conference" or roundtable session to provide an opportunity for them to speak about their submission and is a more informal forum for consideration of issues than a formal hearing process. A formal inquiry will generally conduct public hearings and review public submissions. All three forms of inquiry require a report to be submitted to the Minister, which guides the Minister's formal assessment. It is the latter inquiry process that has exclusively been used in Victoria, although a roundtable conference of experts or stakeholders has sometimes been used to complement a public hearing.

The Minister's assessment of the environmental effects of a proposal is the final step in the EIA process in Victoria. It will provide findings of the potential environmental effects of the project and whether these effects are acceptable. The Minister's assessment is forwarded to the relevant decision-maker(s) and must be considered when decisions are made about the proposal. This assessment is purely advisory and there is no legal obligation on the decision-makers to accept the recommendations of the assessment, as was seen in the case of the proposed Mornington Safe Harbour EES where the Minister's Assessment recommended the project proceed (with conditions), yet the decision-maker (local Council) decided not to allow the development application to proceed to an approval stage.

Ministerial discretion is a key feature of the Victorian process as under the EE Act it is the Minister who decides whether an EES and an Inquiry is required and then provides the advisory assessment.

While EIA occurs mainly under the EE Act, there is some form of a provision for EIA in the *Planning and Environment Act 1987* (PE Act), where environmental, social and economic effects must be considered in relation to both proposed developments and land use changes, and via the *Environment Protection Act 1970* (EP Act). As well as being put through EES procedures, most proposals subject to an EES would require some form of licensing or approval, such as a works approval under the EP Act, or planning scheme amendment to authorise the project under the PE Act. Where the licence/approval process requires public participation, through either the lodging of objections/submissions or a panel hearing, the practice in Victoria has been to link the exhibition period (associated with a planning scheme amendment for example) or inquiry with the submission period (or inquiry) set for the EES procedures. In other words, efforts are made to coordinate the approvals process with the EES process to reduce the need for sequential exhibition periods or inquiries and the delay this would entail as well as reducing the need for community members to unnecessarily participate in more than one process.

Scoping, to specify the matters to be addressed in the EES and the extent of their coverage, is routine for an EES process. Initially (during the 1970 and 1980s) a draft structure or table of contents was made available for public comment (ie the scope), while from 1990-2000 it was normal practice to form consultative committees composed of stakeholder representatives to guide the preparation of the EES. Since 2000, technical reference groups (TRG) composed of agency representatives have been established to provide advice to the proponent and the Department on the relevant matters that need to be addressed in the EES and to provide review and comment on draft EES documentation. The TRG is purely advisory and not a decision-making body.

Following minor amendments to the EE Act in 2006, the Minister now specifies the key matters to be addressed in an EES, as well as the procedures to be followed, as part of the initial triggering decision. The relevant matters are further detailed in scoping requirements for an EES. As with other jurisdictions, scoping requirements outline the studies that should be conducted, although they are not mandatory. Further, if other significant impacts are identified during the EES investigations they should also be addressed. The scoping requirements represent the terms of reference for the EES. The Ministerial Guidelines provide detailed information on what the EES should contain, including assessment of indirect and cumulative effects.

6.10.3 *Major Transport Projects Facilitation Act 2009*

With the introduction of the *Major Transport Projects Facilitation Act 2009* (MTPF Act), a new environmental assessment process was added to Victoria's existing EIA system. This Act enables a "declared project" to be assessed and approved under one of two processes:

- Impact Management Plan process, which according to the DTPLI applies to transport projects where the land is wholly owned by a public authority or reserved for a public purpose and certain approvals are either not required or have been granted; or
- a Comprehensive Impact Statement process.

Unlike the EE Act, an approval decision is also made by the Minister for Planning under the MTPF Act. Guidelines have been prepared under the MTPF Act that provide guidance on the form and content of project proposals that are required to be prepared by project proponents for declared projects under section 19 of the MTPF Act.

The Comprehensive Impact Statement is a rigorous environmental assessment process and projects in Victoria such as the East West Link project have been declared under the MTPF Act for the purpose of applying the assessment, approvals and delivery powers under that Act.

6.10.4 Mines (Aluminium Agreement) Act 1961, amended 2011

This Act was amended in 2011 and now allows for the preparation of an Environmental Impact and Management Report (EIMR) if a proposed mine extension (of relevance under the Act) is likely to have a significant additional environmental impact. If an EIMR is required, the Minister administrating the *Environment Effects Act 1978* must specify the scope for the EIMR having regard to a number of matters in Part 21D(7)(a)-(e) of the *Mines (Aluminium Agreement) Act 1961* (as amended in 2011) and following completion of the EIMR, the same Minister must provide an assessment of the EIMR to the Minister responsible for the *Mines (Aluminium Agreement) Act 1961*. There are no public hearings associated with this process, unlike the EES process under the EE Act.

6.10.5 Environment and Natural Resources Committee

Another process that involved itself with EIA in Victoria was that of the Parliamentary Committee. Before the 1980s, a Parliamentary Public Works Committee had been established to examine projects prepared by State government departments. It seemed that the original intention was to give backbenchers the opportunity to be involved in decisions about major government projects, particularly the financial aspects of those projects. Usually the more major projects, such as power stations, dams and sometimes roads, were referred to the committee.

In the 1970s, when projects were referred to the committee, EESs were sometimes included in the information presented to that committee; for example, Jeeralang 200 megawatt gas turbine power station and the Hazlewood-Cranbourne 500 kilovolt power line. Consequently, the Committee became more deeply involved in examination of environmental effects along with the more usual expenditure and technical details.

With the installation of the Cain Labor government came the introduction of the *Parliamentary Committees (Joint Investigation Committees) Act* in 1982. This Act was responsible for setting up five committees, particularly the Natural Resources and Environment Committee (subsequently Environment and Natural Resources). The functions of this committee were closely related to the more formal EIA procedures under the *Environment Effects Act*, in that the committee was to inquire into, consider and report to parliament on a range of factors that affected environment and conservation.

Under the parliamentary committee system established in 1992, under a new government, the role of the Environment and Natural Resources Committee is to inquire into specific issues affecting the environment, use of natural resources and land use planning, and make recommendations to the parliament. The current powers and responsibility of the Committee would appear to have been broadened compared with the situation in 1982, however, there is now a reduced emphasis on conservation of resources and identification of environmental impacts. The Committee's current responsibilities are to inquire into, consider and report to the parliament on:

- any proposal, matter or thing concerned with the environment;
- any proposal, matter or thing concerned with natural resources;
- any proposal, matter or thing concerned with planning the use, development or protection of land. (Parliament of Victoria (2013))

Importantly, the Environment and Natural Resources Committee does not have legislative or regulatory powers, and can only make recommendations. However, it has been involved in some influential inquiries. It is the Minister's responsibility to address the Committee's recommendations or findings.

During nearly 25 years of operation the committee has been called upon to look at routes for transmission lines, supply and demand for electricity, location of radio masts, diversion of a river, waste management, deposit legislation for beverage containers and most recently on water reform which is a critical issue for Victoria (as with most of Australia) at present.

In 2011, the Victorian Parliament requested an inquiry into the EES process in Victoria. Specifically, to inquire into:

- any weaknesses in the current system, including poor environmental outcomes, excessive costs and unnecessary delays encountered through the process and its mechanisms;
- community and industry consultation under the Act;
- the independence of environmental effects examination when government is the proponent; and
- how better environmental outcomes can be achieved more quickly and precisely and with a reduction in unnecessary costs.

In September 2011, after an extensive public hearing process, the ENRC released its Inquiry Report. The ENRC made fifty recommendations to improve the transparency and rigour of the process and address concerns raised by both industry and the community. The recommendations (in brief) included a number of mechanisms to better administer the Act and aimed at clarifying responsibilities. Other recommendations included enabling any person to have the power to refer a project to the DTPLI as well as appeal rights to VCAT against Minister's decisions.

Importantly, the ENRC recommended that the legislation be amended to confirm ESD principles as the overarching principles underpinning decision-making under the Act (ENRC (2011: 55)). It also recommended three levels for environmental assessment in Victoria:

- Level 1 – assessment on preliminary information
- Level 2 – Public Environment Report; and
- Level 3 – Environmental Impact Statement.

Other recommendations of the ENRC Inquiry into the EES process in Victoria included clarification of opportunities for public comment; building more accountability and transparency into the EIA process; provisions for post-EIA monitoring, auditing, enforcement and evaluation. The ENRC also called for strategic environmental assessment to be incorporated into the EIA legislation.

The Victorian government's response to the recommendations of the ENRC Inquiry was that it intended to proceed with reform of Victoria's EIA framework, including necessary legislative changes, on the basis of the following elements:

- EIA processes will be reformed to provide certainty for all parties about processes, considerations and timeframes.
- The reforms will be grounded in a clear set of objectives to guide their implementation.
- A tiered suite of EIA processes will be developed to enable efficient assessment of project proposals with a variable potential for significant environmental impacts.
- The reforms will embody principles of fairness, including for third party rights, as well as for reviews of key procedural decisions.
- Opportunities will be addressed for establishing clear links between the outcomes of the EIA process and regulatory decision-making under applicable laws, as well as compliance oversight by relevant bodies, to both drive improved environmental outcomes and streamline overall regulatory requirements.
- The reforms will provide an option for enabling the assessment of strategic proposals that are wider in scope than individual projects, where this would offer both efficiency and environmental benefits. (Victorian government response media release (2012))

This type of committee is not unique to Victoria. Other governments have established a variety of processes that provide for a review of public works. These committees, or other processes, frequently enable examination of environmental matters to a degree. As a result, in some situations they could be seen as adjuncts, or alternatives, to the formal EIA process, or in some instances they may even be considered as forms of SEA.

Of interest, the VCEC held an inquiry into environmental regulation in Victoria in 2009. Recommendations from the VCEC inquiry that it considered will improve the current process under the EE Act, and reduce uncertainty, timeframes and costs to business included:

- incorporating clear risk-based criteria in the Environment Effects Act for guiding decisions on the form of environmental assessment, including the provision of an intermediate tier of assessment;
- establishing negotiated, project-specific time limits to each stage of the EES process, and reporting performance against these time limits;
- integrating, at an operational level, the processes of the EES with the environmental aspects of necessary project approvals;

- engendering a stronger 'whole-of-government' approach by departments and agencies by establishing the role of the Co-ordinator General of Major Projects, and ensuring that technical reference groups function more efficiently; and
- promoting the use and recognising the input of strategic assessments in the EES process. (VCEC (2009: 22))

At the time of writing, minimal reforms proposed from either the ENRC or the VCEC inquiries had been implemented. It is also worth noting that although there are opportunities for environmental assessment under these other legislation and inquiry processes mentioned, most environmental impact assessment is undertaken via the *Environment Effects Act* in Victoria.

6.11 Western Australia

Enactment of the *Environmental Protection Act 1971* (EP Act) enabled the setting up of the Environmental Protection Authority (EPA), which is charged with the duty, among others, of enhancing the quality of the environment and conducting EIA in Western Australia. Under this Act, Ministers of the Crown refer matters which may have detrimental effects on the environment to the EPA. The authority can then require the provision of aid, information and facilities to assist in the reporting of environmental aspects of the proposal; in other words, powers for EIA procedures in Western Australia are embodied in the EP Act.

Before 1978, the EPA apparently favoured an ad hoc application of procedural arrangements adjusted to each project (Fowler (1982)). To formalise, and perhaps clarify, these procedures the Department of Conservation and Environment, an agency of the EPA, published the booklet *Procedures for Environmental Assessment of Proposals in Western Australia* in 1978.

Building on this experience the EIA process has evolved formalised administrative procedures (see Environmental Protection Authority (2012)).

A review of the EIA process in 1983 began with a public seminar to discuss working arrangements, and culminated with the enactment of the *Environmental Protection Act 1986*. Through this Act, the EIA process has remained essentially the same, but the role of EIA has been strengthened in several areas: the scope of proposals now includes projects, plans, programs, policies and changes in land use; compulsory referral of proposals of environmental significance operates; there are public rights to appeal advice given and procedures applied, (for both referral and assessment stages) and the Minister sets and enforces conditions, and can reject proposals on environmental grounds.

In 1992, when the Act was reviewed, a key issue concerned the uncertainty of outcome of the EIA process. In conjunction with the state of the environment reporting process, the EPA has worked to establish policies that set the grounds for judging the environmental acceptability of proposals, in advance of their planning and design. Policies have been prepared for several aspects of the environment and represent the Authority's view on particular factors when carrying out an EIA on a proposal; the guidelines for referring planning schemes to the EPA is an example of such a policy (see below).

The Western Australian process does not distinguish between public and private proposals, but is considered to apply to all actions which may have an environmentally significant impact. There are several ways in which a proposal can come to the EPA's attention. Principally, decision-making authorities must refer proposals that appear likely, if implemented, to have a significant effect on the environment. Otherwise a proposal may be referred to the authority by the proponent or a third party, the Minister may ask for an assessment, or the authority itself may call for a review (this may be triggered when a works approval or licence application is lodged). Once referred, all decisions about assessment are notified regularly in the press (see Environmental Protection Authority (2012)).

Proposals deemed not to be environmentally significant are not assessed. For other proposals that may have an impact on the environment, one of several levels of assessment may be undertaken. In-house assessment is reserved for proposals where the EPA is confident that the environmental impact is insufficient to warrant more detailed assessment. A proposal can be a project, plan, program, policy, operation, undertaking or development or change in land use, or amendment of any of these.

The public is given a significant role in the EIA process in Western Australia. All proposals that are referred to the EPA are publicly advertised, thus giving the public the opportunity to have some input into the decision about the level of assessment that will be carried out. Also, a member of the public can appeal to have the level of assessment, decided upon by the EPA, upgraded. After the level of assessment has been set, the proponent summarises the proposal and with the approval of the EPA communicates this information to those who are likely to be affected, or those likely to have an interest in the development. The EPA consults with this group of people when developing a list of key environmental issues that must be covered by the proponent in the environmental review. A summary of the completed review must be distributed to these people. These requirements ensure that those likely to be affected and those with an interest in the proposal are given the opportunity to participate in the EIA process.

The EPA decides not to assess some proposals, but reviews them informally and provides advice to the proponents and government agencies to manage the environmental effects. Although there is no formal public review period, issues raised by the public are taken into account. A review of the EP Act occurred in the late 1990s with changes to the EP Act being introduced in 2003. These changes to assessment procedures means there is an increased level of strategic environmental assessment within EIA. There are two levels of formal assessment in WA, being:

- assessment on referral information (ARI); and
- public environmental review (PER).

An ARI applies to proposals that raise one or a small number of environmental issues that can be readily managed or where it is apparent at the referral stage that the proposal is environmentally unacceptable. Public review is not considered necessary for this level of assessment. There are two categories of ARI – category A and category B (environmentally unacceptable). Category A proposals raise limited number of key environmental issues and these issues can be readily managed. Category B (environmentally unacceptable) are proposals that are inconsistent with established environmental policies, guidelines and standards or that the proposal is likely to have a detrimental impact on the environment (EPA (2012)). A PER level of assessment is used for proposals with significant environmental impacts or where there is major public interest. A scoping document is required for this level of assessment and is prepared by the proponent. Up to an eight-week consultation period is associated with the PER.

In regard to the public, the EPA has sought to increase the level of involvement and awareness in the early stages of the process, by the increased use of public meetings, and through public input in the drafting of the guidelines for the scope of the assessment documents, primarily for PERs.

An extensive range of guidelines have been developed by the EPA to assist those involved in the EIA process, including Guidelines for proponents and decision-makers. These Guidelines can be found at <http://www.epa.wa.gov.au/EIA/assessdev/Pages/EIAAdministrativeProcedures.aspx>.

Amendments made to the Act in 2003 enabled the Western Australian EPA to formally assess strategic proposals likely to have a significant effect on the environment. This means that the EPA can assess the potential impacts of policies, plans and programs. Referral of SEA by proponents is voluntary. This is meant to increase the certainty for proponents and to enable conditions to be set for subsequent proposals ("derived proposals") without assessment. For example, the State government's Department of Resources may refer to the EPA a proposal for a new industrial estate. The EPA would assess the proposal and the Minister would issue a statement of conditions to future industrial development ("derived proposals"). The EPA would decide whether a subsequent proposal was a "derived proposal" or required normal assessment. Amendments in 2010 changed this to assessment of "strategic proposals" rather than SEA, which will be assessed under the PER process.

The Western Australian government announced a further review of critical aspects of the EIA process in February 2008. The outcomes of this review have been implemented recently and include provision of a seven day public comment period before a decision is made on a referral; reducing the formal number of assessment levels from four to two (PER and ARI); EPA prepares the scoping document; and some consultation on draft conditions may occur.

6.12 Comparison of Procedures

There are many levels of comparisons that can be made depending upon the information sought; for instance, comparing the number of EISs per year may indicate differing levels of activity, but the reasons for this would be difficult to uncover. Given that some States/Territories have tiered EIA procedures where EIS is the higher order of EIA, this would not be a true reflection of how many proposals are subject to EIA in Australia.

Overall, the EIA procedures do not differ greatly in practice. What is different is where EIA aligns itself within the system, which is either land use planning (eg, New South Wales, ACT and South Australia), resource management/ environmental protection (Western Australia and Tasmania), a combination of land use planning and environmental protection (Qld) or stand-alone EIA legislation (Victoria and the Northern Territory). All have been subject to some form of review over the past decade to better align with a national approach identified in the IGAE to environmental impact assessment in Australia.

Chapter 7

CONTENTS OF THE EIS

As discussed in previous chapters, EIA is the environmental assessment tool which is used to identify the potential environmental effects of an action (or development proposal). Within the EIA process there are a number of steps: screening, scoping, preparation of the Environmental Impact Statement (EIS), assessment, and post-EIS monitoring and evaluation (although this last step is often not undertaken adequately). The EIS is one of two key products of the EIA process – the assessment report being the other.

In summary:

- Screening – where it is established if an EIA is needed, and hence whether an EIS is to be prepared or whether some other form of reporting would be adequate.
- Scoping – where the likely environmental effects are identified and the extent to which each will be investigated is considered.
- Prediction – where specific environmental impacts are forecast.
- Evaluation – where the significance of the impacts is determined and all the impacts are presented for comparison.
- Reducing impacts – where programs are devised to mitigate unavoidable impacts.
- Monitoring – where monitoring programs are developed to check the effectiveness of the predictions and the mitigation programs.
- Conclusions – where information and results are synthesised to make conclusions about the environmental impacts and make recommendations about the proposal being considered, its alternatives, and the mitigation and monitoring programs.

These last five points are presented as part of the EIS.

This chapter looks at all of these aspects. In addition, the stages of prediction and evaluation are discussed in detail in Chapter 8.

Once the final EIS has been prepared, it forms the basis upon which the assessment report, of the relevant government department and minister, is developed. The EIS is also the document that sets the directions for monitoring the environmental management of the proposal.

7.1 Establishing the Need for an EIA: Screening

Whether a proponent becomes involved in undertaking an EIA, and preparing the EIS, usually depends upon the type of proposal and its magnitude, and sometimes social or political considerations. Also, Porter (1985) points out that issues of timing, data collection and staging of assessment (among others) would affect the EIA process.

To give some direction to proponents, all EIA procedures in Australia have developed mechanisms for the selection of actions requiring EIA. As would be evident from the discussions in Chapters 5 and 6, all EIA procedures have some "trigger" which determines if an EIA is required. This may be fairly vague, such as where significant environmental effects are expected, or may be specifically stated, such as particular types or size of development. In the jargon of EIA, this stage of the process is called screening (also see 4.5.2).

The Australian and New Zealand Environment and Conservation Council's *Guidelines and Criteria for Determining the Need for and Level of Environmental Impact Assessment in Australia* provide a starting point for checking the environmental status of a proposal (ANZECC (1996)). The wording of these guidelines suggests that it is preferable if those administering an EIA process have previously specified the types of proposals which should be subject to an EIA. For example, the European Economic Community (EEC) Directive (see 5.6) lists mandatory projects (eg, oil refineries), and the Australian Capital Territory and New South Wales procedures (see 6.4 and 6.5) specify types of proposals, or the size of proposals which are required to be reviewed through an EIS. This type of designation makes it very clear as to when an EIA is required.

Designation of categories of proposals may be "inclusive" or "exclusive". An inclusive category would be where particular criteria are specified, and where any criterion is met the proposal must automatically be included in the EIA process. For example, it could be specified that an EIA would be required for all quarries that are larger than 2 hectares and are located within 40 metres of a waterbody, or are within 500 metres of a dwelling. Whereas an exclusive approach would be where the criteria for proposals that are to be excluded from the EIA process are specified. If these criteria are met then an EIA is not required, otherwise an EIA is undertaken. For example, public toilet blocks connected to sewerage systems, and quarries in areas zoned for extractive industry could be exempted as it would be assumed that their major potential impacts had already been considered.

Often little direction will be given about the need for an EIA, in which case most procedures rely on the fall-back trigger of whether the proposal is likely to result in significant impacts (see below for a discussion of how significance may be determined). While the concept of significance is the basis of EIA, it is not defined. After considering a range of approaches to assessing significance, Gilpin concludes that significance is "a collective judgment of officers, elected persons, and the public" (1995: 7). However, he also provides direction for that judgment in the form of guidance from the United States Council on Environmental Quality, which highlights the following factors:

- The importance of context – for society as a whole, the region, affected interests and the locality.
- The importance of time – relevance of short-term and long-term effects.
- The importance of intensity – effects on public health and safety; proximity to culturally, historically, ecologically, scenically or recreationally critical areas; controversial nature of effects; uncertainty, or unknown risks; whether precedents would be set; implications for cumulative impacts; effects on nationally registered structures or sites; impact on cultural or historical resources; threats to endangered or threatened species or habitats; and violation of environmental protection laws.

Along with the above guidance, where procedures do not specify the conditions under which an EIA is obligatory, the ANZECC guidelines can provide useful initial direction. The criteria for determining the need for an EIA are reproduced in Table 7.1 (*see opposite*). General questions to be answered under six headings, such as "Character of Receiving Environment", indicate the types of concerns that need to be considered. For each heading there are more specific questions to be considered (these can be found in Appendix A).

Unfortunately, the guidelines provide very little direction about what happens next. The questions can generally be answered yes, no or maybe, but there is no indication of whether a certain percentage of yes answers would mean that an EIA were required. To compound this, in any situation some answers may be more important than others; for example, if the proposal were to impact on a national park, does this mean that an EIA should be considered no matter how many of the other questions resulted in a no answer? Again, the decision is left to the person doing the assessment. Overall, a judgment is made (usually by bureaucrats) as to whether the proposed action will result in a significant impact, taking into account the magnitude of these potential impacts and the location of the development.

The advantage of using the guidelines is that they are generic and can be used for any proposal to give an idea about the types of issues which will have to be considered. A judgment can then be made about the importance of these issues. Also, the information obtained from the questions of the guidelines can be discussed with community groups and with those administrating the EIA process. These discussions can help make a decision about the need for an EIA.

A most important aspect of the guidelines is the reminder they give that assessments are seldom precise. The heading "Confidence of Prediction and Impacts" helps to ensure that the assessor does not take the answers to the questions at face value, but thinks about the assumptions which have been used to derive answers, and remembers that these assumptions will limit the accuracy of the answers.

Table 7.1 Criteria for the determination of the need for and level of EIA in Australia

Character of receiving environment	Potential impacts proposal	Resilience of natural and human environments to cope with change	Confidence of prediction of impacts	Presence of planning, policy framework and other statutory decision-making processes	Degree of public interest
Consider: Is it, or is it likely to be, part of the conservation estate or subject to treaty? Is it an existing or potential environmentally significant area? Is it vulnerable to major natural or induced hazards? Is it a special purpose area? Is it an area where human communities are vulnerable? Does it involve a renewable or a non-renewable resource? Is it a degraded area, subject to significant risk levels, or a potentially contaminated site? NOTE: Offsite as well as on-site characteristics should be considered, where relevant.	Consider: Will construction, operation and/or decommissioning of the proposal have the potential to cause significant changes to the receiving environment? (on site or off site, short term or long term) Could implementation of the proposal give rise to health impacts or unsafe conditions? Will the project significantly divert resources to the detriment of other natural and human communities? NOTE: This should include consideration of the magnitude of the impacts, their spatial extent, the duration and intensity of change, the total product life cycle and whether and how the impacts are manageable.	Consider: Can the receiving environment absorb the level of impact predicted without suffering irreversible change? Can land uses at and around the site be sustained? Can sustainable uses of the site be achieved beyond the project life? Are contingency or emergency plans proposed or in place to deal with accidental events? NOTE: Cumulative as well as individual impacts should be considered in the context of sustainability.	What level of knowledge do we have on the resilience of a given significant ecosystem? Is the project design and technology sufficiently detailed and understood to enable the impacts to be established? Is the level and nature of change on the natural human environment sufficiently understood to allow the impact of the project to be predicted and managed? Is it practicable to monitor predicted effects? Are present community values on land use and resource use likely to change?	Consider: Is the proposal consistent with existing zoning or the long-term policy framework for the area? Do other statutory approval processes exist to adequately assess and manage project impacts? What legislation, standard codes or guidelines are available to properly monitor and control operations on site and the type or quantity of the impacts?	Consider: Is the proposal controversial or could it lead to controversy or concern in the community? Will the amenity, values or lifestyle of the community be adversely affected? Will the proposal result in inequities between sectors of the community?

Source: ANZECC (1996)

Ultimately, the point of screening is to decide if a proposal is required to undergo an EIA. If it has been determined that the decision is "yes", there are a number of considerations to be taken into account so that a satisfactory assessment is produced. In particular, attention must be given to the preparation of a relevant and comprehensive EIS, as this will contain the information upon which the assessment of environmental impact will be based.

As indicated above, the decision whether a proposal requires an EIA will be made through the screening process used for the particular government's EIA process. The remaining sections of this chapter take as the starting point that an EIA is to be undertaken, and focus on the development of the contents of the EIS. Although the emphasis is on the EIS, other levels of reporting may be used, such as public environment reports (as in the Commonwealth and Western Australian procedures). Use of the screening guidelines outlined in Table 7.1 may result in many maybe answers, and so more information would be needed before a final decision about the need for an EIA could be made. In this situation a lesser level (if available) of EIA reporting and assessment would be appropriate. Generally, these levels require much less information to be presented than for a full EIS. They are likely to require a description of the proposal and the environment which could be affected, along with an assessment of possible environmental impacts, but the requirements for this type of reporting should always be discussed with the relevant government agency before any planning for the report is considered. Chapter 6 describes the various types of EIA reporting within Australian EIA procedures.

7.1.2 Determining significance

All approaches to screening and scoping (see 7.4.2) rely on the basic idea that in EIA we need to identify the significant environmental impacts. The concept of significant is the test for the need for an EIA and for what should be included in the assessment process. However, significance can have different interpretations, as pointed out by Beanlands and Duinker (1983):

(a) Statistical significance – relatively value-free approach based on isolating man-induced [sic] effects from natural variation.
(b) Ecological significance – there is little consensus of a definition, but it could include aspects of loss of habitat, extinction of species, reduction of productivity, or similar measures.
(c) Social significance – EIA is inherently an anthropocentric concept centred on the effects of human activities and ultimately involves a value judgment by society of the importance of effects.
(d) Significance to the proposal – effects of any magnitude are insignificant if they are not considered in making decisions about the proposal. This involves the other three concepts of significance in that an effect may be considered significant for planning and/or deciding about the project "if it represents a statistically significant

> change in a socially important environmental attribute, that is either directly or indirectly (through ecological linkages) caused by the project in question". (1983: 45)

Distilling these aspects into something to guide the approach to an EIA, the authors suggest:

> Within specified time and space boundaries a significant impact is the predicted or measured change in an environmental attribute that should be considered in project decisions, depending on the reliability and accuracy of the prediction and the magnitude of the change. (Beanlands and Duinker (1983: 45))

The practical application of this relies on a good deal of subjectivity and the above subsection provides an indication of how the issue can be avoided – by adopting someone else's approach (eg, ANZECC).

Slightly more specific direction is given by the regulations associated with the *National Environmental Policy Act 1970* (US) (NEPA) (see 5.2). Blumm (1988) notes that the six interpretations of significance are:

- A proposal may have significant adverse effects even if in the opinion of the proponent it is considered to have environmental benefits overall.
- Significant effects are more likely to occur if the proposal affects sensitive resource structures or species.
- Significance can be a function of whether a proposal is controversial or sets a future precedent.
- Significance may be a consequence of the proposal's cumulative effects.
- Conflicts with local laws can produce issues that are significant.
- Significant effects may result from uncertainties about a proposal's effects.

However, once again there is obviously a lot of interpretation involved in these directions.

After a substantial review of screening processes, as part of a review of EIA generally, Sadler (1997a) comments that "a 'two track' approach to evaluating significance can be followed, adjusted to the degree of uncertainty and controversy that characterises a specific proposal, and to the changes in understanding that occur in moving from the early to the later phase of environmental assessment". This process is as follows:

- First, apply technical criteria when the likely changes associated with a proposal can be predicted with reasonable accuracy.
- Secondly, use a "negotiating" approach (involving the key stakeholders) when factual information is limited and/or there is a high degree of uncertainty regarding potential impacts (typically associated with the early phases of assessment when activity-environment relationships are only generally understood and there is flexibility in the details of the proposal).

In addition, Sadler (1997b) concludes that as part of the above process, those doing the screening must:

- apply clearly defined criteria to evaluate significance;
- document the reasons for interpretations – this is particularly critical when broadly-based considerations, such as biodiversity, are important;
- describe, as necessary, the confidence levels for impact judgments; and
- provide a non-technical explanation if complex methods are used (eg, for aggregation and weighting).

More particularly, the study concluded that there are some key principles that lead to good practice in determining significance of environmental impacts (see Figure 7.1 *opposite*).

The concept of significance has been tested under the Commonwealth EPBC Act in *Booth v Bosworth* (the Spectacled Flying Fox Case), where it was determined that a "significant impact" is an "impact that is important, notable or of consequence having regard to its context or intensity" (Leshinsky (2010: 1)).

The Spectacled Flying Fox case concerned the significant impact on the heritage values of the Wet Tropics World Heritage Area that could result from destroying large numbers of flying foxes (not themselves listed as endangered), which ventured from the Wet Tropics World Heritage Area in search of food. The species is considered to be important for seed dispersal, evolutionary processes and general ecological function within the Wet Tropics rainforest (McGrath (2005: 2)). The case involved an application for an injunction under the EPBC Act on the practice of a lychee farmer who, to protect crops, had erected electric grids which were killing Spectacled Flying Foxes as they roamed for food.

In this case, the test for "significant impact" proposed by Branson J – "an impact that is important, notable or of consequence having regard to its context or intensity" – is consistent with its plain meaning, case law, extrinsic material and the objects of the EPBC Act, Biodiversity Convention and World Heritage Convention. It also satisfies the constitutional constraints of Commonwealth legislative power. Consequently, it is suggested that the Federal Court will follow this test when considering the EPBC Act in the future and that the test therefore is authoritative.

7.2 Form of an EIS

An EIS is a report which attempts to document the environmental effects that could be expected following implementation of an action. A principal purpose of an EIS is to present information on the environmental effects of actions to decision-makers so that the environmental consequences of the action will be known if a decision is made to proceed with the action. An EIS

Figure 7.1 Good practice principles for determining significance of environmental impacts

1. Requirements:

A statement on the likely significance of environmental impacts should form the basis for judging project acceptability and conditionalities. The evaluation of significance may be made following a four-step methodology of impact analysis to determine:

- the nature and extent of impacts (eg, type, duration)
- likely adverse effects on the receiving environment (eg, sensitive areas, land use, community traditions)
- magnitude of impacts (eg, low, moderate, high)
- options for impact mitigation (eg, reduction, avoidance)

2. Key Actions and Principles

- incorporate tests of significance at various stages of impact analysis
- make a concluding evaluation by reference to several questions:
  - how adverse are the predicted effects (eg, change, loss, foreclosure)?
  - how do these vary in scope and intensity (eg, in their effect on ecological and resource values)?
  - how significant or serious are the impacts (eg, irreversible → inconsequential)?
  - how probable is it that they will occur (eg, high risk → low risk)?
- note, in general, impacts are likely to be significant if they:
  - are extensive over space or time
  - are intensive in concentration or proportion to assimilative capacity
  - exceed environmental standards or thresholds
  - do not comply with environmental policies, land use plans, sustainability strategy
  - adversely and seriously affect ecologically sensitive areas
  - adversely and seriously affect heritage resources, other land uses, community lifestyle and/or indigenous people's traditions and values

Source: Sadler (1997)

will also document the "consequences" of the project not going ahead which is usually in terms of economics, infrastructure requirements, and potential tourism loss to name a few. The other major purpose of an EIS is to disseminate information about the project to the many stakeholders that may be directly or indirectly affected by the proposal.

These points are emphasised in the Victorian Guidelines (2006) (remembering that an EIS and EES are effectively the same):

> *Description of the project*
> A clear and sufficiently detailed description of the proposed project is needed to enable the effective assessment of potential environmental effects. This description should set out:
>
> - Project rationale and objectives;
> - Location, technology and design of project components;
> - Site characteristics and surrounding area;

- Communities, properties and/or residences that may be affected by the proposal, including a description of the way that they may be affected;
- Proposed methods for mitigating adverse environmental effects and risks;
- Proposed program and time schedule for project implementation;
- Proposed method for implementing the project, including responsibility for construction, operation and where relevant, decommissioning.

Description and assessment of relevant alternatives
Alternatives may be:

- Siting and layout alternatives;
- Design or process alternatives;
- Scale, where the magnitude of the project might be varied in response to demand or constraint factors;
- Timing of project activities;
- Staging of project development. (Department of Sustainability and the Environment. (2006: 15))

The Victorian Guidelines go on to detail the relevant environmental effects that the EES should assess, such as:

- physical systems;
- ecological systems;
- human communities;
- land use effects; and
- economic effects. (Department of Sustainability and the Environment (2006: 16-17))

An example of where the specific requirements are included in legislation occurs in the ACT, where specific matters to be addressed in the scoping document are prescribed in s 54 of the *Planning and Development Regulation 2008*. The following includes some of these requirements:

(1) A scoping document for an EIS must contain the following:
  (a) the name, address, telephone number and email address of the people who prepared the document; ...
  (d) each potentially significant environmental impact that must be addressed in the EIS; ...
  (f) if the scoping document relates to a s 125-related EIS – the issues that must be addressed in the EIS in relation to the public health impact of the development proposal to which the EIS relates;
  (g) any current relevant information held by the Territory, of which the planning and land authority is aware, that would be of use in preparing the EIS;
  (h) the requirements for the form and format of the EIS. ...

(4) A scoping document for an EIS may also include any of the following:
  (a) requirements in relation to the methods of assessment to be used in the EIS;
  (b) for each potentially significant environmental impact identified in the scoping document – a requirement that the proponent of the development proposal to which the scoping document relates consider ongoing management, monitoring or reporting regimes;

(c) a list of impacts that are not significant environmental impacts that can be addressed through an altered design or in some other way.

Other States have expressed similar expectations of EISs being essentially the same in concept and direction. They also generally fit in with the more detailed functions which Munn (1975) has proposed. That is, an EIS should:

- describe the proposed action, as well as alternatives;
- predict the nature and magnitude of the environmental effects;
- identify the relevant human concerns;
- list the impact indicators;
- from the predicted values of the environmental effects, determine values of each impact indicator and the total environmental impact;
- make recommendations on acceptance of the project, remedial action, acceptance of one or more alternatives, or rejection; and
- make recommendation for inspection procedures to be followed after the action has been completed.

A critical issue which can affect the form and style of the EIS is whether the EIA is seen to be a process of scientific analysis or a process of negotiation among competing interests. Armour echoes the feelings of many involved in EIA that making and interpreting environmental predictions embodies more political than scientific judgments in the comment that "decision-makers have to be concerned not only with what is 'technically best' but also with what is 'most socially acceptable'" (1983: 28). In this socio-political context EIA becomes a process for negotiation among competing interests. This then changes the emphasis of an EIS from systematic analytical assessment of environmental effects to clarification of environmental values and options to follow. Values cannot be escaped in an EIS; however, EISs typically follow the traditional "scientific analysis format", perhaps with a limited discussion of values, rather than accept their socio-political function.

7.3 Concepts and Definitions

The concept that is basic to an EIS is that:

> Actions initiated by humans lead to environmental effects which lead to environmental impact.

In this context *actions* may be:

> legislative proposals, policies, projects and operational procedures;

while environmental effects represent:

> processes (eg, erosion of soil, dispersion of pollutants, displacement of persons) that are set in motion or accelerated by the actions;

and environmental impact is:

> the net change (good or bad) in human health and well-being (including the well-being of the ecosystems on which human survival depends and

> hence in the health and well-being of humans themselves resulting from an environmental effect) – impact then is related to a difference between the quality of the environment as it would exist with or without the same action.

Consequently an *impact indicator* is an element or parameter that provides an indication (in some qualitative sense at least) of the environmental impact.

7.4 Content of an EIS

7.4.1 General guides to content

In providing information for decision-makers and the public, an EIS typically contains a lot of data. To assist those involved in preparing EISs, government departments frequently provide guides as to what an EIS would be expected to contain. For example, the Western Australia Guidelines for preparing a PER (WA EPA (2012)) note that Public Environment Report document should include:

- the identification of the proponent and proposal location;
- justification and objectives for the proposed development;
- the legal framework, including existing zoning and environmental approvals, and decision making authorities and involved agencies; and
- alternatives considered, including location options. This section should provide analysis of alternatives in the following hierarchy that moves from broad/strategic to increasingly narrow/project specific in nature:
  - Need/meeting needs – is this development needed? Consider no-action alternative.
  - Mode/meeting general goals – is this development proposal the best way to meet the general goal? Consider alternative technologies or options.
  - Location/meeting project objectives spatially – what is the best location for the project. Consider alternative locations with a view to minimising environmental impacts.
  - Timing/meeting project objectives temporally – what is the best sequence of development for components of the project?
  - Implementation mechanisms/designing project – what is the best way to optimise the project so as to minimise environmental impacts? Consider detailed site design, layout, technologies and mitigation strategies. (WA EPA (2012: 4))

Australian States, and others, have been guided by the approach adopted under the NEPA (see 5.2). For this Act the Council on Environmental Quality published guidelines, which noted:

> The following points are to be covered:

(a) a description of the proposed action, a statement of its purposes, and a description of the environment affected, including information, summary technical data, maps and diagrams where relevant; all adequate to permit an assessment of potential environmental impact by commenting agencies and the public;
(b) the relationship of the proposed action to land-use plans, policies and controls for the affected area;
(c) the probable impact of the proposed action on the environment (including positive and negative effects for both the national and international environment, together with secondary or indirect, as well as primary or direct effects);
(d) alternatives to the proposed action, including where relevant those not within the existing authority of the responsible agency. A rigorous exploration and objective evaluation of the environmental impacts of all reasonable alternative actions, particularly those which might enhance environmental quality or avoid some or all of the adverse environmental effects, is essential;
(e) any probable adverse environmental effects which cannot be avoided (eg, water or air pollution, threats to health);
(f) the relationship between short-term uses of man's [sic] environment and the maintenance and enhancement of long-term productivity;
(g) any irreversible and irretrievable commitments of resources that would be involved in the proposed action;
(h) an indication of what other interests and considerations of federal policy are thought to offset the adverse environmental effects of the proposed action. (Canter (1977: 258))

It is interesting that under NEPA there was precise direction as to what an EIS should contain, whereas other guidelines are just that – a guide to the contents.

Guidelines such as the above provide a general indication of the broad material to be covered and are relevant to all EISs. However, they do not provide much information as to what should be covered under each heading for a particular proposal, for example a power station. At this stage, assistance can be gained from previous investigations (EISs) and from generalised methods (such as those discussed below). However, the technique of scoping is being increasingly relied upon.

All this may give the impression that preparation of an EIS is a step-by-step, linear process. However, Sadar (1994) reminds us that EIA is an iterative process; as indicated in Figure 7.2 *over page*. It may be necessary to return to previous stages and reconsider previous conclusions because:

- new or unforeseen issues may arise, including changes in public attitude or project design;
- environmental and social conditions may change, especially for a multi-year project; and/or
- knowledge about the environmental interactions may improve and lead to better identification of issues, prediction of impacts or their significance. (Sadar (1994: 14))

Figure 7.2 The iterative nature of EIA

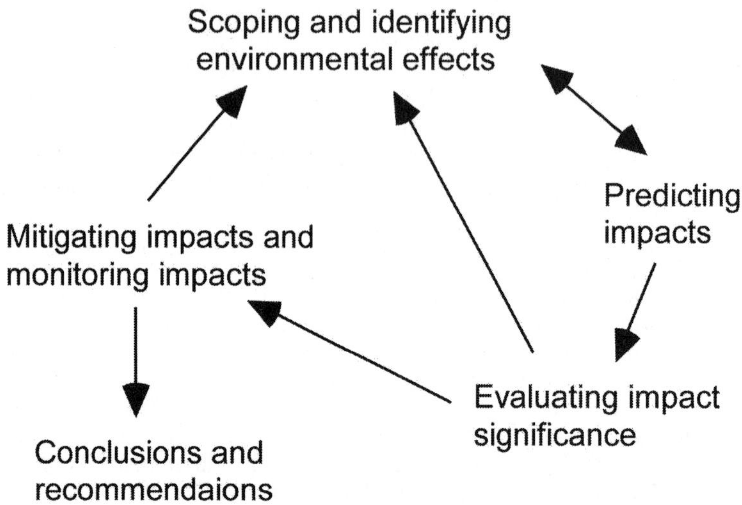

Source: Sadar (1994)

7.4.2 Scoping the contents of a particular EIS

Scoping is a method finding favour to determine the issues to be investigated in the EIS. This process has the ability to identify what the EIS should contain, and the extent to which each issue requires attention. A judgment about significance (see Figure 7.1) is frequently required to decide whether an EIA is undertaken; for example, when "there is a likelihood of regionally or State significant adverse effects on the environment" (Department of Sustainability and the Environment (2006: 2))

The point of the scoping stage is to identify the important items to be included in the EIS, as most attention is given to those environmental effects that may be significantly affected. Tomlinson lists the aims of scoping as:

(i) to identify concerns and issues requiring consideration;
(ii) to facilitate an efficient EIS preparation process;
(iii) to enable those responsible for EIA to properly brief the study team on the alternatives and impacts to be considered at different depths of analysis;
(iv) to provide an opportunity for public involvement;
(v) to save time. (1984: 186)

Simply stated, scoping involves the bringing together of the ideas, for the contents of the EIS, held by a variety of people in the community – the proponent, government, non-government organisations and interested individuals for a particular development proposal, action or policy. Tomlinson (1984) discusses the concept of scoping in terms of some methods for developing the scope, or terms of reference of the assessment, and the responsibilities of those involved in the EIA process. Scoping is evolving and there is little documentation about it, but Tomlinson points out that with involvement of the public, acceptable terms of reference can be developed and the likelihood of major controversy about the EIS is reduced. Scoping relies on the exchange of information and concerns between the interested parties through an appropriate organisation. Scoping is often the first formal opportunity for the public to participate in the EIA process, through providing input into the scope of the EIS.

According to United Nations Environment Programme (2002), the key objectives of scoping are to:

- inform the public about the proposal;
- identify the main stakeholders and their concerns and values;
- define the reasonable and practical alternatives to the proposal;
- focus the important issues and significant impacts to be addressed by an EIA;
- define the boundaries for an EIA in time, space and subject matter;
- set requirements for the collection of baseline and other information; and
- establish the Terms of Reference for an EIA study. (2002: 228)

The Terms of Reference for an EIS are broadly the outcome that would be sought from the scoping phase. These should clearly identify the work to be done, and who is responsible for undertaking the stages of the EIA. Nonetheless, flexibility has to be built into the stages to allow the study to adapt to changes if necessary.

Tomlinson (1984) provides an overview of scoping procedures, many of which rely on some form of checklist. An example of this sort of list is ANZECC's (1996) *Guidelines and Criteria for Determining the Need for and Level of Environmental Impact Assessment in Australia.* (Table 7.1 on *page 182* and Appendix A reproduce these guidelines.) In essence, the guidelines provide a reminder to the person doing the scoping that there are many elements of the environment that potentially may be impacted upon. Answering the questions of the checklist with yes, no or maybe identifies those parts of the environment where there are likely to be impacts, thereby identifying those environmental effects which have to be covered in the EIS.

In 8.3, examples of general checklists are presented which can provide a starting point for scoping, or thinking about the important elements of the EIS. More specific checklists have been developed for application to particular types of proposals, such as hydro-electric power plants, main roads and community facilities. Gilpin (1995) provides a comprehensive overview of the range of these checklists and their contents.

Clearly decisions about what is valued will depend on the experiences of those involved. In general, experience is at the base of most of the techniques that have been developed to assist with scoping. The personal and professional experience of the person planning the EIS is always going to be a starting point for development of the scope. This experience can be expanded by including the input of colleagues, and by calling on the experience of relevant experts. The content of previous EISs, for similar proposals, is another source of guidance along with that from the checklists discussed earlier.

Another approach, which may be used in conjunction with experience and checklists, has been used in several EIA procedures across Australia since the early 1980s. Preparation of EISs have frequently begun by the proponent publicising the intended contents of the study (and document), seeking suggestions from interested parties. Comment from relevant government agencies and community and conservation organisations has also been sought to help determine the range of issues. Scoping is also flexible – unknown factors/ issues may arise during the course of preparing the EIS and the proponent will need to investigate these emerging issues. In these cases, scoping guidelines may be amended.

Boundaries are also an essential element of scoping. Sadar (1994) comments that this entails setting limits to what is to be included in subsequent steps in the assessment. Common sense and practical considerations guide the process which should:

- focus on the most important issues;
- limit the amount of information to be gathered to manageable levels; and
- be able to propose realistic mitigative and monitoring measures.

He also suggests that the relevant boundaries relate to spatial, temporal and jurisdictional limits. As importantly, the ecosystems and social components to be excluded should be identified along with those that will be included in the EIS.

7.5 Comments on Contents

The following headings are used as an example of the structure for an EIS. Under these headings the material that would typically be covered is outlined.

7.5.1 Objectives

These should be objectives of the proposal, just what it is intended to do, not objectives of the EIS itself. Some examples will help to illustrate this.

Alcoa in its Portland aluminium smelter EES (1980) notes in the objectives section, after giving a brief history of Alcoa in Victoria, that "Alcoa's proposal is to construct a primary aluminium smelter, with an ultimate production capacity of about 530,000 tons per year on a site at South Portland, Victoria" (1980: 107). The section goes on to discuss the various approvals the project required. However, in "Need for the Proposal" it is noted that aluminium smelting will expand in countries where there are domestic energy, and bauxite surpluses (eg, Australia) and that "because of problems ... in other countries with surplus energy and bauxite, Australia is well placed, in the short term, to develop a competitive new aluminium smelting capacity to supply world markets".

Hence, the objective is to build a smelter to produce aluminium for export. However, the point that the aluminium is for export is important and could have easily been stated as part of the objectives; rather than the reader having to ferret it out.

A good example of clearly stated objectives is the Victorian Country Roads Board's EES for the Omeo-Mitta Mitta Road (1980: 6). Following an outline of the board's aim to provide safe, efficient and environmentally acceptable roads, five specific objectives are listed:

(1) To provide an all weather, all season road for the safe and efficient movement of people and goods between Omeo and Mitta Mitta.
(2) To develop a route which meets and enhances the social and economic need of the people and areas it serves.
(3) To develop a route which will exploit the tourist potential of the area without detriment to the environment.
(4) To design and construct a route commensurate with the best engineering practice, minimising its effect on the physical environment.
(5) To plan and construct a route with a funding program compatible with the board's overall financial resources and commitments.

It is notable that the intention to minimise environmental effects was actually considered part of the proposal and is clearly stated as such.

The Ryan Corner Wind Farm EES (EES Summary (2006: 4)) project objectives include:

- contributing to State and Federal government policy objectives to maintain a secure, efficient and affordable supply of energy in Victoria while reducing the intensity of greenhouse gas emissions from the energy sector;
- minimising the adverse impacts of the facility on visual landscape values;
- ensuring that the construction and operation of the facility does not unduly affect the amenity of nearby houses;
- avoiding or minimising impacts on species and communities listed under the *Flora and Fauna Guarantee Act 1988*;
- avoiding or minimising impacts on places with Aboriginal and non-Aboriginal cultural heritage values; and
- minimising any physical impacts from the construction and operation of the facility, such as erosion, sedimentation, road damage and traffic hazards.

Another example of a project's objectives is the Bootu Creek Manganese Project in the Northern Territory. The Public Environment Report (PER) stated that the objectives of the project are to:

- produce a manganese product for export to an OM Holdings owned Ferro Alloy plant;
- produce manganese product at a low cost whilst maintaining high levels of performance in terms of quality, safety and environmental production; and
- ensure benefits of the project are shared with members of the Northern Territory and local communities. (Martinick Bosch Sell Pty Ltd (2004: 23))

In summary, the clearer the objectives, the easier it is to determine exactly what the project is supposed to do, and what potential impacts the project is aiming to avoid or minimise. As importantly, objectives provide a basis on which to evaluate the performance of the proposal, and its alternatives. For example, by taking the Omeo-Mitta Mitta Road, will it provide a safe all-weather road and will it exploit the tourist potential without detriment to the environment, and will the Ryan Corner wind farm unduly affect the amenity of nearby dwellings?

Clear objectives also help in choosing between alternatives. For instance, it may become clear that one alternative could fulfil the most important objective and fail the rest, while another alternative scores reasonably on all. In such a case the trade-offs will be clearer to the decision-makers. Further, a straightforward statement of objectives will help to indicate whether or not environmental considerations are part of the proposal's design.

An emphasis on articulating clear objectives is promoted by the Environmental Protection Authority of Western Australia. Morrison-Saunders and Bailey (2000) note that this clarity is sought to promote a transparency in decision-making associated with EIA. Under the Authority's processes, the screening and scoping stages are used to identify the key environmental factors that will be investigated in the EIA, and to link these factors to environmental objectives established by the EPA. Highlighting these objectives makes all the interested parties aware of the issues against which the impacts of the proposal will be assessed. The EPA prepares guidelines for the EIA, including coverage of the objectives that have been identified, then the proponent prepares the EIA documentation in accordance with the guidelines. During the latter stages of the EIA process the documentation (and proposal generally) is evaluated by the EPA using the objectives established for the particular environmental factors. This evaluation becomes the basis for the EPA's assessment report to the minister.

7.5.2 Description of proposal and main alternatives

There are no hard and fast rules here, but enough information is needed to evaluate how the most important environmental factors would be affected. Text, diagram, maps and examples of similar actions are all useful. Discussion of alternatives should be objective, and where some are not to be investigated fully, reasons should be given for their deletion.

The *do nothing* alternative should always be included, with a discussion of it being adopted; that is, what would the future look like without the proposal? The do nothing (or no build) alternative is always feasible and gives a "base case" against which the performance of other alternatives can be compared in terms of environmental impact, economic effects and other performance measures indicated by the objectives.

In addition, there is likely to be a wide range of alternatives to the proposal. The *scope* of the alternatives may be substantial. For example, if the proposal is construction of a road, the alternatives could reasonably be expected to include the broad transport alternatives, such as rail transport and demand management (to reduce the need for more road space).

In many instances there will be alternative *locations* for the project. However, with private projects this alternative may not be possible if the property has been purchased, or if discussion of alternative sites would lead to land speculation and raise the cost of the possible sites. In these cases, alternatives are discussed in terms of alternative *layouts or designs*.

On any site there will be a variety of alternatives relating to the *scale* of the project. Frequently projects comprise a number of parts, some of which may be deleted to reduce the size of the project. The scale of a power station could be reduced by constructing fewer generation units or an office building could be built with four stories rather than seven. Rather than the project having discrete pieces, it may be more of a continuous nature; so the hole for a quarry need not extend to the geological limits of the material being mined, but could stop after a proportion had been taken.

During the design and planning of any project there will be thousands of opportunities to consider alternatives for the *site* details. Water tanks are available in a variety of shapes, and colour is another choice. Likewise the location of plant, equipment and buildings associated with a large project will have many possible variations, which may reduce the environmental concerns without affecting technical efficiency.

Finally, when considering the *processes and operation* of the project there will again be alternatives. Water quality may be achieved using different processes (eg, dosing with chlorine, ozone treatment), while noise may be controlled through the use of ear-protectors or by using silenced equipment.

The range of categories of alternatives has been summarised by United Nations Environment Programme (1997) as:

- Demand alternatives (eg, using energy more efficiently rather than building more generating capacity).

- Activity alternatives (eg, providing public transport rather than increasing road capacity).
- Location alternatives, either for the entire proposal or for components (eg, the location of the processing plant for a mine).
- Process alternatives (eg, the re-use of process water in an industrial plant, waste-minimising or energy efficient technology, different mining methods).
- Scheduling alternatives (where a number of measures might play a part in an overall program, but the order in which they are scheduled will contribute to the overall effectiveness of the end result).
- Input alternatives (eg, raw materials, energy sources).

7.5.3 Description of existing environment

The existing environment includes not only the physical environment, but also the built, social, cultural and economic environment. As Canter (1977) suggests, this description provides the baseline information necessary to predict the environmental effects of the various alternatives. However, there is no single list of environmental items that is appropriate for every EIS, as each project will have its own potential impacts. Rather, a separate list should be considered for each EIS.

Some guidelines for EIS preparation provide descriptions of items to be included; most do not. A flexible approach means that an EIS which embodies a range of actions can be produced with the least irrelevant data. Data for data's sake does not encourage a sympathetic review of the EIS. However, it is important not to leave out something that is relevant. Past EISs for similar actions provide one of the best guides for what would be included. For consistency and clarity, it is worth grouping the data items in categories of socioeconomic, physical, chemical and biological. Munn (1975: Table 3.1) gives examples of socioeconomic items.

Clearly the scoping exercise (see 7.4.2) undertaken for the EIS will provide specific direction for which parts of the environment should be described. Having identified the potentially significant issues, or areas where there may be impacts, it will be important to describe the existing situation, and then investigate the extent of those impacts.

7.5.4 Description of likely environmental impacts

This is a difficult but critical area. The accuracy with which we can assess physical impacts is dependent on our understanding of how the environment functions. Given perfect understanding, it should be possible to predict exactly the direction, degree and rate of change of the physical environment as it responds to a given stimulus. As our knowledge is often less than perfect, Hey (1976) cautions that the accuracy of any predictions will be reduced. Nevertheless, attempts must be made to predict the effects of alterations on the socioeconomic, physical, chemical and biological items previously described in the EIS.

There are many scientific approaches and models that can be used to predict impacts on the air, water and noise environments. Less quantitative methods are available for predicting the impacts on the biological environment. The prediction of the impacts on the socioeconomic environment vary from quantitative (eg, increased income) to qualitative (eg, whether alcohol addiction may increase if houses are isolated by a road proposal).

Accurate predictions of ecological change are difficult. Generally the only predictions that can be made about ecosystems are broad ones based on previous experience. That which is not expected cannot be predicted and certain changes are unpredictable for they are based on random events. So, as Moss (1976) suggests, it may be relatively simple to make predictions about change, but it is impossible to guarantee correct predictions.

However, methods have been developed to aid the process of predicting environmental changes. For instance, to help identification of important items and outline their relationship to environmental changes, checklists, matrices or flow diagrams (discussed in 8.3.2) could be used.

Checklists are comprehensive lists of environmental effects and impact indicators designed to stimulate the analyst to think broadly about the possible consequences of actions. A word of caution is needed, though, as relying upon the checklist could lead the analyst to ignore factors that are not on the list.

Matrices typically employ a list of human actions in addition to a list of indicators. The two are related in a matrix, which can be used to identify cause and effect relationships.

Flow diagrams can be used to help the analyst visualise the connection between action and impact. While the flow chart could become complex for other than simple actions, its use in the initial stages could help to clarify some relationships.

Scoping (see 7.4.2) has the ability to take the information and methods which are seen to be most relevant to a particular proposal, rather than rely on generalised processes that have been developed for similar situations; general methods can never be expected to be accurate when applied to specific situations. With scoping, relevant individuals and/or organisations are asked to list the important factors, based on their experience and knowledge, which they consider should be included in the assessment.

Methods for prediction cover a wide spectrum and cannot be easily categorised. All predictions are based on conceptual models of how the universe functions. They range in complexity from those that are totally intuitive to those based on explicit assumptions concerning the nature of environmental processes. Provided that the problem is well-formulated and not too complex, scientific methods (mathematical models) can sometimes be used to obtain predictions (eg, air pollution estimation).

However, methods for predicting the behaviour of qualitative variables are difficult to find or to validate. In many cases, the prediction consists of indicating merely whether there will be a degradation, no change or enhancement of environmental quality; in other cases qualitative ranking scales are used.

It is essential to note that the environment is never as well-behaved as assumed in models. Anyone making predictions is to be discouraged from accepting off-the-shelf formulae without an understanding of how the formulae work and whether they are in fact relevant to the situation under study. Generally, scoping does not usually involve prescribing the methodologies to be used.

As a final point, the description of environmental impacts should also cover all the environmental effects which had not been identified as important, in sufficient depth to demonstrate why they could be classed as unimportant. The description should also consider effects in terms of immediate short-term effects, long-term effects, adverse effects, beneficial effects, and irreversible and irretrievable commitment of resources.

7.5.5 Safeguards to avoid, minimise or ameliorate adverse impacts

This section could come either here, or after the evaluation section. The point of considering safeguards here is to encourage the analyst to think about the "extras" that may have to be added onto the basic proposal, before the evaluation of the proposal is undertaken. A crucial point to remember is that after safeguards are added to a proposal, the alternatives may become more attractive (eg, cheaper).

The sorts of things covered in this section are fairly obvious; for example, soil conservation measures that would be needed to stabilise road batters in dunes, landscaping that would be needed to hide a car park, or "scrubbers" that would be needed to remove high levels of sulphur from air emissions.

Any safeguards and mitigation proposals should be reported in sufficient detail to indicate that they are feasible, and that there would be a commitment to carrying them out.

7.5.6 Evaluation and discussion of alternatives

This is one of the more difficult sections. There are any number of books and papers that have been written on the topic of evaluation, so evaluation techniques will be covered in detail in Chapter 8 and the current discussion will be confined to a brief summary.

The evaluation section can be expected to cover the basis for the final recommendations of the EIS. This is the section which brings it all together. Ideally, there should be an assessment of how each alternative measures up to the objectives that were outlined at the start of the EIS, plus the evaluation of how each alternative would affect each of the environmental items that were previously identified. In practice, usually only the evaluation of impacts is undertaken. There are a number of ways of determining the importance of each "impact indicator" and displaying the evaluation. The three most popular types are:

1. *Display of sets of values of individual impact indicators* – All the raw data for each alternative are listed for the reader to decide their relative importance.
2. *Ranking of effects within impact categories* – In this case, the raw data is interpreted for the reader, to the extent that some form of designation is placed on each effect to show the extent/magnitude/direction of the effects; for example, ranked from 1 (most desirable) to 5 (least desirable).
3. *Normalisation and mathematical weighting* – This basically means putting all the environmental effects into comparable units and having an "objective" method for assigning numerical weights to each factor. Benefit-cost analysis is a special case of normalisation. (Crudely put, economic values are assigned to each factor, then all the benefits are summed and divided by the sum of the costs. An overall benefit would occur when the ratio was greater than one.)

The assessor has a wide range of methods to choose from. In terms of the resources required to use any method, its ability to replicate results and the degree of detail displayed, it may be fairly easy to choose which method best suits a particular project. However, each method needs to be regarded with caution; not so much for what it shows, but for what is hidden by virtue of the method's limitations and the simplifying assumptions which are always involved (see 8.2 for further discussion).

Another aspect of evaluation is the value system which is inherent in the collection, analysis and reporting of data. Throughout the preparation of an EIS, numerous objective and subjective judgments must be made; about which environmental features to study, how to study them, how to compare changes in them (impacts) and how to summarise these findings for the EIS reader. Matthews (1975) suggests that both subjective and objective judgments will be found in an EIS, but it is important that the two types are not confused. These judgments rely on the values of the person making them, so it is necessary for the analyst, and EIS reviewer, to recognise the value system they work under. Preferably, the EIS should indicate where value judgments are made. While such clarity is advisable throughout the EIS, it is particularly important where evaluations are made, so that the logic behind comments made about impacts and alternatives is clear.

For the Stockland Point Lonsdale Residential and Waterways Project EES the proponent prepared evaluation criteria (see Table 7.2 *over page*) for its project, which were used to assess the proposal as well as alternatives put forward (Stockland (2007)).

Table 7.2 Evaluation criteria for the Point Lonsdale Residential and Waterways Project EES

Evaluation Objectives	Assessment Criteria
Development is integrated with Point Lonsdale	Connectivity with Point Lonsdale and the existing Lonsdale Lakes Estates residential area. Provision of a formal edge to extent of development within the Point Lonsdale township. Minimisation of impact on the local road network and surrounding intersections, notably access from Bellarine Highway. Ability to cater for future provision of public transport.
Development with good internal layout and linkages	Internal connectivity and linkages between land uses and open space networks.
	Ease of movement within the site for vehicles, pedestrians and cyclists.
Sustainable use of infrastructure and resources	Ability for fill material to be sourced within the site to create development platforms.
	Ability to provide services and infrastructure.
Conservation and enhancement of sensitive environmental areas on site	Minimisation of impacts on Aboriginal or post-settlement cultural heritage.
	Retention of significant native vegetation and consistency with principles of Victoria's Native Vegetation Framework.
	Minimisation of impacts on migratory waterbirds and other species of significance.
	Potential for creation of new habitat.
Creation of a variety of residential opportunities	Opportunities for creation of a variety of residential development forms across the site.
Maintenance of water quality and environmental conditions	Maintenance or improvement of surface water flows between Lake Victoria and Lakers Cutting.
	Maintenance of on-site water quality.
	Ensurance of water quality and ecological character of Swan Bay Ramsar area and the Marine National Park.
Opportunities for environmentally sensitive design outcomes	Provision of opportunities for optimisation of solar orientation.
	Provision of opportunities for minimisation of potable water demand.
	Provision of opportunities for water sensitive urban design for storm water management.
Optimisation of use of low lying land for open space and surface water management	Extent of opportunities to utilise low lying areas required for sourcing of fill to be used for open space and surface water management.
Provision of an overall societal benefit, taking into account economic benefits, social outcomes and residual environmental impacts.	Enhancement of employment and local business opportunities.
	Enhancement community services networks, community planning, and recreation and conservation value.
	Establishment of framework for long-term, sustainable management of the site through Environment Management Plans and formalising environmental commitments through approval conditions.

Evaluation Objectives	Assessment Criteria
Provision of commercial and community facilities for residents.	Ability of any commercial/retail facilities to service local and wider community.
	Ability of community facilities to service residents within the development and wider community.
Consistency with local and State planning policies	Meeting of provisions of local and State planning policy frameworks, specifically in relation to public transport, urban settlement and growth, and environmental management.

Source: Adapted from Stockland & Maunsell Australia Pty Ltd (2007: 3:17)

7.5.7 Conclusions/recommendations

EIS guidelines may or may not include "Conclusions" in their suggestions for contents. Nonetheless, sections are usually identified which provide the opportunity for "tying up loose ends" and making concluding assessments of potential impacts and any mitigation and monitoring proposed for the project. These sections could equally well be written into a single chapter if it is decided that the material does not warrant extensive coverage. Conclusions could cover:

- comments from government departments, conservation groups and the public;
- discussion of how the proposed, or favoured, alternative involves compromises between short- and long-term effects, and the extent to which the proposed action forecloses future actions;
- comments from the consultant, if appropriate; and/or
- reasons the proposed action has been endorsed.

Recommendations would probably make the obvious point that some action should be undertaken: whether as proposed or with modifications. In some situations the recommendation may be that the proposal should not go ahead; but given all the work that would have been put into it by the proponent at this stage, and the opportunities for modification, such a recommendation would be a rare occurrence.

Recommendations may go on to suggest how certain departments or decision-making groups should proceed; for example, how a parcel of land should be zoned, or how discharge licence applications should be considered. Recommendations could also comment on the need for involvement of certain organisations (government or private) in parts of the project. Importantly, recommendations could emphasise the role and responsibilities of a monitoring program to review the prediction and mitigation of environmental impacts.

To give this a practical focus, Sadar reminds us that conclusions and recommendation should be "guided by accepted principles of objectivity, fairness and, above all, common-sense. It should also be remembered that EIA is only an advisory process and that the final decisions most often rest with others and not those who conduct the assessment" (1994: 61). As a result he suggests taking into consideration:

- the mandate of the proponent and the importance of the proposal to the objectives of the organisation and the public interest;
- the nature and basis of the opposition or concerns of the public or pressure groups;
- the issues and concerns that have been identified and which cannot be dealt with by the proponent alone;
- whether there are issues in addition to the environmental concern about the proposal which may affect approval of the proposal;
- whether there are any legal implications from not dealing with some issues;
- whether other spheres of government, departments or agencies, or specialists need to be consulted before a final decision is made;
- if some issues can be dealt with through short-term commitments, or if long-term allocations of resources will be required;
- the resource implications (both money and people) for implementing the proposed mitigation and monitoring programs;
- the nature and extent of public participation, of other spheres of government, departments and other stake-holders in the proposed mitigation programs;
- whether there are obstacles (such as resource restrictions) which could affect the implementation of the proposal, or the mitigation or monitoring programs; and
- if there may be public or political reaction, or other costs, as a result of the proposed recommendation.

7.6 Building in Bias

Hollick states, "The EIS must be objective and unbiased if it is to aid objective decision-making" (1986: 114).

This may be the desirable goal, but even Hollick recognises that an unbiased EIS is not possible. He sees that bias can creep in through attempts to promote a particular outcome, or the subconscious intrusion of personal values. There can also be distortion from ignoring or playing down impacts, neglecting some phases of the proposal, omitting certain information, being overly optimistic about the effectiveness of safeguards, and using value judgments about the importance of particular factors or impacts. To add to this general list of possible biases, Beder (1990) points to the use of language and the favourable interpretation of data.

The way in which information is presented in an EIS, or any report, can have a strong influence on the reader. Careful choice of words and the order in which data are discussed can draw out reactions wanted by the author. Distortion, exaggeration and omissions help to make a stronger argument. This type of bias is not necessarily intentional; unless authors are aware of their own values, they may not appreciate that what they have written is not the "whole truth".

The possibilities for misdirection are noted by Spry:

> A large EIS may provide overwhelming reasons why a development is benign, but a single impact which is obscured by the bulk may be so critical as to require modification or termination of the project. Most such impacts are highly technical and require both a specialised knowledge of the subject and an appreciation of environmental values. Such obscured impacts may escape detection by both the public and the approving authority. (1976: 253)

These types of bias can be tolerated if the role of the EIS is understood. Spry helps to clarify what can be expected from an EIS:

> The traditional view of the EIS since its inception is that it must be comprehensive, objective and, above all, impartial. One might, however, consider the possible advantages of an alternative explicitly partisan EIS. The "impartial" EIS is expected to consider all alternatives and consequences and, by itself, allow an authority to make a well-informed decision. No significant impact may be omitted and no impact obscured. The "partisan" EIS on the other hand would present only the view favourable to the developer, on the basis that he [sic] is entitled to present his own case in the best possible light. The advantage of the partisan EIS is that the assessor would be aware that he [sic] must investigate the problem himself and challenge every fact and opinion before accepting it. Thus he is less likely to overlook critical points and obscured impacts. (1976: 225)

It is reasonable to assume that all EISs fall into the "partisan" category, whether by design or not.

A particular example of bias has been presented by Lyon (1989) in connection with economic analyses, in which a value may be assumed for the discount rate to give an unrealistically good picture of the proposal's economic performance. He suggests that the bias occurs when there is an unwillingness to incur costs to receive future benefits, but a willingness to accept present benefits while discounting future costs. Researching the factors influencing public policy-makers, he found: "Participants [in the study] were less likely to perceive health impacts as being important relative to the financial benefits, and were more likely to vote for projects when the health impacts were not apparent for years" (1989: 255).

Proponent bias is an issue that repeatedly comes up during the EIA process, especially regarding the EIS and information presented within it. This is largely due to an EIS being prepared and funded by the proponent. Beder (1993) suggests that the EIS is increasingly perceived as a biased public relations document and can never be truly objective since they are often undertaken late in the project planning process when the proponent will almost certainly have committed considerable financial resources to a particular option at a particular site. Morgan (1998) suggests that while most stakeholders participating in the EIA process recognise the benefits of EIA, the assessment authorities need to ensure bias has been considered. He suggests that one mechanism of overcoming proponent bias is to charge the proponent of the impact assessment study and then contract an independent group to carry it out. Although a good idea in terms of potentially eliminating bias, as Morgan suggests we are unlikely to see a major shift away from proponent funded and managed EIA studies in the near future, and a formal quality control step is important. There are more instances now of particular studies within an EIS being peer reviewed or independently reviewed by another specialist. This certainly allows for an increase in the robustness of the assessment, however it may not resolve the issues around bias.

By giving attention to the points discussed above, and the many other areas where bias may creep into an assessment, most of the obvious areas of "inaccuracy" may be kept under control. To assist, Gilpin (1995) provides a checklist of points to consider in an EIS to reduce subjectivity and increase objectivity. However, there will always be some inherent biases related to culture and philosophy of the author(s) and the time and context in which the EIS is prepared. An awareness of the possible biases should always be in the mind of the EIS author and the reviewer.

7.7 Monitoring, Surveillance and Auditing – Checks on the EIS and EIA Process

Of the many aspects of EIA, monitoring usually receives some attention, surveillance receives negligible comment, while auditing is rarely considered. Monitoring is usually spoken of in terms of systems for observing, measuring and recording information about specific aspects of a project (eg, levels of pollutants in air emissions). Discussion of safeguards for a proposal (see 7.5.5) frequently includes suggestions for monitoring programs. Sors (1984) provides a general discussion of monitoring, while ANZECC (1991a, 1991b) outlines the role of monitoring in the Australian context.

Surveillance can be taken to mean a system designed to look out for trends. This can be a monitoring of the monitoring programs, or a "watchdog" role assumed by a government agency or community group to ensure that recommendations about safeguards (or agreed actions) are being complied with.

Auditing is effectively an evaluation of the EIA process. It investigates whether or not predicted impacts have actually occurred, whether methods used to make these predictions were reliable, whether recommendations were followed, whether safeguards were effective. There is considerable interest in auditing worldwide. As Gilpin (1995) discusses, this form of post-project analysis has been the subject of a United Nations Economic Commission for Europe task force report, with the conclusion that it is a necessary part of EIA. However, in most countries post-project analysis is still rarely undertaken and indeed it is seen as a crucial deficiency in EIA at present.

One of the few studies to look at the accuracy of predictions is that of Sewell and Korrick (1984), who also monitored a sample of EISs for science content and usefulness. More generally, the quality of an EIS can be assessed from a variety of points of view; for example, whether it is comprehensive, or clear for non-technical readers. A systematic approach for reviewing EISs has been developed by Colley and Lee (1990). As well as providing a means of comparing the adequacy of an EIS, the review provides a checklist, for anyone preparing an EIS, of the important aspects of the document. The review involves four levels of factors, an overall assessment and specific criteria. More particularly, the review areas focus on the extent of the description of the proposal and baseline conditions, identification of key impacts, identification of alternatives and mitigation of impacts, and communication of results.

Following this work, Hickie and Wade (1998) report the results of another British study to improve the effectiveness and efficiency of environmental assessment. The criteria for the review of the EISs came from nineteen specific questions, see Figure 7.3 *over page*, and provide guidance what can be done to ensure that an EIS provides clear and useful information.

This comprehensive, but possibly pedantic, list of criteria provided several proposals for the improvement of EISs. The chief proposals included: the need to improve the communication of information (to be less technocratic and to give greater consideration to readers); and development of standardised EIA procedures and formats to ensure that all aspects would be covered in a structured and logical manner. Hickie and Wade (1998) also commented on the desirability of developing environmental action plans to manage the EIA process. Such a plan would have four parts: the management system for the environmental assessment process from the release of the EIS to the completion of the project; the inclusion of objectives targets for each of the environmental "constraints" identified in the EIS; a summary list of environmental specifications for inclusion in the engineering and other contract documents for the project (to ensure that the environmental measures were incorporated in the construction/implementation stages; and drawings showing all the objectives and targets in plan form.

While the criteria we have looked at so far focus on the detail of EISs, Raff (1997) has considered the "bigger picture" issues. His list of ten "principles of quality in environmental assessment" (see Figure 7.4 on *page 207*) are largely based on the standards and practice that have come from the US experience with NEPA. Clearly, these principles have much in common with the experiences and reflections from other countries.

Figure 7.3 Criteria and questions for the evaluation of the effectiveness of EISs

The following questions were developed to relate to both the content and the presentation of the EIS:

- The project – were sufficient data provided to enable a non-specialist to visualise the project?
- Site and local environment factors – were sufficient data provided to enable a non-specialist to visualise the site and local environment?
- Maps – was the provision of maps satisfactory?
- Areas affected – was sufficient information on the affected areas provided?
- Photographs – was the provision of photographs satisfactory?
- Adjacent land use, site designations and legal rights – was the information provided satisfactory?
- Baseline conditions – were the baseline surveys sufficient?
- Consultation – were the ranges of consultation sufficient?
- Impacts – were the full ranges of impacts sufficiently identified?
- Effect prediction, magnitude, and significance – were the indications sufficient?
- Alternatives – were all alternatives considered and adequately assessed?
- Mitigation measures – was the information provided satisfactory?
- Enhancement (similar to mitigation measures, but may include specific positive environmental outcomes from the proposal) – was the information provided satisfactory?
- Monitoring program – was an adequate monitoring program provided?
- EIS layout and presentation – was the layout sufficiently clear and logical?
- Non-technical summary – did it provide an adequate summary in a succinct form?
- Emphasis – was the information provided in an unbiased manner?
- Key issues – was it clear to the reader what the key issues were?
- Overall impression of the EIS – what was the impression given?

Source: Hickie and Wade (1998)

In Australia there has been plenty of informal discussion about the quality of EISs, but there is not much accessible written discussion. However, Gilpin (1995) discusses three reviews of New South Wales EISs which provide guidance for anyone preparing an EIS and wanting to produce a quality document.

Some of the concerns raised by the three studies include:

- technical and complex presentation in the EISs;
- insufficient, or omission of, information relevant to the decision-making process (such as coverage of environmental impacts);
- extraneous or non-essential information;
- the offering of a justification for a prior decision;
- a narrow focus on aspects of the proposal (eg, engineering detail) rather than a broader environmental planning approach.

Figure 7.4 Ten principles for assessing the quality of an EIS

The quality of an environment assessment and its EIS is reflected in the following:

- an assessment of environmental impacts cannot be restricted to "site specific" environmental effects – but should consider the effects that the proposal may cause "downstream" and to neighbouring areas and communities;
- the assessment must contain a statement of alternative courses of action and their environmental significance;
- bigger/greater plans or policies must be assessed in addition to single phases in their execution – elements of cumulative impact assessment and strategic environmental assessment may be required;
- segmentation of the project should not occur – the focus of the EIA of a large project should not be narrowed into smaller phases by segmenting it;
- viable and realistic time-frames are needed – aspects of science fiction and crystal ball inquiries should not be included in the consideration of alternatives and major shifts in community norms in the short term;
- alternative courses of action must include the option of doing nothing;
- the assessment must engage in a real inquiry and not merely dispose of alternatives in favour of a decision that has already been arrived at;
- environmental effects are not to be disregarded just because they are difficult to identify of quantify;
- the EIS must take a "hard look" at the environmental consequences of the proposal and what it entails – a hard look includes applying acceptable standards of reasonableness and good faith, considering alternatives as realistic possibilities, and alerting decision-makers to inherent problems with the proposal; and
- the findings of the EIS must be presented in clear language and the methodologies employed to arrive at them must be explained.

Source: Raff (1997)

A more recent instance of EIA review was the audit conducted of the EIS and its supplement for the proposed second Sydney airport (SMEC (1999)). This extensive audit concluded, in summary, that the methods used in the assessment of specific issues were generally adequate, but there were deficiencies in the areas of noise assessment, water quality and cumulative impacts. Also, the documents did not examine consequences of not constructing the proposal, nor did they provide a clear description of the proposed action. Specifically regarding the (draft), the audit commented that EIS lacked sufficient detail for a project of this size. So the overall review was fairly mixed, with some key deficiencies being identified.

A general approach to the review of an EIS has been used in Canada for some time. Ross (1987) notes that once an EIS has been prepared under the federal process, it is evaluated by the panel set up under the Environmental Assessment Review Process for the proposal. This evaluation focuses on three aspects:

(1) Is the EIS suitably focused on the key questions which need to be answered to make a decision about the proposed action?
(2) Is the EIS scientifically and technically sound?
(3) Is the EIS clearly and coherently organised and presented so that it can be understood? (1987: 140)

With some three decades of experience with the production of EISs we would expect that their quality should be showing signs of improvement. However, Tzoumis and Finegold (2000) have found that in the US there apparently has been a decline. They point out that under the processes associated with NEPA (see 5.2), all EISs (particularly the draft EISs that are commented on by the public) have been rated for both information accuracy (using a three level scale), and for the environmental impacts of the preferred alternative (using a four level scale). Based on the sample of over 19,000 EISs, their study concludes that over time it has been more difficult for an EIS to achieve the top rating on both scales, and EISs have been getting lower ratings. They note that the "top" agencies, those that have been submitting the bulk of the EISs, show a lack of consistent performance. While this may be the case, another explanation of the results could be that expectations of reviewers have increased over the years and with the subjectivity built into the ranking scales, recent reviewers may be harder in their judgments. Nonetheless, the authors suggest that the ratings can be valuable means of assisting the agencies preparing EISs to look into ways of improving performance. Also, the information about the ratings of EISs should be public information so that those preparing EISs would be able to identify those that are "good performers" and be able to use them as models.

The European Union issued a *Guidance on EIA: EIS Review* in 2001 which "aims to help developers ad their consultants prepare better quality EISs and competent authorities and other interested parties to review them more effectively" (2001: 5).

Overall, it is clear that as with other aspects of EIA there is no one way of checking on the quality of EISs. However, the types of evaluation discussed above could help to improve the prediction capability and acceptability of EIA. Inevitably they will be used to contribute to reviews of EIA procedures, and they may guide performance improvement in the future.

7.8 Final Comments

Jain, Urban and Stacey (1981) present criteria for the review of documents which are most relevant to EIS. These criteria are reproduced in Table 7.3 (see *page 210*) and are useful for reviewing both the EIS which has been written (to see if it has the opportunity of being reasonably intelligible to non-experts) and that which is being examined (to remind the reviewer of where unclear interpretations may occur). As with all checklists, it covers most general cases rather than every situation, but the general approach can be adapted for most purposes.

Further guidance in writing a "neutral" EIS is provided by Bendix (1984). However, perhaps there is a role for more imaginative writing styles that provide the opportunity to engage the reader's interest so that reading the EIS need not be tiresome.

No matter what the writing style, fundamental bias inevitably occurs when data to be included in the EIS is selected and interpreted. It comes as a shock to most people that data, particularly that relating to environmental issues, is seldom objective; that is, it can be interpreted to have different meanings. In a very readable article, Savan (1986) gives examples of how scientists can be "coopted" to present a particular view, and how a set of data can be given more than one interpretation. More specifically related to EIA, Nelkin (1975) uses proposals to develop a nuclear power plant and a new airport runway to examine the involvement of experts, and the lay public, in the technical and political debate that surrounded the proposals. She notes that these analyses emphasise the politicisation of expertise; that is:

> [T]he way in which clients ... direct and use the work of experts embodies their subjective construction of reality – their judgments, for example, about public priorities or about the level of acceptable risk or discomfort. When there is conflict in such judgments, it is bound to be reflected in a biased use of technical knowledge, in which the value of scientific work depends less on its merits than on its utility. (1975: 54)

It is well to be wary about where data has come from, and understand the motives of whoever is presenting it. To add to the complexity of this, however, the form of reporting information can consolidate biases.

A further comment on the content of an EIS is that over the past 40 or more years that EIA has been operating, the general content has not changed; rather it is the presentation, complexity, community and regulators expectations and delivery of the content that has changed. The earlier EISs were mostly a single document focused on key issues relevant to the proposal whereas now there are multiple volumes of the EIS and technical reports covering key issues as well as many "potential effects". There are a number of reasons for this, including the introduction of numerous regulations and policies that require attention in the EIS, as well as the fact that the way we do science and report findings may have changed over the same period. More critically, communities, government, NGOs and private sector expectations of the EIS (and the EIA process) have also changed from the early 1970s, when the US NEPA was introduced, to the present. We have certainly seen more recourse to the courts in Australia since the introduction of the EPBC Act where various decisions and assessments have been challenged (discussed in 6.3). Issues to be considered when writing or reviewing any document:

Table 7.3 General document review criteria

Issue	**Criteria**
A. Quantification	1. Use well-defined, acceptable qualitative terms. 2. Quantify factors, effects, uses and activities that can be quantified.
B. Data	1. Identify all sources. 2. Use up-to-date data. 3. Use field-data-collection programs as necessary. 4. Use technically approved data-collection procedures. 5. Give reasons for use of unofficial data.
C. Methods and procedures	1. Use quantitative estimation procedures, techniques and models for arrival at the best estimates. 2. Identify and describe all procedures and models used. 3. Identify sources of all judgments and assumptions. 4. Use procedures and models acceptable by professional standards.
D. Interpretation of findings	1. Consider and discuss all environmental effects or factors before any are dismissed as not applicable. 2. Give thorough treatment to all controversial issues, and discuss the implications of all results. 3. Consider the implications for each area of a range of outcomes having significant uncertainty. 4. Analyse each alternative in detail and give reasons for not selecting it. 5. Scrutinise and justify all interpretations, procedures and findings that must stand up under expert professional scrutiny.
E. Flavour and focus	1. Do not slant or misinterpret findings. 2. Avoid use of value-imparting adjectives or phrases. 3. Avoid confusion or mix-up among economic, environmental and ecological impacts and productivity. 4. Avoid unsubstantiated generalities. 5. Avoid conflicting statements.
F. Presentation	1. Use consistent format. 2. Use tables, maps and diagrams to best advantage and link these to the text. 3. Avoid mistakes in spelling, grammar and punctuation.
G. Readability	1. Write clearly. 2. Remove all ambiguities. 3. Avoid the use of technical jargon all technical terms should be clearly explained.

Source: Based on Jain, Urban & Stacey (1981)

Chapter 8

DETERMINING IMPACTS FOR THE EIS

8.1 Role of Impacts

Selection of the environmental issues to be addressed in the EIS was discussed in Chapter 7, as were the points to be covered in the objectives and description of the proposal. Preparation of the EIS continues with the important steps of describing the existing environment and assessing the environmental impacts, while identifying how significant these impacts may be. As a result, the environmental impacts can be assessed relative to each other, and relative to technical, economic and political factors that may be involved.

Initially though, the impacts have to be predicted. Sadar (1994) suggests the need to consider:

- Whether each impact can be ascribed to the proposal.
- How a particular element of the proposal may give rise to an impact.
- The probability of each impact actually occurring.
- The magnitude of each impact and their extent over space and time.

Whether related to environmental matters or not, most evaluations are undertaken by comparing one factor (often expressed as some number) to another factor (number). Hence, the first step is to establish a basis upon which comparisons can be made, or attempt to give the varied factors a standard point of reference; this is usually done by quantifying the factor.

This chapter raises issues to be considered in attempts to quantify impacts, and specifically reviews the general methods that can be used to estimate impacts. This review also considers the methods' utility for evaluation of alternatives to the proposal and evaluation of the variety of impacts associated with a proposal.

8.2 Quantifying Effects

Previously it may have been good enough merely to say that air pollution would increase to "astronomical levels" without saying what this meant. It is difficult to consider the importance of this statement compared with, say, "many families will have to be relocated". The problem with using general or vague statements is that the reader is forced to assume the impacts. For example, an obvious question would be: what is the relationship between air pollution levels and human health?

The "tidy" way to compare things is to use numbers. Particularly with environmental effects, it makes life much simpler if the person who has to make a decision about the proposal can simply compare the number that has been derived to represent the environmental impact of the proposal with other numbers, such as the cost of the proposal or the number of hours of travel saved. Leopold et al (1971) began looking at quantification procedures for environmental effects in the early 1970s. The cornerstone of their procedure was to assign figures to the magnitude and importance of effects and then to sum the results.

In summary, the approach is as follows:

- Decide which environmental factors may be affected (ie, the environmental effects, or the processes altered by the proposed action) – for example, air pollution, aesthetics or social interaction.
- Determine the magnitude of the change in each of these factors – for example, small or large change – and (if possible) assign a figure to this change.
- Determine the importance of the change – for example, not important or very important, and (if possible) assign a figure to this importance.
- Combine the magnitude and importance measures for each environmental factor, and (if of the same units) combine all the "scores" together.

Uncertainty is a major concern for the prediction of environmental changes, and hence the assessment of impacts. As De Jough (1988) notes, uncertainty occurs in prediction, but also in the expression of values and in decision-making. To cope with uncertainty he suggests several approaches. In particular, in the case of model development, checks can be made through the evaluation of output, residual error analysis, block and recursive algorithms, or speculative simulation modelling. For handling uncertainty in input data, Smyth (1990) proposes the use of sample program design, sensitivity analysis or Monte Carlo error analysis; while for prediction, the techniques include scenario building, Monte Carlo simulation, speculative simulation modelling and expert systems.

However, enthusiasm for quantifying effects needs to be kept in context. The "inescapability of value judgments, and therefore the importance of pluralistic participation", has been highlighted by Barbour (1980: 201). He has noted the results of a study on technology assessments (and these results are equally appropriate to other forms of assessment) which show that value-laden decisions occur at a number of points in assessment procedures. For example, there are influences associated with the biases of assessors, as a result of their cultural backgrounds, and the political environment in which they operate.

8.2.1 A word about assigning figures

Assigning values to impacts is sometimes difficult but not impossible. At the very least, it forces the assessor to think quantitatively and produce numbers that can be debated. This is the rub, however. What has led to assigning that particular number? Why not another number? Is the abstract/real model (relationship) being used to derive the number infallible, or are there approximations in it? Would someone else doing the work arrive at the same number?

It may well be that there is no definitive theory or even a reasonable model to help derive these numbers. It may be that this analysis is as good as anyone else's. Even scores which come from a recognised model (or formula) depend not only on the model being an accurate reflection of the situations but also on the quality of the data which is fed into it; error and distortion can creep into both areas. In other words, the numbers are not fixed, but depend to an extent upon the person doing the work (and on the person's knowledge, background, experience and values).

With particular reference to matrices, Barbour (1980) discusses some of the issues regarding the assigning of numerical scores. Administrative personnel lacking depth in environmental and resource studies can use matrices or other guides to organise terms of reference for ecological studies, and to help ensure that the study design is comprehensive. However, Barbour cautions that matrices cannot provide quantitative answers to technical and judgmental questions, and, if used for these purposes, can damage the effectiveness of the study:

> The greatest danger in the matrix method is that it is often prepared by a person from a single-discipline background who feels compelled to fill in every blank space, bypassing years of professional experience available in a region ... (1980: 110)

Further, the use of numerical scores, whether associated with matrices or some other quantifying approach, tends to gloss over the systems (or interconnecting) aspects of the analysis.

Rather than relying solely on numerical methods, Barbour suggests that it is preferable for an environmental manager to use concurrent numerical methods and judgment. Sensitivity analysis can help. If numerical scores are used, controlled variation of the numbers can determine the critical factors that could push the analysis in one direction or another; that is, identify how sensitive the analysis is to certain inputs.

The point is that it is important to remember that the vast majority (if not all) the numbers assigned will be open to debate. In this situation numbers provide a guide only and cannot be regarded as sacrosanct.

8.2.2 A word about magnitude and importance

There may be those who will argue that an EIS should be a totally objective statement of likely effects, and that it should contain no value judgments. Cullen (1975) comments that this view proposes that the EIS should be a factual catalogue of impacts and their consequences, and that it is up to the decision-makers to decide on the importance of each effect. However, this view is simplistic (see 7.6; Matthews (1975)).

The person preparing the EIS must make some value judgments, even if it is just to decide which effects are trivial and can be omitted. If you try to avoid this decision and include every conceivable effect, not only will the EIS be impossibly costly in money and time, but it will probably be useless, as readers will not be able to identify the important issues from the mass of trivia. Trivia can help to "window-dress" the EIS and suggest an aura of validity, but it will not aid the decision-maker.

Once the decision has been taken as to which environmental effects will be included in the study (scoping stage of EIA), the next stage is to determine the magnitude and importance of the changes in the environmental effects.

Deciding on the importance of the many effects that may be predicted requires the assigning of utility, or worth, to each. What is the cost of destroying some species in an area? The costs of saving it are probably clear (ie, the benefits foregone by not proceeding with the proposal). The cost of destroying the species is harder to assess. For instance, Australian society's attitudes to kangaroos are different from its attitudes to mosquitoes, and if we were about to destroy the last breeding ground for some strain of mosquito it is unlikely that opposition to the proposal would be based on this effect. The same could not be said for kangaroos.

Leopold et al define importance (of an effect) as "the weighting of the degree of importance of the particular action on the environmental action" and magnitude (of the effect) as "the degree, extensiveness or scale of an interaction" (1971: 2).

So, *importance* of an effect refers basically to the significance of the effect to the community. We need to distinguish between effects that will be of most concern, and those that are considered to be less important. Care must be taken to distinguish between "unit importance" (the importance to an individual) and "total importance" (importance to the whole community). Short-term importance may also be different from long-term importance.

"Importance" depends on the effect of the change that may take place, and this has to be distinguished from the effect of the action (eg, the kangaroo-mosquito case). The importance of an effect is difficult to define and evaluate, but it is necessary to attempt an estimate, as it will be necessary to compare quantitative values of the magnitude and importance of many effects – such as comparing a large number of small adverse effects with a single large beneficial effect.

As a guide, Sadar (1994) suggests that the elements of significance include:

- determining which environmental factors are most at risk from the proposal or its alternatives;
- determining the social, ecological and aesthetic significance of the identified factors;
- prioritising the issues raised by the public;
- identifying legal requirements, standards, guidelines and the like that need to be met; and
- ranking predicted impacts in order of priority for avoidance, mitigation, compensation (if applicable) and monitoring.

In summary, he sees that the main elements for assessing the significance of impacts will be the level of public concern, scientific and professional judgment, the extent of disturbance to valued ecological systems, and the degree of impact on social value and quality of life. In this context he considers that:

- Ecological significance relates to aspects of the environment critical for ecosystem functions as well as those valued for aesthetic, sentimental, or for other human centred reasons, such as:
  - effects on plant and animal behaviour;
  - rare and endangered species;
  - ecosystem resilience, sensitivity, biodiversity and carrying capacity; and
  - variability of local species population.
- Social significance includes:
  - effects on human health and safety;
  - potential loss of species or resources (such as farm land) with current or potential value;
  - recreational or aesthetic value;
  - demands on public resources and services;
  - demands on public infrastructure (eg, transport, schools); and
  - demographic effects.

The *magnitude* of an effect depends on the extent of the action and the significance of the environmental effect included. As an example, consider hill-face quarrying. The magnitude of the visual effects of a quarry depends on the viewing distance, aspect and the time the quarry is seen. To a person whose lounge-room window faces the nearby quarry, the magnitude of the visual effect would be large. However, for a person driving past at some distance from the quarry, the magnitude of the visual effect would be small.

Of course, the issues surrounding the determination of the magnitude of an effect are more complex. The importance of the visual effect depends on the degree of visibility of the quarry (its contrast to the background) and its aesthetic effect. For a person distant from the quarry, the importance of the visual effect would probably be small. A more sensitive/concerned/ involved/ knowledgeable observer, however, may consider the hill face as a whole to have been ruined, and the visual effect to have a large importance. If an observer never sees the quarry, then it may be thought that they would consider the visual effect to be zero. However, this is not necessarily so, as people are concerned about the effects of some actions even if they cannot see them (eg, Franklin River, whaling in the Antarctic).

It is important to remember that, like beauty, the importance of environmental effects is in the "eye of the beholder". Using the example of the quarry, at one extreme there would be observers who would consider that the quarry's impact is beneficial because it helps to relieve the monotony of the hill face. At the other extreme, there would be observers who would consider the effect to be disastrous.

Hence, depending on personal value judgments and trade-offs made by individuals, effects may be considered good through to bad and important through to minor. Consensus, as to whether an effect is beneficial or not, may be easy to obtain for some environmental effects; for example, reduction in the number of fatalities on the road. For other factors, such as the visual effect of hill-face quarries, a general opinion may be difficult to obtain. Wind farm proposals are a more recent example in Australia of the different magnitude of impacts to varying observers when weighing up visual and landscape impacts against the use of renewable sources of energy instead of non-renewable, such as coal.

Hrezo and Hrezo (1984) provide a background to the links between values, attitudes and opinions, and hence how opinions need to be interpreted in connection with environmental issues. Recognition and interpretation of these opinions can provide a means of presenting arguments so that they are more likely to be accepted.

8.2.3 A word on units

When the magnitude of an effect is determined it is usually done so by measurement. Choosing the appropriate units for the measurement can be a difficult decision, as you want something that is relevant, representative, consistent and feasibly done. Consider the example of filling in swamplands. Such filling could be described in several ways:

- Volume of swamp filled as a proportion of the swamp system.
- Number of species lost.
- Reduction in the area of mosquito breeding habitat.
- Changes in peak water flows.

Selection of units requires yet another judgment by the analyst. In many cases it will not matter which units are used. The important thing is to be consistent (do not swap units around).

8.2.4 Methods and models for predicting impacts

Prediction is of concern because once the important environmental factors and effects have been identified the next step is to decide how much change would occur if the proposal (or an alternative) were undertaken. Prediction involves a number of steps, as illustrated by Sadar (1994):

- Make the understanding of environmental factors focused and as precise as possible.
- Identify links between the proposal's components and these environmental factors.
- Identify *direct* impacts, where the proposal's elements interact with the social and biophysical environments and there is an immediate cause-effect consequence from an activity of the proposal.
- Identify *indirect* impacts that are at least one step removed from the project activity in terms of cause-effect links (eg, reduced employment opportunities resulting from the acquisition of shops for a road proposal).
- Identify *cumulative* impacts which come from the interaction of elements of the proposal or other activities occurring simultaneously, or subsequently.
- Predict *residual* impacts, if any, that may result from impacts that cannot be avoided or mitigated.
- Predict the probability, magnitude, distribution and timing of expected impacts.
- Forecast what will happen to the affected components of the environment if the proposal goes ahead.

Direct impacts, Sadar suggests, can be identified from the assessment of which components of the proposal will interact with the range of environmental factors. Also, networks of ecosystems and social interconnections can be used to identify indirect impacts, which may be as important as direct impacts. See 6.3.4 for a discussion of the importance of identifying indirect impacts under the Commonwealth's EPBC Act in the context of the Nathan Dam example where it was determined that the potential indirect impacts of this proposed dam, including the potential use of the insecticide endosulfan being able to flow (and therefore potentially polluting) the Great Barrier Reef World Heritage Area (GBRWHA) as well as increased sediment and nutrient pollution to the GBRWHA, should have been a consideration of the Commonwealth Minister for Environment's decision to approve the project.

Identification of cumulative impacts is likely to be more difficult as these are the potential additive or combined effects of past, existing and proposed activities, including the proposal under assessment. Cumulative impacts may have to be examined under a full Cumulative Impact Assessment (CIA) (see 3.9).

Looking at a variety of environmental factors in the physical and social environments, and some issues of special concern (such as chemical hazards), Erikson (1994) discusses predictive approaches appropriate to each factor. In this review he considers direct and indirect impacts, and how quantitative or qualitative methods may be available for the assessment. Sadar (1994) also comments on the use of quantitative methods (such as models, "worst-case" calculations where the results of modelling are limited, laboratory or field experiments, and case studies). He suggests that case studies could be used for identifying qualitative thresholds, and that trend extrapolation and scenario development may be useful in some situations.

More generally, Glasson, Thérivel and Chadwick (1994) suggest that predictive methods can be classified in many ways. In terms of the scope of the methods, they can be divided into *holistic* and *partial* methods. Partial methods are subdivided into the type of project (eg, retail impact assessment) and the type of impact (eg, wider economic impacts).

Some methods are *extrapolative*, where predictions are made on the basis of past and present data (eg, trend analysis, scenario generation, analogies and intuitive forecasting). Otherwise, there are *normative* methods, which work backwards from the desired final state of the project to assess whether the project can be achieved without causing environmental problems.

Mathematical and mechanistic models can be used for prediction. These describe cause and effect relationships which are assumed to be correct, and therefore the output of the model is taken as being correct; these issues have been discussed previously. Glasson, Thérivel and Chadwick (1994) note that the general range of flow charts and mathematical functions used in EIA work can be classified as:

- mass balance models;
- statistical models;
- physical, image or architectural models;
- field and laboratory experimental models; or
- analogue models.

However, in practice, they comment, there has been a tendency to use less formal predictive methods, especially expert opinion, and even where the more formal approaches have been used they have generally been of a simple form. Munn (1975) has also discussed the use of the formal mathematical models, particularly identifying some of the difficulties in their use.

As discussed by Duncan (2008), the use of computer modelling in impact assessment is quite common and allows "an analyst to experiment with a mode to test a system's behaviour under varying conditions and with alternative interventions" (2008: 54), although the author acknowledges the issue of values assigned to the models, as decisions always have to be made as to what data goes into the model.

This pessimism is not to say that these types of predictive approaches should not be used. Rather, the point is that they may be very useful in helping to identify differences between alternatives, including the "do nothing" alternative. However, when they are used, the uncertainty in the model and the data has to be recognised. One way of doing this is to use sensitivity analysis in conjunction with the prediction. In this analysis, a comparison is made between the results of the model using the data which is thought to be correct, and the results when a small difference, or possible error, is inserted into the data. If there is a substantial difference in the two results, the results coming from the model should be considered with caution.

8.3 Evaluation Techniques

8.3.1 Assessment criteria

An indication of the issues which should be considered in assessments can be gained from the following list of criteria suggested by Canter (2000). He considers a methodology should aim to include:

- *Comprehensiveness* – The environment contains intricate systems of living and non-living elements bound together by complex interrelationships. An adequate methodology must consider effects on all these systems.
- *Flexibility* – Sufficient flexibility must be contained in the methodology, as proposals of different size and scale result in different types of impact.
- *Detection of true effects* – The actual effects are the changes in environmental conditions resulting from a proposal, as opposed to the changes that would naturally occur from existing conditions. Moreover, both short- and long-term changes must be reported.
- *Objectivity* – The methodology must provide impersonal, unbiased and constant measurements, immune to tampering by political and other outside forces. An objective and consistent procedure provides a firm foundation, which can be periodically updated, refined and modified, thereby incorporating the experience gained through practical application. To be effective as an aid to decision-making, environmental assessments should be repeatable by different analysts and able to withstand scrutiny by various interest groups.
- *Ensured input of required expertise* – Sound, experienced, professional judgments must be assured by a methodology, especially as subjectivity remains inherent in many aspects of environmental evaluation. Input of the necessary expertise can be achieved either through the design of the methodology itself or through the rules governing its use.
- *Utilisation of state of the art* – Maximum possible use of state of the art techniques or approaches should be made, drawing on the best available techniques.
- *Employment of explicitly defined criteria* – Evaluation criteria, especially any quantified values, employed to assess the magnitude or importance of environmental effects should not be arbitrarily assigned. The methodology should provide explicitly defined criteria and explicitly stated procedures regarding the use of these criteria, with the rationale behind such criteria being documented.

- *Assessment of actual magnitude of effects* – Means must be provided for an assessment based on specific levels of effects for each environmental factor, in the terms established for describing that factor. Assessment of magnitude based on generalities or relatives (qualitative comparisons between alternatives) is usually considered to be inadequate.
- *Provision for overall assessment of total effect* – A means for aggregating multiple individual effects is necessary to provide an evaluation of overall total environmental impact.
- *Pinpointed critical effects* – The methodology should provide a warning system to pinpoint and emphasise particularly "bad" effects. In some cases, the sheer intensity or magnitude of effect may justify special attention in the planning process, regardless of how narrowly the effect may be noticed.

8.3.2 Evaluation methods

Evaluation methods have been developed to bring together the various environmental effects (and impacts) associated with a proposal and its alternatives. The methods enable comparisons of impacts and alternatives to be made. They also lead directly to an output which summarises all the environmental impacts and provides the basis for the evaluation of the proposal (and alternatives) by decision-makers. Methods can be classified into eight main types, although this is obviously arbitrary and not necessarily consistent for different researchers. The categories which cover the vast majority of methods are:

1. Ad hoc methods
2. Checklists
3. Matrices
4. Overlays
5. Systems diagrams
6. Networks
7. Quantitative or index methods
8. Mathematical models

Shopley and Fuggle (1984), in their comprehensive review of evaluation methods, also comment on evaluation and adaptive techniques, which combine aspects of the major groupings. The demise of checklists and numerical data manipulation methods, according to Bisset (1988), has been predicted for years, but they continue to fulfil a role in the technocratic "objective" decision-making processes used by most proponents.

The methods for assessing impacts have been categorised by Bisset (1988), and Wathern (1988a), as follows:

- for assessing direct impacts – checklists, matrices; and

- for assessing indirect impacts – interaction matrices, flow diagrams and network analysis, overlays.

Direct impacts are where there is an explicit cause and effect relationship (such as removal of vegetation leading to erosion and loss of habitat), while indirect impacts have less obvious links (such as dredging leading to reduced fish breeding).

Whether for investigation of direct or indirect effects, when choosing a method it is important to remember what the evaluation is for. The final product you are hoping to achieve is the presentation of information that will help the decision-maker come to an informed decision about the proposal. Consequently, it is necessary to present the effects in a meaningful way so that the trade-offs between positive and negative effects can be made; and so that these environmental effects can be considered alongside the economic and technical information. This usually means that the assessments present the effects in one of three ways:

1. Simple display of information; or
2. Tabular comparison; or
3. Aggregation of effects.

The output from the following types of methods falls into one of these groupings.

Ad hoc methods

These methods typically present the environmental effects in a descriptive form. As there is usually a large amount of information of a varied nature associated with an EIS, such lists of information may become rambling catalogues of environmental concern.

An improvement on this is where a summary is provided of the key issues. The result of this is a list or table comprised of numerical data and descriptions such as "good", "bad", "dangerous", "polluting" or "minimal environmental impact". These terms might not be defined or put into context, so it is then up to the decision-maker to interpret the effects relative to each other, and relative to the other information. Table 8.1 (*see over page*) provides an example of an ad hoc tabular approach developed for a transport proposal where three alternatives were to be compared, and evaluated (Department of Infrastructure and VicRoads (1998)).

This table illustrates the typical combination of quantitative information (things that can be measured, like capital cost) and qualitative (those issues that cannot be directly measured, like open space and recreation). It provides the reader with an overall summary of the issues that were of concern, or the indicators on which the alternatives can be judged, and an indication of how each alternative "performs". The information is reasonably "transparent" in that the information associated with each indicator is presented for interpretation by the reader. However, we would have to go to

Table 8.1 Example of an ad hoc tabular presentation pf predicted impacts and their comparison for evaluation – Scoresby Transport Corridor EES

INDICATOR	STRATEGY OPTION 2	STRATEGY OPTION 3	STRATEGY OPTION 4
Operational Performance			
Average vehicle speed in the Corridor (kph)	51.4kph	53.2kph	54.2kph
Percentage of congested road links in the Corridor	19%	18%	15%
Public transport mode share in Melbourne	6.4%	6.3%	6.4%
Percentage increase in total vehicle kilometres of travel in the Corridor compared with current	23%	27%	30%
Accessibility			
Percentage of the Corridor with improved accessibility to work by car compared with 2011 Basecase	10.3%	Between Options 2 and 4	79.6%
Percentage of the Corridor with improved accessibility to work by public transport with 2011 Basecase	14.7%	Between Options 2 and 4	36.6%
Capital Cost			
Road projects	$274M	$532M	$775M
Public transport projects	$209M	$234M	$239M
Total	$483M	$766M	$1014M
Return on Investment			
Benefit cost ratio	1.2	1.2	1.5
Business benefits of Victoria in 2011	$10M	$15M	$31M
Impact on Victorian Gross State Project in 2025	$55M	Between Option 2 and 4	$200M
Impact on annual National Gross Domestic Product in 2025	Not calculated	Not calculated	$400M
Benefit cost ratio with indirect business benefits	1.4	1.5	2.0
Benefit cost ratio with Enhanced Dandenong landuse	-	-	-
Benefit cost ratio with Urban Village landuse	1.2	-	-
Safety			
Reduction in road casualty accidents in the Corridor compared with 2011 Basecase	9	48	98
Pedestrian and cyclist safety compared with 2011 Basecase	No net improvement	Improved conditions in north	Improved conditions overall

Based on Department of Infrastructure and VicRoads (1998)

INDICATOR	STRATEGY OPTION 2	STRATEGY OPTION 3	STRATEGY OPTION 4
Air Quality	• Carbon monoxide and nitrogen dioxide objectives will be met by a considerable margin; • PM10 concentrations unlikely to change and will just comply with proposed NEPC objectives; • Ozone concentrations will decrease slightly but still exceed EPA objectives		
Noise Proportion of the community likely to be highly annoyed due to road traffic noise, compared with 2011 Basecase Change in the number of residences exposed to noise levels of more than 68 dB(A) L10 18hr compared with 2011	Minimal change Minimal change	Low negative change Slight decrease	Medium negative change Slight decrease
Landscape Character - Visual impacts along Scoresby Reservation	3kms of moderate impacts	13kms of moderate/high impacts	19kms of moderate/high impacts
Open Space and Recreation - Change in the quality of recreational experiences	Minimal change	Impact on valley parklands	As for Option 3
Property Requirement Number of properties to be acquired	Springvale Road 20 Within Reservation 21 Outside Reservation 8	Springvale Road Nil Within Reservation 40 Outside Reservation 5	Springvale Road Nil Within Reservation 102 Outside Reservation 37
Community Effects - Social dislocation and psychological severance	Some change	Moderate effects	As for Option 3
Urban Form - Additional pressure for development on urban fringe	Low	Potential increase	Potential increase
Greenhouse Gas Emissions - Percentage increase in CO_2 emissions in the Corridor, compared with current	14.0%	16.7%	18.5%
Water Resources and Quality Change in existing drainage patterns Change in surface water and groundwater quality	Potential for improvement Potential for impacts	Potential for improvement Potential impacts	Potential for minimal change Potential for impacts
Native Flora and Fauna Potential impact on areas of high conservation value Area of native vegetation/fauna habitat depleted (ha) Number of species that are rare or threatened at the National, State or Regional level potentially affected	Nil 3.3ha 7	Nil 7.8ha 14	High 18.2ha 38

the relevant sections of the EIS to see how these summaries were derived. If an explanation is not available, we would have to question the reliability of the information in the table.

Another example is shown in Table 8.2 (*see opposite*) where the qualitative information is presented pictorially. This example also comes from the transport area and the intensity of shading indicates the different levels of impact (VicRoads (2000)).

Checklists

Checklists represent one of the basic methods used in environmental assessment. According to Sadar (1994), checklists:

- are a structured approach for the identification of the relevant environmental factors to be included in the EIA;
- encourage discussion during the early stages of the assessment process;
- may range from a simple listing of environmental factors to a listing that incorporates mathematical modelling to describe environmental conditions;
- represent the collective knowledge and judgement of those who have developed them; and
- should be exhaustive to ensure that nothing has been left out.

Importantly, Sadar notes that "[c]hecklists *cannot* represent the interdependence, connectivity, or synergism between interacting environmental components, nor are they able to describe variation of environmental conditions with time" (1994: 19).

Canter (2000) suggests that there are five broad categories of checklists:

1. Simple checklists

These are sometimes classed as ad hoc methods. A simple checklist is a straightforward list of relevant parameters with no guidelines being provided about how the environmental effects are to be measured or interpreted. An example would be short descriptive statements used to outline the effects associated with a proposal. For instance, with a transportation project descriptive comments could be made as follows:

- Planning and design phase
  - impact on land use through speculation
  - acquisition of property for the project
- Construction phase
  - noise
  - displacement of people

Another approach would be to list the potential effects and indicate the phase(s) in which the effects would be expected. In either case, while the

Table 8.2 Ad hoc and pictorial presentation of information

ASSESSMENT CRITERIA	AlternativeKF1	AlternativeKF2	AlternativeKF2 ALT	AlternativeKF3
TRANSPORT ECONOMIC EVALUATION (based on initial 4 lane construction)*				
Project Total Estimated Cost ($ Million)	103	104	105	115
Benefit Cost Ration (BCR)	0.75	1.11	0.86	0.78
Net Present Value (NPV) ($ Million)	-18.9	8.2	-10.4	-17.7
IMPACTS				
Land Required (ha)	299	265	305	290
No of Houses Acquired (within right-of-way)	2	6	3	6
Road Safety (reduction in accidents in first year of operation)	2.7	3.4	3.0	2.9
Business and Tourism				
Agriculture				
Social				
Traffic Noise				
Land Use Planning				
Flora and Fauna				
Exotic Vegetation				
Landscape				
Archaeology and Heritage				

Least impact
↓
Most Impact

Source: VicRoads (2000)

Figure 8.1 Example of a simple checklist

	No effect	Some effects								
		Positive	Negative	Beneficial	Adverse	Problematic	Short-term	Long-term	Reversible	Irreversible
Flora			X			X	X			
Fauna	X									
Soil characteristics	X									
Natural drainage	X									
Groundwater		X		X						
Noise			X				X			
Recreation	X									
Air quality			X		X			X		X
Visual intrusion	X									
Health and safety	X									
Economic factors		X		X				X		
Infrastructure	X									

Source: Based on Rau & Wooten (1980)

effects have been identified, there is no attempt to measure or interpret the effects; for all the reader knows, the effects may be positive or negative.

Simple checklists may provide some indication of whether there will be changes associated with each environmental effect that had been identified (see Figure 8.1 *opposite*).

2. Descriptive checklists

These include an identification of the relevant environmental effects and guidance on how to measure the parameters. As such, they provide slightly more information than simple checklists.

3. Threshold of concern checklists

In these cases a list of environmental effects is again presented, and alongside each factor is a threshold which indicates a level at which the assessor should become concerned about the impact. See Table 8.3 for an example.

Table 8.3 Example of a threshold of concern checklist

Environmental component	Criterion	Threshold of concern (TOC)	Impact (Alternative 1)	Impact > TOC? (Alternative 1)
Air quality	emission standards	1	2C	Yes
Economics	benefit-cost ratio	1:1	3:1	No
Endangered species	number of pairs breeding spotted owls	35	50D	No
Water quality	water quality standards	1	1C	No
Recreation	number of camping sites	5000	2800C	Yes

Source: Glasson, Thérivel & Chadwick (1994)

On the basis of the above analysis, air quality and recreation would be unacceptably impacted upon. Impacts can also be rated for their duration: in the above, A indicates a duration of the impact of 1 year or less; B indicates 1-10 years; C is for 10-50 years; and D indicates an irreversible impact. Additional columns can be added to assess the effects of other alternatives.

4. Scaling checklists

Similar to descriptive checklists, these include additional information which is basic to the scaling of parameter values; that is, they use scaling techniques.

For example, with a transport project the environmental effects may be outlined as follows on *page 229*:

Figure 8.2 A scaling checklist

Key

–	= minor negative impact	+ +	= major negative impact
– –	= major negative impact	x	= undermined impact
+	= minor positive impact	0	= no appreciable impact

	Elements	Impacting actions									
		During project				After completion					
		Relocation of business	Demolition	Construction	Wildlife habitat	Residential buildings	Commercial buildings	Recreation	Historical conservation	Modifications to street system	Cultural protection
BIOPHYSICAL	Soil and geology	0	0	0	0	0	0	0	+	0	0
	Hydrology	0	0	–	–				0	0	+
	Flora	0	0	–		– –	+	+	0	0	+
	Fauna	0	0	0	0	0	0	0	–	0	0
	Air quality	0	0	–	0	0	– –	– –	– –	+	0
	Land use	0	0	– –	– –	0	0	0	– –	– –	0
trans-portation system	Roads	0	–	– –	–	+	+	+	– –	0	+
	Public transport	0	0	–		0	X	X	X	0	X
	Cyclists	–	–	– –	–	0	– –	– –	+	– –	x
	Open space	0	0	00	0	0	– –	–	–	– –	x
SOCIO-ECONOMIC	Demand for ancillary services	+	+	+	–	0			0	0	+
	Health and safety	0	0	–	–	– –			0		+
	Industry	–	– –	–	–	0	– –	+	+	– –	x
	Commercial	– –	– –	–	–	0	– –	+	+	– –	x
	Tourism	– –	– –			0	– –	+	+	– –	x
	Infra-strucure	0	0	–	–	+	+	0	0	+	+
	Landscape	0	0	–	– –	0	+	+	–	+	0
CULTURAL	Historic structures	0	0	–		+	0	0	x	+	– –
	Ethnic groups				+	+	+	–	–	–	–

Source: Based on Rau & Wooten (1980)

	Alternative			
Environmental effect	**A**	**B**	**C**	**Comments**
Noise effect	- 2	- 1	0	reduction of local traffic
Air pollution	+ 5	+ 2	+ 4	improved traffic flow
Open space	+ 3	+ 1	- 6	some structures removed etc

In this case, the scaling values have been assigned from a range of – 5 to + 5 using subjective evaluation. From this list, or table, an assessment of three alternatives could be made by:

- comparing the number of plus and minus ratings;
- taking the ratio of plus to minus ratings;
- taking the algebraic sum of the ratings;
- comparing the average of the ratings; and so on.

Figure 8.2 (*see opposite*) provides another example of this type of checklist. The scaling may also be presented in a diagrammatic or pictorial form, as was illustrated in Table 8.2 where the shading, or size of a symbol, has specific meaning.

5. Scaling-weighting checklists

These checklists are essentially scaling checklists with information provided to enable the subjective evaluation of each parameter with respect to every other parameter.

An example of this type of checklist is that developed by Battelle Laboratories in 1972 for the United States Bureau of Reclamation (see Munn (1975)). The Battelle system (or Environmental Evaluation System) comprises a list of 78 environmental factors (see Table 8.4 *on page 231*) and a scale of environmental quality for each factor, scaled between 0 (very bad) and 1 (very good). Figure 8.3 (*see over page*) provides an example of environmental quality (EQ) scales. Qualitative variables such as aesthetics and quantitative variables such as water quality are all present, within the range of 0 to 1, even though they may have been assessed in the field using ranges of, say, 0 to 5 or 10. This obviously involves value judgments, but a common scale has been selected for what 0 and 1 represent. Explicit weighting factors are then used on the different environmental measures (these are indicated in Table 8.4 in parentheses). Weighting factors have been developed, and published for use in assessing water-development projects, using the DELPHI technique. (DELPHI is an interactive process where experts are asked for their opinions, an average position is established then returned to the experts, who modify their positions to give a new average; the cycle is continued until the positions reach a consensus.) Canter (2000) explains that using the Battelle method involves the following steps:

Figure 8.3 The Battelle system's environmental quality

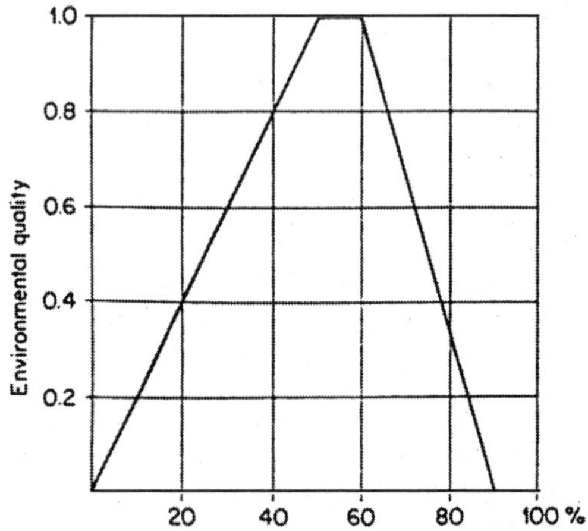

Per cent of carrying capacity based on animal units of browsers and grazers

Source: Munn (1975)

- Obtain existing-conditions data for each of the 78 environmental factors. Convert these parameter data to EQ scale values. Multiply the scale values by the weighting factor for each environmental factor to develop a composite score for the environment without the project.
- For each alternative, predict the change in the environmental parameters.
- Using these predicted changes, determine the EQ scale value for each parameter and for each alternative.
- Multiply the EQ values by the relevant weighting factor and aggregate the information for a total composite score for each alternative.

The Battelle system is a highly organised method and as such it helps to ensure a systematic and comprehensive approach to identifying critical changes. There is no passing or failing score, as the numerical evaluation has to be interpreted by the reader. As in the case with other methods, very little emphasis is given to socioeconomic factors. Other criticisms have been noted by both Cullen (1975) and Munn (1975):

- The method is inflexible in terms of application to different types of projects.

Table 8.4 The Battelle system – example of a scaling weighting checklist

The numbers in brackets are the weights which are to be applied to the environmental quality results obtained (eg, the environmental quality for browsers and grazers comes from Figure 8.3 on page 232)

Ecology	Aesthetics	Physical/Chemical	Human Interest/ Social
Terrestrial species and populations • Browsers and grazers (14) • Crops (14) • Natural vegetation (14) • Pest species (14) • Upland game birds (14) Aquatic species and populations • Commercial fisheries (14) • Natural vegetation (14) • Pest species (14) • Sport fish (14) • Waterfowl (14) Terrestrial habitats and communities • Food web index (12) • Land use (12) • Rare & endangered species (12) • River characteristics (12) • Species diversity (14) Ecosystems	Land • Geologic surface material (6) • Relief and topographic character (16) • Width and alignment (10) Air • Odour and visual (3) • Sounds (2) Water • Appearance of water (10) • Land and water interface (16) • Odour and floating material (6) • Water surface area (10) • Wooded and geologic shoreline (10) Biota • Animals – domestic (5) • Animals – wild (5) • Diversity of vegetation types (9) • Variety within vegetation types (5) Human-made objects • Human-made objects (10) • Composition • Composite effect (15) • Unique composition (15)	Water quality • Basin hydrological loss (20) • Biochemical oxygen demand (25) • Dissolved oxygen (31) • Faecal coliforms (18) • Inorganic carbon (22) • Inorganic nitrogen (25) • Inorganic phosphate (28) • Pesticides (16) • pH (18) • Streamflow variation (28) • Temperature (28) • Total dissolved solids (25) • Turbidity (20) Air quality • Carbon monoxide (5) • Hydrocarbons (5) • Nitrogen oxides (10) • Particulate matter (12) • Photochemical oxidants (5) • Sulphur oxides (10) • Other (5) Land pollution • Land use (14) • Soil erosion (14) Noise pollution • Noise (4)	Education/Scientific • Archaeological (13) • Ecological (13) • Geological (11) • Hydrological (11) • Historical • Architecture and styles (11) • Events (11) • Persons (11) • Religions and cultures (11) • "Western Frontier" (11) • Cultures • Indians (14) • Other ethnic groups (7) • Religious groups (7) Mood/Atmosphere • Awe/Inspiration (11) • Isolation/Solitude (11) • Mystery (4) •"Oneness" with nature (11) Life patterns • Employment opportunities (13) • Housing (13) • Social interactions (11)

Source: Munn (1975)

- Although the method appears to be scientific in its development of quantitative measures of environmental quality and the weighting factors, in reality it entails little more than summing the value judgments of the analysts (the method makes these judgments reasonably explicit so that they may be questioned and perhaps changed).
- Impacts that cannot be readily quantified are likely to be distorted or masked in such an analysis.
- Aggregation to a single index is undesirable in that when alternatives are being considered it may be possible to consider modifications that would make an alternative more acceptable.

- There is no effective mechanism for estimating or displaying interactions between the environmental effects.
- The method is not strictly mutually exclusive. While effects are not counted twice, the same effect may appear in different parts of the method; for example, water quality appears in the physical/chemical section and may come into the section on aesthetics (if the water is turbid).

However, the method provides a good checklist for factors likely to be affected by water-resource developments, even if it is not used actually to evaluate these effects.

Matrices

In many respects, the term "matrices" is a misnomer. It gives the impression that the data in a matrix has some mathematical derivation, or at least the data can be manipulated mathematically, whereas this is seldom the case. It would often be better simply to use the term "tables".

Nonetheless, the term has been widely associated with the methodology outlined in the Leopold matrix method. Leopold et al (1971) developed the method for the US Geological Survey. The matrix was designed for the assessment of almost any type of construction project. Its main strength is as a checklist that incorporates qualitative information about cause and effect relationships, but the matrix can also communicate results.

The system is an open-cell matrix containing 100 project actions along the horizontal axis and 88 environmental "characteristics" and "conditions" along the vertical axis (see Figure 8.4 *opposite*). The matrix is comprehensive in covering both the physical-biological and socioeconomic environments. However, Munn (1975) notes that the list of environmental characteristics is weak from the point of view of balance; for example, swimming (activity) and water temperature (indicator of state) are both included. Also, the list is biased towards the physical-biological environment.

Use of the Leopold matrix involves:

- identifying all the actions associated with the project; that is, along the horizontal axis;
- placing a slash in each matrix cell where there is likely to be an environmental effect;
- in the upper left-hand corner of each cell with a slash placing a number from 1 to 10 to represent the magnitude of the possible effect (using a plus sign if the effect is beneficial); and
- in the lower right-hand corner placing a number from 1 to 10 to represent the importance of the effect.

A further step could involve the summation of the values in the cells, by row and/or column, to obtain some idea of overall effect and help the person trying to interpret the mass of data.

Figure 8.4 The Leopold matrix

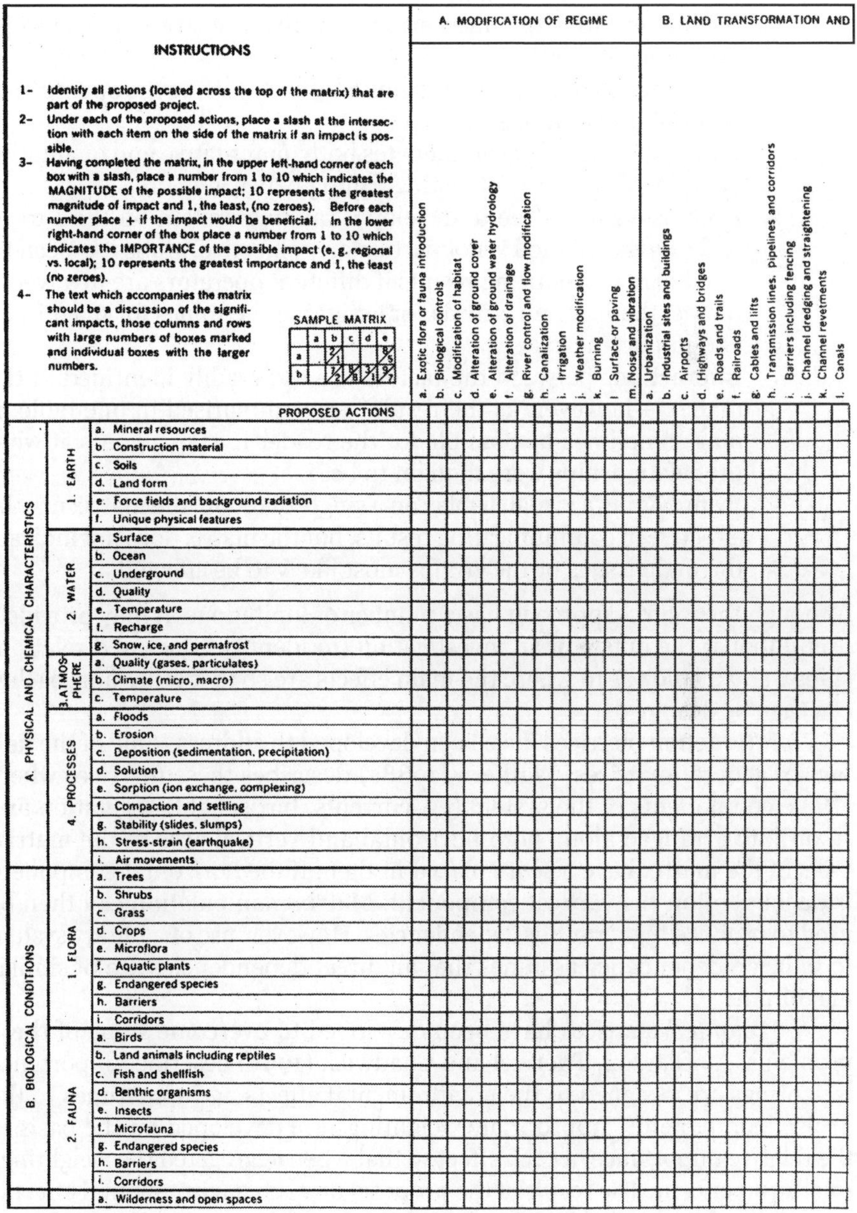

Source: Leopold et al (1971)

Problems with the method include:

- The information would be difficult to absorb let alone interpret without some summations being done.
- Each alternative or time period requires a separate matrix to be produced.
- Aspects of the matrix are not mutually exclusive, so the opportunity for double counting exists.
- While the method accommodates both quantitative and qualitative data, there is no way of distinguishing between them.
- Each assessor is free to develop a ranking system on the 1 to 10 scale, so the method is not particularly objective, nor is it especially reliable in reproducing results if different operators are involved – that is, there are no criteria for assigning the values for magnitude and importance.
- Interactions between the factors are not readily identified in the matrix – however, as the results are summarised in one table, it may sometimes be possible for the reader to take a guess at what the relationships were thought to be.
- By providing a visual display on a single diagram, the matrix may be effective in communicating results, but the matrix does not indicate the main issues or the groups most likely to be affected.

Although the matrix approach has a number of limitations, it may provide a useful tool at the outset of an investigation (to identify issues and relationships), and a guide as to where the main effects are so that other approaches can then be designed.

Interaction matrices have been developed to address some of the failings of general matrices. Wathern (1988a) describes these matrices where all the components of the system (eg, currents, birds, light, vegetation and so on) are arranged along both horizontal and vertical axes of the matrix. Cells of the matrix have a score placed in them if the horizontal component is dependent on the vertical component. Matrix manipulation can then be used to produce higher order dependencies. However, use of a binary system to indicate dependency assumes that all direct dependencies are of similar magnitude.

Weighted matrices have been developed to overcome some of these problems. As Glasson, Thérivel and Chadwick (1994) describe, importance weightings are assigned to the environmental effects and sometimes to the project components. Appropriate weightings are developed on the basis of the relative importance of each effect so that when aggregated the weightings total 100 per cent. The result of the proposal on each environmental effect is assessed and multiplied by its weighting to gauge the impact on each factor. These impacts are then aggregated to provide the overall impact. However, these matrices still suffer from all the restrictions of numerical representation of environmental effects, and the numeric manipulation of the data.

Another style of matrix is discussed by Canter (2000), and titled the *stepped* matrix. In effect, a stepped matrix has a similar approach to the networks (discussed below) in that a logical progression from action to effect is documented. The sequence for completing such a matrix involves:

- identifying discrete parts of the action (future improvements) and where these interact with "casual factors" (ie, physical outcomes of the action);
- identifying aspects of the environment which are likely to be altered by the actions and the extent of the alteration by a symbol (to indicate the direction and extent of the change); and
- providing relevant description under the appropriate column headings.

This type of matrix is similar to the checklists discussed above and exhibits similar problems. See 3.5 for a case study that illustrates the use of a risk assessment approach in EIA for Shell Australia's Darwin Joint Terminal project.

Overlays

The overlay technique was first suggested by McHarg in the late 1960s. With this method, a set of transparencies is used to identify, predict, assign relative significance to environmental effects and communicate environmental impacts. The method relies on the mapping of environmental characteristics (such as biodiversity, aesthetic value, social status) to describe the area surrounding the proposal. Impacts are then identified, as Sadar comments, "by noting the affected environmental characteristics within the project area boundaries. This presents a graphical display of the types of impacts, the impacted areas, and their respective geographical location. The result is to identify areas which are compatible with the proposed action" (1994: 24).

A simple overlay is to map the extent of a particular environmental impact, such as noise contours resulting from a road, or the ground-level concentrations of particular air emissions. Transparencies can then be used to represent alternatives (eg, road design). These are then laid over a base map of the proposal's area to indicate the extent of impact, or to show how the proposal may be designed to avoid sensitive geographical locations.

More complex approaches using up to ten transparent sheets are also possible. Here the method relies upon displaying the most important effects. One sheet is used for each environmental effect, and usually the sheet indicates the degree of environmental impact (often by shading). A geographical-style map provides the base upon which the "coloured-up" transparent sheets are put. After all the sheets have been placed, the density of colour (or shading) indicates the relative impact that the project would have on a particular geographical area.

Computer programs can be written to perform the task of aggregating the predicted impacts. This is done by coding the geographical area in terms of small zones, assigning a numerical weighting to each zone to represent the impact for each environmental effect, and summing all the numbers allocated to each particular zone. This is the basis of Geographic Information Systems (GIS) as discussed by Sadar (1994), who sees that they have the potential for storing and accessing very large data sets assembled form diverse sources. Such systems are efficient at performing multiple map overlays, allowing a number of different scenarios to be investigated quickly, by varying the parameters for successive analysis runs. The generation of descriptive statistics may also be useful for some assessments.

Problems with the overlay method as seen by Munn (1975) include:

- There is restriction on the number of factors that can be considered.
- Objectivity is low – except with respect to the spatial positioning of effects and impacts (eg, flooded land), interactions are not displayed.
- Extreme impacts with small probabilities of occurrence are not considered.
- Each alternative would require a separate set of overlays to be produced, and less than gross differences between alternatives may be difficult for the decision-maker to detect.

The overlay method cannot be considered ideal. Despite its limitations, however, it is useful for illuminating complex spatial relations (in a greatly simplified fashion). It is relevant to large regional developments and corridor selection problems, but the assessor should still view the analysis with at least a degree of caution.

Systems diagrams

Bisset (1988) explains that a systems diagram consists of a chart detailing the socio-environmental components linked together by lines. These lines indicate the direction and sometimes the amount of energy flow between the components. Here energy is used as the common unit to measure impacts. Activities associated with a proposal and likely to result in impacts can be included in the systems diagram.

The advantages of systems diagrams are the acknowledgment of the complexity in the interactions of environmental parameters, and that impacts have a common basis for comparison (eg, energy units). Disadvantages are primarily that construction of the diagrams are expensive and take time, not all ecological relations may appear (eg, habitat links) and socio-environmental impacts are difficult to incorporate.

Networks

Networks are the documentation of the logical steps the analyst would take in thinking about the environmental impacts of a proposal. Like the checklists and matrix approaches, the network method identifies the major effects by considering how the proposal would change various environmental effects.

Rau and Wooten (1980) provide a generalised framework for construction of a network.

The method is started by listing the most significant environmental effects that would be affected by the proposal, followed by a listing of the aspects of the proposal that would influence the environmental effects. This is similar to the Leopold matrix. The next stage is to identify all the environmental impacts that result from the interaction of the proposal and environmental effects. Munn (1975) explains that these impacts are presented in a flow diagram to indicate how one impact can lead to another impact. Using the example of dredging, presented in Figure 8.5 *over page*, and following one of the possible paths:

Removal of bottom material:

- leads to changed bottom topography;
- leads to increased water depth;
- leads to inhibited eelgrass growth;
- leads to destroyed fish habitat; and
- leads to reduced sporting and commercial fishing.

At this stage the network is in a purely descriptive form. The analyst could then derive values for the magnitude and importance of each environmental effect, multiply them together and add the resulting numbers to obtain a total environmental impact.

Problems with the method are similar to those noted for other methods, that is:

- lack of objectivity makes the analysis difficult for reproduction of interactions that are not explicit;
- each alternative would require a separate network to be made; and
- with more than a limited number of environmental effects being considered, presentation of the network becomes cumbersome.

The method is probably most applicable to the first stage of an analysis, visually presenting the interactions so that they can be quantified in later stages. It may also be useful in illustrating the interactions to the decision-maker.

Quantitative or index methods

These methods are based on a list of environmental effects thought to be relevant to a particular proposal and which are differently weighted. Likely effects are first identified and then assessed for magnitude and importance. These effects are transformed into a common measurement unit; for example, a score on a scale of environmental quality. The scores and factor weightings are multiplied and the resulting scores added to provide an aggregate effect score.

By this method beneficial and harmful effects can be summed and total scores compared. Alternatively, all total effect scores for all alternatives can be aggregated and compared, the alternative giving the "best" score being the preferred option.

Figure 8.5 Examples of a network for dredging

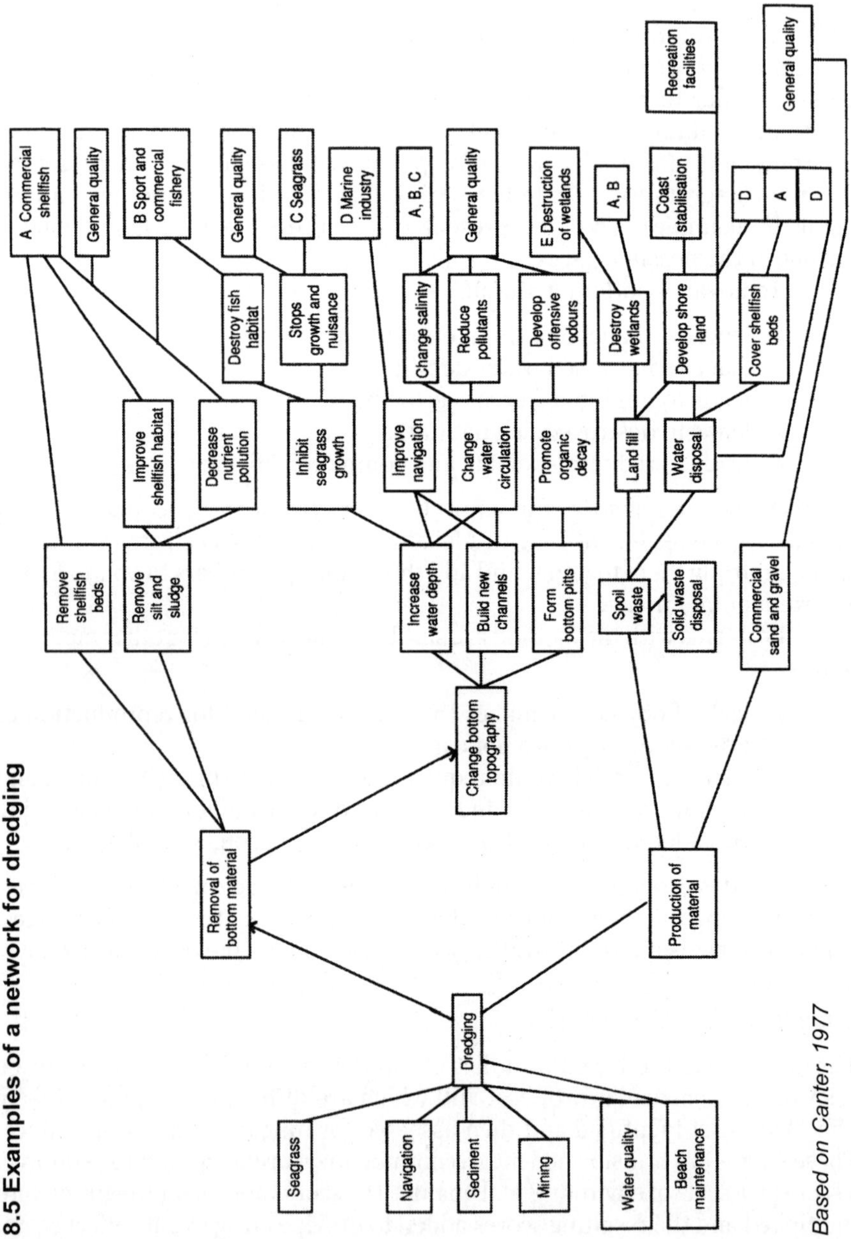

Source: Based on Canter, 1977

Figure 8.6 Example of an index method

The following model of residential environmental quality is designed to end up with a number which is intended to represent an index of quality.

$$\Delta RAj = a\sum_{i}^{N} Ri\,(NOij - NO_{iE}) + \text{ß}\sum_{i}^{N} Ri + (VBij - VB_{iE}) + \text{¥}\sum_{i}^{N} R_{GC}\,(VIj - VI_{E})$$

where	ΔRAj	=	change in residential amenity relative to existing state
	RGC	=	number of residences
	NOij	=	noise impact in area i due to alternative j
	NOiE	=	existing noise impact in area i
	VBij	=	vibration impact in area i due to alternative j
	VBiE	=	existing vibration impact in area i
	VIj	=	visual/intrusive impact in area due to alternative
	VIE	=	existing visual/intrusive impact within area
	aß_¥	=	weighting functions
	N	=	number of potential areas affected by alternatives

Source: Centre for Environmental Studies (1976)

This seems to be the ideal method. At last we are reaching a point where the environmental impacts are being neatly aggregated and comparisons between alternatives are possible. The SAGE method described by Hyman, Moreau and Stiftel (1984) is a form of index method designed to assist in evaluation of alternative proposals. They explain that this approach involves predicting the effects of the proposal and alternatives, eliciting weights to be applied to the weights, and combining the weights and effects in a form useful to the decision-makers. The combination of weights and effects is achieved through the development of a mathematical formula, which results in a numerical value ascribed to the particular alternative. These values can then be compared and used to guide the choice of the preferred alternative; depending on the way the formula has been developed the "best" value may be that which is the smallest, or the largest. Figure 8.6 *above* provides an example of a mathematical formula, which is the key element of the index method.

Quantitative methods certainly overcome some of the problems associated with the other methods, but in so doing introduce weighty problems of their own (but which are also evident in other methods):

- There is frequent lack of indication of the relationships between and interaction of environmental factors and effects.
- Opportunities exist for lack of objectivity in predicting environmental effects and the importance of these.
- Biases can result from the analysts deriving relationships and weights.

- Assumptions are made that accurate data can be obtained and which will be of equal quality to the sophistication of the formulation used.
- There is a false security in arriving at a figure for the environmental impact of the proposal.
- The methods frequently run into trouble in whether the numbers can correctly (in a mathematical sense) be combined in the way suggested; that is, whether the numbers are cardinal or ordinal and whether combining these through multiplication, addition and so on is correct.

As long as the limitations of the methods are realised and allowed for, they can be used to give an indication of environmental impact. However, as indicated, they may suffer from severe conceptual and data-availability problems.

Mathematical models

Most of the other methods also use mathematical models in some form. Wherever "importance" is derived using a formula and where "magnitude" and "importance" are combined according to some rule (formula), a model is being used. Models in the environmental assessment sense usually mean a formula that has been derived to combine the magnitude and importance of effects together and to arrive at some figure of total environmental effect. (In this respect there are many similarities with index methods.)

Benefit-cost analysis like cost-benefit analysis (CBA) is an example of a model used in this way (see 8.4 for additional discussion). In a benefit-cost analysis the model is:

$$\frac{\text{sum of environmental benefits (ie, magnitude x importance)}}{\text{sum of environmental costs (ie, magnitude x importance)}}$$

where "importance" is designated by the dollar value that is assigned to the unit of change for the particular environmental factor.

The advantage of this formulation is that other factors such as travel-time savings/increases and capital/maintenance costs can be added to the appropriate line. The reason for doing this is so that all factors that would be included in assessment of the proposal can be considered at once. It also makes it very easy to compare alternatives with one another.

Mathematical models in general offer a number of perceived advantages for decision-makers:

- A total environmental impact is usually produced.
- The interrelationships of factors and effects with magnitude and importance are usually stated explicitly in mathematical terms.

- As the analysis involves mathematics, many would say it has a "ring of respectability".
- The method is easily handled by computer programs.

The problems are closely related to the "advantages", especially:

- Possible (probable) errors arise in the mathematical relations for environmental effects – and there is the difficulty of quantifying qualitative information.
- The use of numbers, particularly aggregation, tends to obscure important points in the assessment.

Hobbs et al (1984) provide a valuable discussion of the problems of adopting particular mathematical functions and weighting procedures for the models. In general, mathematical models are tools that are available to the analyst and can serve a useful purpose. However, their limitations have to be realised, and the results obtained treated accordingly.

Expert systems

These systems "incorporate the knowledge and experience of experts from relevant disciplines into a structured *decision-making* analytical tool" (Sadar (1994: 28)). Processes and interrelationships need not be precisely defined, and "value-judgements" and "best-guesses" about the likely outcome are basic ingredients of expert systems. So expert systems reflect to typical decision-making processes where decisions are made on the basis of incomplete information and incorporate values.

8.3.3 Combinations of methods

The assessor has a wide range of methods from which to choose. Sometimes it may be decided to use more than one method, either:

- consecutively at different stages or levels of detail of the assessment; or
- concurrently, at the same stage.

An example of the latter would be in something like "multi-objective planning" as used by water-resource bodies. In this case the final evaluation is done via a table that mixes descriptions of effects and numbers that describe the effects. As discussed in 8.2.1, there can be advantages in combining different types of assessment methods so as to overcome some of the biases of the methods.

Table 8.5 (*see over page*) presents an evaluation of how the methods rate for the various stages of an EIS. Atkins (1984) also gives an evaluation of the utility of six typical methods. In practice though, it appears that the "simple" approaches are most favoured. Reporting on a survey of EIA practitioners Cooper and Canter (1997) found that professional knowledge and judgement was used in 40 per cent of EIA cases, while simple checklists were

Table 8.5 Overview of Methods and their Applicability to Stages of an EIA

Method	Scoping	Impact identification	Describe effected environment	Impact prediction	Impact assessment	Decision-making	Communi-cating results	Relative extent of usage
Analogs (analogous proposals, case studies)	√	√		√	√			W
Checklists (simple, descriptive)		√	√				√	W
Decision-focused checklists (for conducting trade-off analyses or alternatives)					√	√	√	S
Environmental cost-benefit analysis				√	√	√		S
Expert opinion		√		√	√			W
Expert systems	√	√	√	√	√	√		S
Indices or indicators	√		√	√	√		√	M
Laboratory testing and scale models		√		√	√			S
Landscape evaluation (for aesthetic or visual resource assessment, usually expressed as an index/indicator)			√	√	√			S
Literature reviews		√		√	√			M
Mass balance calculations (for comparison of existing to proposed conditions for air, water, etc)				√	√		√	W
Matrices	√	√		√	√	√	√	W
Monitoring (baseline information)			√	√	√			S
Monitoring (field studies of similar proposals)			√	√	√			S
Networks		√	√	√	√			M
Overlay mapping (probably using Geographic Information Systems)			√	√	√		√	S
Photographs/montages			√	√			√	M
Qualitative modelling (conceptual)			√	√				M
Quantitative modelling			√	√				S
Risk assessment	√	√	√	√	√			S
Scenario building (considering alternative futures)				√		√		S
Trend extrapolation (extending historical trends)			√					S

After Canter (2000)

used 24 per cent of the time. Models, used to generate quantitative information of impact evaluation, qualitative case studies and detailed checklists were the next most used, while index methods, networks and matrix approaches were seldom mentioned.

Table 8.6 (*see over page*) provides a comparison of methods with specific information about the potential for particular methods to be used at the different stages of an EIA. Data for the table have come from the International Study of the Effectiveness of Environmental Assessment (see 10.1). Broadly speaking, the stages, in the table, of impact assessment and decision-making, are similar to what we have called the evaluation of impacts, which has been the focus of much of this chapter. The table also gives an assessment of the extent to which the methods have been used. Compared with the earlier work of Atkins (1984), it appears that checklist and matrix approaches have gained some greater acceptance, while quantitative models have become less important.

8.4 Economic Evaluations and EIA

EIA has traditionally been interpreted to be the process by which physical environmental effects of an action are assessed. Taking a broad perspective of "environment", however, means the consideration of social effects and, as social and economic aspects are highly interrelated, the economic implications of actions/projects. The links between economic analysis and environmental issues are a major feature of the process of economic impact assessment discussed in Chapter 3.

Even without a conceptual reasoning to include economic issues, an EIS will inevitably include the costing of the project (and alternatives) and frequently estimates of the direct monetary benefits (eg, value of timber products, travel cost savings). There are also occasional complaints that unless EIA includes an economic assessment, the environmental effects are being considered without regard for the (presumably more important) economic factors.

Usually economics are covered by the listing of the costs and income (or benefits) that are directly attributable to the project; for example, capital and maintenance costs, costs of safeguards, value of export earnings, reduced accident costs. These costs and benefits are then compared against all the other environmental factors. However, in some EIAs an attempt has been made to handle the whole assessment using benefit-cost (or cost-benefit) analysis. This approach is promoted by Ahmad (1983), Hufschmidt et al (1983), and Department of Arts, Heritage and Environment (1985a). Ahmad (1984), and Department of Arts, Heritage and Environment (1985b) illustrate how it has been used to varying degrees for some projects.

Much has been written on the advantages of using this type of analysis; for example, it enables quantification of effects (in dollars), brings all factors to the same base to allow comparison, and makes comparison between

Table 8.6 Comparison of methods

Type of method	Addresses impact								Ease of application			Resource requirement			
	Identi-fication		Measure-ments		Interpret-ation		Evaluation					Staff needed		Computer needed	
	Yes	No	Yes	No	Yes	No	Yes	No	Difficulty	Moder-ately difficult	Not difficult	Highly skilled	Moder-ately skilled	Desired	Not needed
Ad hoc	x			x		x		x			x		x		x
Overlays	xx			x		x		x		x		x			x
Checklists	xxx			x		x		x			x		x		x
Matrices	xxx		x		x		xx			x			x		x
Networks	xxx		xxx		xxx		xxx		x			x		x	

Source: Based on Rau & Wooten

Key
x = Fair
xx = Good
xxx = Excellent

alternatives straightforward. However, there are many difficulties which are not always noted, or recognised; principally, the problems of determining "shadow prices" for environmental factors, ensuring all the costs and benefits are included and taking account of the redistribution effects (who/what gains or loses). Cooper (1981) provides additional insight to these difficulties.

The infrequent use of benefit-cost (or cost-benefit) analysis in EIA suggests that most analysts find other ways of assessing environmental effects to be more satisfactory.

8.5 A Risk-based Approach

We are seeing the increase in use of a risk-based approach to EIA in more recent years by both proponents in preparing their EIS's (or their consultants on their behalf) and regulators. For example, the Victorian *Ministerial Guidelines for assessment of environmental effects under the EE Act 1978* state that in preparing an EES, the proponent must incorporate the following specific principles of best practice:

- A systems approach to identifying, assessing and managing potential environmental effects to ensure that relevant effects and responses are considered;
- A risk-based approach to ensure that required assessment, including the extent of investigations, is proportionate to the risk of adverse effects;
- An integrated perspective of the relationship between and significance of different effects to inform decision-making; and
- The need to assess the consistency of proposed works with principles and objectives of ecologically sustainable development. (1978: 3)

The Australian/New Zealand and International Standard for Risk Management is AS/NZS ISO 31000:2009, and is the standard that some consultants/proponents are adopting to use in their EIA. This standard provides a structured approach and is widely recognised and routinely used as a basis for EIA and other risk assessments.

The risk management process outlined in AS/NZS ISO 31000 is iterative and can be applied to specific projects and activities. The main elements are:

- **Communicate and Consult** – communicate and consult with internal and external stakeholders at each stage of the risk management process.
- **Establish the Context** – establish the external, internal and risk management context in which the rest of the process will take place.
- **Identify Risks** – identify where, when, why and how events could prevent, degrade, delay or enhance the achievement of the objectives.

Figure 8.7 Risk management process

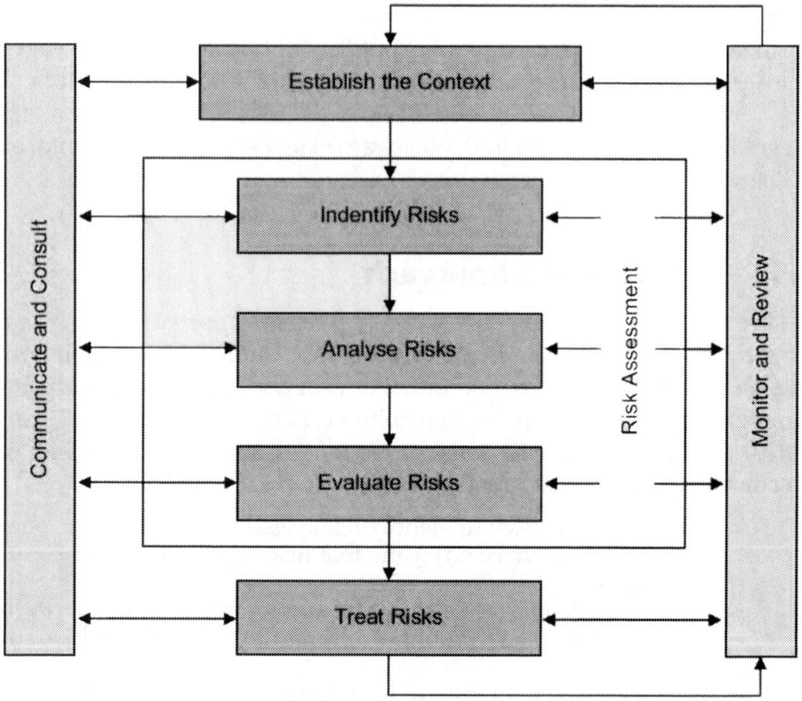

Adapted from AS/NZS ISO 31000:2009 (GHD (November 2012))

- **Analyse Risks** – identify existing controls, determine likelihood and consequences and determine the level of risk.
- **Evaluate Risks** – compare estimated levels of risk against the criteria and consider the balance between potential benefits and adverse outcomes.
- **Treat Risks** – develop and implement specific cost-effective strategies and action plans for increasing potential benefits and reducing potential costs.
- **Monitor and Review** – monitor the effectiveness of all steps of the risk management process. (GHD (November 2012: 4))

GHD, in its preparation of an EES for the *Western Highway Project – Section 3: Ararat to Stawell* in Victoria, provides an illustration of how the risk assessment process will be undertaken for the project to enable the proponent to identify risks and therefore aim to mitigate these.

8.6 Winding Up

Evaluation of environmental effects is not easy, that is, if the job is tackled honestly. It can be made easy by blithely adopting one of the evaluation methods and plucking some weighting technique out of some other study. Given the difficulty in measuring environmental systems and their values, it does not seem that the "easy" approach to evaluation would be a professional approach (ie, doing the best, socially responsible job with the resources available). There are difficulties in:

- determining the relevant environmental effects;
- determining the relationships between these effects and the environmental impacts;
- determining the extent of changes resulting from an action;
- determining the importance of these changes;
- comparing the magnitudes and importance of a variety of changes; and
- comparing environmental impacts of different proposals.

The evaluation methods have been able to cope with these problems with varying degrees of success and can be used to provide an indication of the extent of environmental impacts that may result from a proposal. The important thing is to acknowledge and document the limitations of the method being used, and to use the results with due regard.

It should be noted, however, that the theoretical approaches to evaluation discussed in 8.3 seldom appear in published EISs, but they may have been used as part of the EIS preparation. In practice, evaluation tends to be mainly descriptive (ad hoc or simple checklists), suggesting that the difficulties of the methods preclude their general use.

Ideally, the purpose of using any of the evaluation approaches is to draw together all the potential environmental impacts and to provide an assessment of the overall effect of the impacts. In other words, we look to the evaluation to indicate whether the total effect will have a severe effect on the environment or whether it is acceptable. As a result there is a preference for evaluation approaches which simplify all the information to present an "answer", usually a number. In general terms, the purpose of EIA is to assess the risk to the environment resulting from some proposal, so the comments of Stern and Fineberg (1996) regarding the analysis (or evaluation) of risks are equally relevant to EIA. They note that analysis is "often used to reduce inherently multidimensional risks to a single dimension so as to facilitate decision making" (1996: 116). However, there are dangers in attempting to reduce everything to a single item. As Stern and Fineberg comment:

> Analysis conducted to simplify the multidimensionality of risk or to make sense of uncertainty can be misleading or inappropriate, can create more confusion than it removes, and can even exacerbate the conflicts it may have been undertaken to reduce. Because of the power of analytical techniques to shape understanding, decisions about using them for these purposes

> and about interpreting their results should not be left to analysts alone, but should be made as part of an appropriately broad-based ... process. (1996: 117)

Likewise, with the evaluation approaches used in EIA we need to be careful that attempts to simplify the information presented to decision-makers do not result in important elements being hidden or ignored. No evaluation approach has been able to resolve this situation, so there will inevitably be a tension between seeking simplicity and displaying the critical information.

When involved at the other end of the EIS – that is, reviewing it – awareness of the limitations of the assessment is needed. It is important to keep a wary eye for assumptions, their effect on the results, and how the results have been interpreted for the reader. Also, it is well to consider what has not been included or said, as well as what has been.

Finally, it is important to understand the numbers used. If indices (ordinals) for noise, view, pollution and so on are derived, can they be added together as if they were cardinal numbers? Equally so, can they be multiplied? It is enough to say that, however much people trust or mistrust them, numbers will usually be the basis of environmental evaluation. However, they will have to be numbers which the majority of people who are numerate can understand and support.

Chapter 9

STEPPING THROUGH EIA

The theory and practice of EIA have been covered in previous chapters. While EIA may appear a fairly straightforward and simple process, it is no doubt apparent that EIA can be very complex. If the procedures are taken seriously and an honest attempt at a "good" job is attempted, then EIA is a challenge. One of the great difficulties, and perhaps strengths, of EIA is the extent of interrelationships between seemingly separate facets of the process; for example, the levels of public participation provided for, and the style in which the information in the EIS is presented.

The degree and depth to which the subject of EIA has been considered so far may discourage anyone from attempting to incorporate all the many important aspects. Such a situation would be an unacceptable result of this discussion of EIA. Consequently, to make the process clearer, the following sections present the key points that should be appreciated and taken into account when an EIA is undertaken.

What follows is essentially a summary of the main points concerning the practice of EIA. These are dealt with generally in the same order as the preceding chapters. As a result, this chapter provides a step-by-step guide to EIA which is appropriate to most situations. However, anyone involved in the preparation or review of an EIA should acquaint themselves with the local procedures and legislation with which their EIA will be required to comply.

9.1 Environmental Attitudes (see 1.1)

Although people give some credence to the idea that they should act as stewards towards nature, in our society there is a strong emphasis on the use (or exploitation) of natural resources. There is an emphasis on development, with some consideration for the ways to reduce the adverse effects of such development. This generally means that environmental or conservation issues tend to be given lower priorities compared with other aspects of society; for instance, the economy and employment. As a result, EIA sometimes appears as a rearguard action to ensure that environmental aspects are at least taken account of.

9.2 Technology (see 1.2)

The speed, scope and magnitude of effect of human actions on the natural environment are determined by the level of technology available to the community. With present technology, society can make significant changes to the environment in a short time. Hence, to avoid irreversible environmental damage, society needs to be aware of the potential environmental impacts of proposals, before the proposals are undertaken.

9.3 Background to EIA Procedures (see 1.3)

Community concern over environmental issues has been indicated by the many examples of legislation dealing with pollution, public health and nature/national parks. While this legislation was effective in ensuring that specific sections of the environment were considered, and possibly protected, no legislation or procedures were available to consider all the environmental impacts together. Further, because of a lack of strong environmental attitudes, other factors, such as economic issues, have tended to be given more weight. EIA procedures are an attempt to redress these imbalances.

Formal EIA procedures were first introduced into the USA with the passage of the *National Environmental Protection Act 1970*. Australia soon followed with the introduction of the *Environment Protection (Impact of Proposals) Act 1974*; now superseded by the *Environmental Protection and Biodiversity Conservation Act 1999*.

9.4 Issues for EIA (see Chapter 2)

EIA may seem to be a purely technical process for bringing environmental considerations into decision-making. However, its underlying political nature must be remembered. Wherever it is practised, EIA provides only *advice* to decision-makers, and cannot veto a proposal. Nonetheless, the outcomes of EIA can have a definite role in shaping environmental approvals and policies. An aspect of this role is the place seen for EIA as part of ecologically sustainable development and the resultant environmental quality, and social equity issues.

9.5 Features of EIA (see 2.2 and 2.3)

In summary, EIA is a process which:

- identifies the potential environmental (ie, environmental, social and economic) effects of undertaking a proposal;
- presents these environmental effects alongside the other advantages and disadvantages of the proposal, for the information of the decision-makers;

- requires proponents involved with the proposal, as well as decision-makers, to think about the environmental effects (and thereby avoid negative effects), and enables the public to become aware of the details of a proposal;
- does not provide a veto over a project; and
- still enables decisions which are unsatisfactory from an environmental point of view to be made, but these decisions are made with full knowledge of the environmental consequences.

The EIA process normally entails four key steps:

1. Screening to determine if a proposed action requires environmental assessment under the relevant legislation, which is often determined by the potential for it to have significant impacts on the environment.
2. Scoping to determine what matters require assessment as part of the *environmental impact statement* (EIS). Essentially, the scoping step of EIA is formalising a terms of reference for the EIS and will often include some form of community and/or government agency input before the scoping document is finalised.
3. Preparation of a document which provides information on the existing environment and predictions about the environmental effects which could flow from a proposal. This document is usually called an *environmental impact statement* (EIS), but is also known in Australia as an environment effects statement (EES) or public environment report (PER).
4. Review of the EIS by the public and government officers to consider the accuracy of the EIS and, in view of the predicted effects, whether or how the proposal should proceed. This review is reported to whoever makes a decision about the proposal in some form of an *assessment report*.

Hence, the key outputs of EIA = EIS + Assessment Report.

The combination of these two components makes up the majority of the EIA process as the screening and scoping stages of EIA generally occur over shorter timeframes. With recent moves to "decentralise" EIA, by incorporating environmental assessment into various levels of decision-making, formal documentation through an EIS and assessment report may not always be apparent. For instance the environmental assessment may be undertaken through an informal process set up under an organisation's Environmental Management System (see 10.5), or similar arrangement, or be incorporated into local council planning processes. However, no matter what terms are used, the two stages will be apparent in some form; that is, there will be a bringing together of information about the impacts of the proposal, and there will be some form of report about the acceptability of the impacts and what is recommended for the proposal. Again, these two elements combine to make the overall EIA.

Whenever we are discussing EISs, it is worth remembering that in some EIA procedures, such as that of the Australian Commonwealth, there are two parts to what becomes the final EIS. The initial document produced by the proponent is called the *draft EIS*, which is made available for public comment. The proponent responds to the public's comments in one of two ways. The draft is revised and released as the final EIS. Once the final EIS has been prepared, it forms the basis upon which the assessment report, of the relevant government department and/or minister, is developed. The final EIS is also the document that sets the directions for monitoring the environmental management of the proposal. In some EIA procedures, there may be a form of independent review or inquiry process, such as in Victoria under the *Environment Effects Act 1978* or under the Canadian EIA system (see 5.3).

9.6 Contributions to EIA (see 2.4)

Usually the proponent of the action prepares the documentation for the EIA; that is, they (or often their consultants) prepare the EIS. Public comment on the EIS is normally called for, either via submissions to the proponent or relevant government agency, or through a public inquiry. An assessment report is then prepared by the relevant government department, based on the EIS and the public comment.

9.7 Options for Impact Assessment (see Chapter 3)

EIA is arguably the best known form of assessment process. However, some other forms of assessment have predated it (such as economic assessment), while others have been proposed because of perceived deficiencies of EIA.

- *Economic impact assessment* (3.1) – Specific approaches have been developed to consider aspects of the economy; the basic purpose of these is to assess changes for government policy. Project-specific approaches, such as benefit-cost ratio, have also been developed.
- *Energy analysis and greenhouse assessment* (3.2) – Introduction of changes is considered in terms of the various forms of energy used or conserved. More recent EISs (certainly where the proposed development is within a coastal area) are also including an assessment of potential sea level rises and potential climate change effects on the proposed development.
- *Health impact assessment* (3.3) – Because EIA has typically not included health as part of the environment, HIA focuses on the impacts of a proposal on public health; while environmental health impact assessment has been developed to include both fields of concern. HIA usually assesses the effects of pollution on human health, as well as the impacts of siting facilities within particular areas and whether adequate services and infrastructure exist for the community.

- *Regulatory impact assessment* (3.4) – This process looks at the potential effects, on identified parts of the community, of changes in government rules, policies, regulations and legislation.
- *Risk analysis* (3.5) – This has been used in industrial contexts for many years to provide a measure of the chance of "things going wrong", and of the potential consequences and likelihood of such occurrences.
- *Social impact assessment* (3.6) – Social factors were generally not considered part of early EIAs, so SIA was developed to focus on the social consequences of proposals; it is also a vehicle for ensuring that the community is involved in setting the scope of an impact assessment. SIA is also important for determining how the community values an area or site, which can then be incorporated into the EIA.
- *Species impact assessment* (3.7) – This is an element of EIA to ensure that impacts on biodiversity are recognised and assessed.
- *Technology assessment* (3.8) – The social and economic effects of introducing new technology are the main concerns of this process, either to highlight how parts of society would be affected, or to consider how effects can be avoided.
- *Cumulative impact assessment* (3.9) – A series of separate insignificant decisions can lead to an overall significant effect, so this assessment process aims to take the longer and broader view to identify the possible total effect of decisions about a number of seemingly isolated projects.
- *Strategic environmental assessment* (3.10) – Whereas the practice of EIA and other assessment approaches has been to investigate the effects of specific projects or actions, SEA has developed to assess the environmental effects of broad policies, plans, and programs (such as regional land-use plans, and energy policies). While specific projects will be part of these policies, plans and programs, and may be assessed separately through EIA or the other approaches, SEA aims to identify the broader scale effects and has much in common with cumulative impact assessment.
- *Integrated impact assessment* (3.11) – This approach proposes the consideration of all possible impacts at the one time rather than having to conduct several separate impact assessments.
- *Sustainability impact assessment* (3.12) – A relatively recent form of impact assessment is sustainability assessment which is the assessment of likely economic, social and environmental impacts of policies, plans and programs and strategic projects. (Arbter (2008))

Depending upon definitions and approaches, procedures such as SIA or HIA can be used as a part of EIA to provide information for EIA. Alternatively, EIA could be conducted within the structure of a social or health assessment. Generally, all forms of assessment provide insight into EIA.

9.8 Public Participation (see Chapter 4)

9.8.1 Value of public participation (see 4.2)

The principal value attributed to public participation is that it results in better decision-making. More particularly, participation programs enable:

- engagement with key stakeholders and the community;
- dissemination of information;
- identification of relevant issues (and perhaps values); and
- a more open decision-making process.

9.8.2 Objectives of public participation (see 4.4.1)

Many things are wanted or expected of participation programs. A range of objectives is apparent, although most objectives relate to:

- communication – letting the public know what is going on; or
- providing an opportunity to present the public's views; or
- (rarely) enabling the public to be involved in making decisions.

The objectives for a public participation program will depend upon the political and social context of the proposal. A comprehensive set of objectives may be difficult to define; however, it is important to attempt a definition. Only when the objectives are clearly expressed is it feasible to determine appropriate techniques to arrive at the objectives, and then to evaluate the effectiveness of the participation program.

Links with an SIA may be a particular objective (see 4.4.2).

9.8.3 Levels of participation (see 4.4.3)

Levels of participation are many and varied. The most concise groupings suggest three broad categories:

1. Informing the public.
2. Soliciting input.
3. Public representation on decision-making bodies.

There is a strong relationship between levels of participation and the objectives of a program. Specification of the level of participation will largely determine the objectives that could be expected to be met; an objective of "providing useful additional information to decision-makers" will not be met if the level of participation is restricted to "informing the public".

9.8.4 Timing (see 4.4.4)

Most objectives of participation will be enhanced by devising and implementing a participation program as early as possible in the planning phase.

However, there are some factors that need to be remembered when designing a participation program; that is:

- Allow sufficient time to achieve the determined levels and objectives of the participation.
- Start early enough in the planning/design process to enable the levels of objectives to be achieved.
- Have ideas worked through sufficiently to give the public some information to respond to.
- Have a sufficiently short participation phase to minimise uncertainty and "planning blight".
- Fit in with various requirements of governments (eg, meetings or elections).

9.8.5 Who participates? (see 4.4.5)

Those sections of the community which would be likely to be involved in a participation program are:

- people directly affected by the proposal;
- people or organisations expressing a public interest in the proposal; and
- relevant government departments.

The people who are most likely to want to be involved in expressing an opinion about a proposal tend to be those who have the ability to find out about a proposal and have the access to participation mechanisms. These people are likely to be well educated and have a strong interest in their community. Initially, at least, they are likely to be opposed to the proposal (wanting to object). People who would be sought out to provide the opportunity to be involved include those who are in support of the proposal, and special interest groups (eg, ethnic or disadvantaged) who would be adversely affected, but may not have the ability to present their concerns. For some projects, during the formal public review phase of an EIS, there may even be as many supporters of the project as objectors, although it may be argued that this could be a result of a good public relations exercise by the proponent, or simply because the community was divided on the project due to the potential jobs that it may create. An example of where submissions on a project were very similar in numbers for and against the project was the Portland Wind Energy Project in south-west Victoria where there were more than 500 formal public submissions received and many of these were in favour of the project.

9.8.6 Public participation in EIA (see 4.5)

Participation under EIA procedures typically takes the form of public (written) submissions to the EIS (or draft EIS). These comments are used by the proponent, or relevant government department, in conjunction with the

EIS to provide an assessment of the environmental impacts of the proposal. Alternatively, a public inquiry may be convened to which the public contributes comment and after which the assessment report is typically produced by either the relevant government department or the inquiry panel.

In Victoria, for example, the *Environment Effects Act 1978* notes that the minister may invite public comment of the final EES, and the guidelines for implementation of the Act note that the EES will be exhibited from a period of between 20 and 30 business days. During this time, the public can make submissions to the minister. Normally, however, once an EIS is prepared, its availability for public comment is made known through newspaper advertisements calling for submissions. Written comments are received by the relevant government department, and taken into account in the preparation of the assessment report. Public inquiries have been held in Victoria, sometimes in conjunction with procedures under Acts other than the *Environment Effects Act* (eg, *Planning and Environment Act 1987*).

Screening (see 4.5.2) is the stage at which a decision is made as to whether an EIA is required for a particular proposal. Various triggers are used to help this process, to which the public rarely has access. However, making the decision about the scope (and any associated conditions) available on the Internet, which most States/Territories and the Commonwealth do, can provide the public with some information about the process.

Scoping (see 4.5.3) has been employed to involve the public in advising on the issues to be covered in an EIA, and the depth of coverage. In most jurisdictions, a draft scoping document (or assessment guidelines) are advertised for a specific period, somewhere between two and four weeks, to enable public comment on the draft scope for the EIS.

9.8.7 Techniques for obtaining public participation (see 4.6)

The methods by which participation has been sought are many and varied. For each participation program the techniques used to provide opportunities for participation are likely to be different from previous programs. In particular, the techniques used will depend upon the objectives of the program. Table 4.2 lists a number of techniques and provides an assessment of when a particular method would be effective.

9.8.8 Evaluation of public participation (see 4.7)

The evaluation of public participation programs is important so that some guidance is available as to how each program achieved its objectives, and whether there are any lessons to be learned for future programs. Because evaluation can only take place after a program has been completed, there is a tendency to forget about it. However, even a limited evaluation is likely to provide useful information which would assist in the design of future participation programs.

9.9 Environmental Impact Legislation (see Chapter 5)

The first EIA legislation was passed in the United States in 1970. This was the *National Environmental Policy Act* (NEPA). Since then most Western countries have:

- passed similar types of legislation; or
- brought in administrative arrangements which have a similar effect to this legislation; or
- introduced some form of EIA procedures through other legislation.

The introduction and operation of EIA in developing countries requires special consideration to ensure that the context and the resources for EIA exist (see 5.7.1). Many developing countries do now have some form of EIA procedures or are captured under foreign "donors" who require EIA to be undertaken when funding projects in developing countries (see 5.8).

9.10 EIA Procedures in Australia (see Chapter 6)

At the Commonwealth level of government, EIA legislation came into being with the *Environment Protection (Impact of Proposals) Act 1974*. This Act has been superseded by the *Environment Protection and Biodiversity Conservation Act 1999*. Under this Act, the Commonwealth Environment Minister determines if an action is "controlled", and therefore if the action will be subject to EIA (ie, to be controlled by the Act). Actions include projects, developments, any undertaking, and an activity or series of activities. Within these, a controlled action is one that either:

- is likely to have a significant impact on a matter of national environmental significance, or
- is likely to have a significant impact on the environment associated with Commonwealth land.

If the proposal is a controlled action, the Minister selects one of five levels of investigation and assessment; ranging from preliminary documentation to an inquiry (see 6.3.4).

All the Australian States and Territories have introduced some form of EIA procedures, although only Victoria and the Northern Territory have passed legislation that deals exclusively with EIA. Western Australia has included EIA procedures in the *Environmental Protection Act*. The Australian Capital Territory, New South Wales, South Australia and Tasmania have included EIA procedures in land-use planning legislation, while Queensland has a relatively dispersed EIA procedure focused on land-use planning, although there are some EIAs undertaken for mining and other major projects under relevant Acts within these jurisdictions. There are also provisions for States and Territories to make "special" legislation or conduct EIA through a major project type of legislation, for example the Tasmanian *Pulp Mill Act 2007* and the earlier example of the South Australian *Roxby Downs (Indenture*

Ratification) Act 1982 which provided for an override of existing Aboriginal and environmental legislation relating to the Olympic Dam site.

As an example, the Victorian *Environment Effects Act* (and associated 2006 guidelines) sets out a process for undertaking EIA which is fairly typical. The Act relates to all government proposals, and private proposals which require some form of approval or licensing from a government department, and where there is likely to be a significant environmental effect as a result of the proposal. The proponent refers the proposal to the minister and the minister determines if an EES is required (that is, if there are likely significant State or regional environmental impacts).

As with all EIA procedures in Australia, the proponent is responsible for the preparation of the EES, which is advertised to enable public comment to be obtained. Once the submission period, or public inquiry, is completed, an assessment report is prepared by the minister. The minister then forwards the assessment to the relevant decision-makers. The decision-maker(s) is obliged to consider this assessment report together with the technical, economic and other information in relation to the project before making the final decision on the project. There is no obligation on the decision-maker(s) to abide by the recommendations or conclusions of the Minister's Assessment.

The process described for Victoria above is similar to other Australian EIA procedures. In 1991 the Australian and New Zealand Environment and Conservation Council adopted a set of principles to establish a national approach for EIA in Australia (see 6.2). Also aimed at developing consistency is Schedule (3), dealing with EIA, in the InterGovernmental Agreement on the Environment. This Schedule sets out a common set of principles which will achieve greater certainty about the application of EIA throughout Australia, and avoid duplication and delays in the process. Reviews of EIA procedures followed with the aim of establishing a consistent Australia-wide approach. The assessment bilateral agreements associated with the Commonwealth *Environment Protection and Biodiversity Conservation Act 1999* are also intended to develop a degree of consistency in the procedures across Australia (see 6.3.4). All States and Territories have an assessment bilateral agreement with the Commonwealth.

9.11 Determining the Need for an EIS (see Chapter 7.1 and descriptions of specific State and national procedures)

Whether a proposal goes through an EIA procedure depends upon the particular requirements of the procedure. These requirements are determined from the "screening" process associated with the EIA procedure. If the proposal is not screened out, the next stage is preparation of an EIS, or some less detailed form of EIA documentation (eg, preliminary documentation under the Commonwealth's *Environment Protection and Biodiversity*

Conservation Act 1999), if provided for in the procedures. The basic "trigger" for the screening process is whether a proposal may have significant environmental impacts, so 7.1 contains a discussion of how significance may be determined. In some EIA procedures, there may be lists of projects that require EIA, for example Tasmania.

9.12 Purpose of an EIS (see 7.2 and 7.3)

The principal purpose of an EIS is to present the environmental impacts of actions to the decision-makers so that the environmental consequences of the action will be known if a decision is made to proceed with the action. The other major aspect of an EIS is to disseminate information about the project.

The concept which underpins an EIS is that human-initiated *action* leads to *environmental effects* (such as air pollution) which lead to *environmental impact*. The extent of all environmental impacts of an action may not be fully known at the EIS (and assessment) stage, so it is important that the EIS also presents a mitigation and monitoring program that specifies how environmental impacts will be prevented, addressed, managed and what ongoing monitoring is to occur to ensure these impacts are minimised.

9.13 Content of an EIS (see 7.4)

Most guidelines for the preparation of an EIS are similar in that they require:

- full description of the proposed activity;
- statement of the objectives of the proposed activity;
- full description of the existing environment likely to be affected by the proposed activity;
- indication and analysis of the likely interactions between the proposed activity and the environment;
- analysis of the likely environmental impacts or consequence of carrying out the proposed activity;
- clear justification of the proposed activity;
- safeguards to be taken in conjunction with the proposed activity to protect the environment and an assessment of the likely effectiveness of those measures; and
- consideration of viable alternatives to the proposal.

The detailed requirements for the appropriate form of EIA documentation should always be ascertained from the relevant government organisation.

9.14 Comments on Contents (see 7.5)

- *Objectives* – These should be objectives of the proposal, not merely objectives of the EIS itself.

- *Description of proposal and main alternatives* – Enough information is needed to evaluate how the most important environmental factors would be affected. The "do nothing" alternative should always be included. Alternatives can mean alternative sites and/or alternative technologies. For example, wind farm EISs would include alternative types of both non-renewable and renewable energies to meet the same objectives (that is to produce electricity).
- *Description of existing environment* – This provides the baseline information necessary to predict the environmental effects of the various alternatives. There is no single list of environmental items that is appropriate for every EIS; rather, a separate list should be developed in each case. Scoping is used to determine the list of relevant (and often site-specific) issues.
- *Description of likely environmental impacts* – The accuracy with which we can assess physical impacts is dependent upon our understanding of how the environment functions. As our knowledge is often less than perfect, the accuracy of any predictions will be reduced. It may be relatively simple to make predictions about change, but it is impossible to guarantee accurate predictions. Nevertheless, attempts must be made to predict the impacts of actions in the socio-economic, physical, chemical and biological items previously identified as being the important environmental effects. Checklists, matrices or flow diagrams can assist, and in some cases mathematical models can provide an indication of the degree of impact.
- *Safeguards to avoid, minimise or ameliorate adverse effects* – Safeguards (or mitigation measures) such as soil conservation measures, landscaping and measures to maintain air quality should be reported in sufficient detail to indicate that the safeguards would be feasible, and that there would be a commitment to carrying them out. Safeguards must be considered as part of the proposal (eg, included in the overall cost) so that the proposal can be equally evaluated against the alternatives.
- *Evaluation and discussion of alternatives* – This section provides the basis for the final conclusions and recommendations of the EIS. Ideally, there should be an assessment of how each alternative measures up to the objectives plus the evaluation of how each alternative would affect each of the environmental items previously identified. This evaluation is usually presented as:
  - a display of sets of values of individual impact indicators; or
  - a ranking of effects within impact categories; or
  - normalisation or mathematical weighting of the impact indicators.
- *Conclusions/Recommendations* – To bring elements of the EIS together this section could cover comment on public submissions,

consultants' views (if appropriate), reasons for endorsing the proposal, and outlines of monitoring and safeguard programs.

9.15 Quantifying Environmental Effects (see 8.2)

Quantification involves:

- deciding which environmental effects may be affected (eg, air pollution, aesthetics or social interaction);
- determining the magnitudes of the changes in these;
- determining the importance of the changes;
- combining the magnitude and importance of measures for each environmental effect (if of the same units) and combining all "scores" together. It should be noted that the numbers which are assigned to magnitude or importance (for example) are not sacrosanct. The reliability and validity of the number, and the relationship from which the number derived, should all be considered with caution.

The magnitude of an effect depends on the magnitude of the action and the magnitude of the environmental effect affected. Importance of an effect refers basically to the significance of the effect to the community. Care must be taken to distinguish between unit importance (the importance to an individual) and total importance (importance to the whole community). Importance depends heavily upon the values of the person considering the effects; it may be that one person considers an effect to be good while another would consider it to be bad.

9.16 Evaluation Techniques (see 8.3)

The typical methods used to evaluate environmental effects are:

- ad hoc methods;
- checklists;
- matrices (tables);
- overlays;
- systems diagrams;
- networks;
- quantitative or index methods; and
- mathematical models.

These usually present the information in terms of:

- simple displayed information; or
- tabular comparison; or
- aggregation of effects.

At one extreme (ad hoc methods), the evaluation is presented in generalised descriptive terms and much of the analysis is left to the reader. At the other extreme (quantitative or index methods, mathematical models), predetermined relationships are used to derive some index for the effects. Hence, moving from one end of the spectrum of methods to the other, the degree of analysis increases and thereby the complexity for the reader is reduced; however, assumptions and value judgments tend to be hidden from the reader.

All methods have advantages and disadvantages. The difficulty for the person preparing an EIS is to adopt the method(s) which presents the evaluation in the most suitable form for the reader, and decision-maker. It is important to be aware of the efficiencies of the particular methods, and to make the reader aware of these. Evaluation of environmental effects presents difficulties in:

- determining the relevant environmental effects to be considered;
- determining the relationships between these effects and the environmental impacts;
- determining the extent of changes resulting from an action;
- determining the importance of these changes;
- comparing the magnitude and importance of a variety of changes; and
- comparing environmental impacts of the alternatives.

The use of risk-based assessments in evaluating the environmental effects of a proposal is becoming more popular in EISs (see 8.5).

9.17 Assessment of the EIS

Each of the EIA procedures in Australia has a different mechanism for assessing the EISs produced. These assessment mechanisms are discussed in Chapter 6 in relation to the particular EIA procedure under discussion. Typically, preparation of an assessment report is undertaken by the relevant government department. This report is based on:

- the EIS;
- comment from the public;
- comment from government departments and statutory authorities (eg, water authorities); and
- information additional to the EIS, provided by the proponent, either voluntarily or upon request.

The assessment is sent by the minister responsible for the EIA process (or, in some cases, the head of the department) to those ministers and government departments responsible for making a decision about the proposal (ie, granting approvals), and to the proponent. Recommendations contained in the assessment report provide *advice* to the decision-makers. Under some procedures the assessment report is published. Under the Commonwealth *Environmental Protection and Biodiversity Conservation Act*, the Commonwealth Minister for the Environment prepares an assessment and also makes a decision to approve (or not) an action (such as a development proposal).

9.18 Finally, What Hasn't Been Said

The above are points to remember when thinking about EIA. They by no means provide all the details necessary to assist in undertaking an EIA and the assessment of environmental impact; however, they do provide a checklist from which an EIA can be designed. The preceding chapters provide more detail on the particular aspects of an assessment process.

It can be tempting to follow a process (such as that set out in this chapter) to produce an EIS, and to think that this is all that has to be considered. However, it would be naive and short-sighted to be involved in an EIA without appreciating and making attempts to overcome its limitations. Section 2.4 and Chapter 10 explore some of the issues which are hidden in the EIA legislation and processes, particularly:

- It is largely the technically literate who become involved in the EIA process, but the proposals frequently affect those who are not technically literate (disenfranchised groups of people, and plants and animals).
- EIA has a tendency to be used as a vehicle for assisting development to occur, and may not give adequate attention to alternatives (such as do nothing).
- Frequently there are attempts to quantify and place a value on environmental impacts where such attempts have little objective basis (eg, the value of a view in dollar terms).
- EIA might be thought of as the only technique for reviewing what might result from going ahead with a proposal, when in reality many processes are used by people to arrive at a judgment.
- Some may see EIA as providing "the objective information" upon which decisions are made; however, because all decisions about resource use are social by nature, EIAs are also based on value judgments and are political decisions. Indeed, EIA is both a technical and a political process.
- There is little, if any, post-EIA follow up after the EIA process is complete in terms of monitoring and managing potential environmental impacts. Such post-EIA monitoring is important to determine if the potential environmental impacts occur as predicted in the EIS and that environmental safeguards are implemented as described.

Chapter 10

THE FUTURE OF EIA

What does history tell us? Over four decades of experience with the operation of EIA or forms of environmental assessment (EA) have provided opportunities to reflect on the worth and practice of the procedure. Sadler (1997b) identified two clear trends in the late 1990s which are still relevant today:

> First is the widespread establishment of EA systems by many developing countries and by countries in transition. Second, is the emergence in several industrialized nations of a second-generation, integrated, strategic EA process more closely linked to national planning and decision-making processes.

Adoption of the EIA process by governments worldwide suggests that EIA is seen to fulfil a valuable role for the decision-makers at least. This degree of acceptance is related to the way in which EIA has been used to assess proposals; that is, the practice of EIA which has developed, and continues to evolve. While EIA was initially seen as a tool of the conservationists, a threat to development, and a time delay, it is now becoming an accepted part of the project development process as an important tool for identifying relevant environmental, social and economic effects of a proposal. Being part of the development process is the substance of Sadler's second point, that assessment of environmental aspects is being introduced at the early stages of policy-making and planning, as we discussed in relation to Strategic Environmental Assessment (section 3.10) and also in trade negotiations with the introduction of the EU's Trade Sustainability Impact Assessments.

Drawing on a brief review of the evolving support basis for EIA, this chapter reflects on our four decades of experience with EIA, considers its future directions, and comments on the understanding of EIA's social context which has emerged.

In the literature that has evolved to report and critique EIA, there is a continuing discussion of EIA's objectives, role, and its future directions and use. The theory that supports EIA seldom receives scrutiny. This issue has been emphasised by Lawrence who points out that EIA practice has "evolved without a sound conceptual foundation" (1997: 79) and that a more coherent theory is required. He suggests that building this theory is not an end unto itself, but is a means to achieve better decision-making and improved practice. He also sees that this theory will need to draw on a wide range of concepts, including: directed; rigorous; practical; evolving; heuristic; collaborative; risky; uncertain; complex; critical; reflective; designed; value-full; ethical; political; pluralistic; contingent; boundary-spanning; constrained; opportunistic. Lawrence's thorough discussion of theory will provide directions for the refinement of EIA, and comes at an appropriate juncture in the evolution of EIA. As in other fields of human endeavour, practice usually precedes the theory, which serves to document the practice, and provide a vehicle for its critical analysis. From this comes the "directions" (theories and models) for how to proceed to achieve the best results.

While Lawrence has provided a framework for the discussion of theory, explicit discussion of theoretical concepts is less common than that which gives emphasis to the practice of EIA. However, an instance of consideration of the "theory of EIA practice" has been provided by Suter and Cormier (2008). They "propose a theory that derives 19 principles from 3 axioms: 1) assessments inform environmental management decisions, 2) assessments are science based, and 3) management decisions accommodate multiple goals and constraints" (2008: 478). These axioms tend towards the promotion of a scientific focus in EIA, however, the third allows for integration across disciplines, scales, and types of evidence. Nonetheless, the focus provides little room for consideration of the political aspects of decision-making.

There is a growing literature related to EIA. While much of this focuses on improvements to practice, there is growth in discussion from academics and theoreticians. The theoretical coverage outlines the philosophy behind EIA, its context and role, the "rules" that have been developed to guide practice, and EIA's relationship to planning and environmental management (such as the relationship to Ecologically Sustainable Development (ESD)).

Equally important for the day-to-day operation of EIA is the review of its administration and practice, in that this provides us with ideas for how the tool can be better used. However, the key issue is whether we want to continue to use the tool in the same way, or not use it at all. These points are the basis of our consideration of the broader issues.

10.1 Reflections on Practice

In several of the previous sections there has been the opportunity to discuss points that have been learnt from the experiences of applying the principles and concepts of EIA, and associated assessment approaches (especially section 7.7). Generally the literature abounds with reflections on the effectiveness of EIA. Even if the title and focus of the material is not on assessing or evaluating EIA, or EISs, there are inevitably insights that can be used to lead to improvements.

Importantly, now that EIA is a mature area of practice and its use has been demonstrated in developing countries, worldwide experiences are now available to guide improved practice. For example, in Pakistan, Nadeem and Hameed (2008) found that EIA has been used mainly as a project justification tool, rather than as a project planning tool to contribute to achieving sustainable development. They identified the main deficiencies to be: inadequate capacity of EIA approval authorities, deficiencies in screening and scoping, poor EIS quality, inadequate public participation and weak monitoring. These are problems that have been raised in relation to other countries, as is their view that needs to be a stronger political will and institutional capacity to undertake EIA. Political will and the values that underlie it are matters to which we will return.

In 1993, at the Shanghai meeting of the International Association for Impact Assessment (IAIA), the initial International Study of the Effectiveness of Environmental Assessment was launched. Participants in the meeting were asked whether environmental assessment had achieved its goal of helping us reach better decisions. As a result, a three-year study was generated involving 12 international partners, including government agencies and professional associations (see Sadler (1997a)). The study was led by the Canadian Environmental Assessment Agency in collaboration with the IAIA with contributions from many organisations and individuals; the vast majority being from "developed" nations. Overseen by Barry Sadler, the study focused on four issues:

- Foundations of EA, focusing on guiding values and principles.
- New dimensions in EA, focusing on the application of sustainability concepts, strategic environmental assessment, and cumulative and large-scale effects.
- Process strengthening, focusing on the relationship of EA to decision making.
- Capacity building, with particular reference to needs of developing countries.

The focus of the study was to assess the effectiveness of environmental assessment. Since EIA began there has been much debate about its scope and role (see Chapter 2). This has led to the development of a range of assessment approaches (see Chapter 3) so that specific aspects of the environment (eg, social, health) were given particular attention. In recognition of this range of approaches, the generic term Environmental Assessment (EA) has been adopted to encompass the variety of approaches that assess the likelihood that some proposed action will have environmental effects. Clearly, EIA is the key assessment approach that comes under the banner of EA.

In summary, the results from the survey of the status of EA practice, that sought the views of practitioners and managers, are an important insight into how well EA has been operating. Generally, the results indicated that:

- EA has been broadly successful in identifying appropriate mitigation measures and in providing clear information on the potential consequences of proposals;
- there has been only limited success in making verifiable predictions, in specifying the significance of residual impacts, and in providing advice on alternatives; and
- in addition to informing decision-makers, EA promotes greater awareness of environmental and social concerns, expands professional capabilities, and promotes public involvement in decision-making.

Given the issues associated with predicting environmental change (see Chapter 8) it is not surprising that predictions have been found to be wanting. Clearly the successes of EIA have been associated with the broader picture of environmental awareness, and general understanding of activity-environment interactions, rather than at the level of detail.

We have referred to the conclusions and proposals coming from the study in several sections where they relate to specific aspects of EIA. However, in addition there are some key proposals that may help to shape the direction of EA generally, and EIA in particular. The proposals have been presented under the headings of:

- Going Back to Basics – including clarifying the relationship of EA to decision-making and reducing duplication of assessments.
- Upgrading EIA Processes and Activities – such as improving quality control and public involvement.
- Extending Strategic Environmental Assessment (SEA) as an Integral Part of Policy Making – including developing flexible approaches that can be related to the varied situations of policy making.
- Sharpening EA as a Sustainability Instrument – particularly incorporating relevant indicators in screening, significance, and other "checklist" processes.
- New Opportunities and Challenges for EA – especially the expansion of EA into SEA and sustainability areas.

A review of the IAIA Study of the Effectiveness of Environmental Assessment was launched at the IAIA 28th Annual Conference in Perth in 2008. Delegates participated in a discussion of impact assessment effectiveness, what it means and how practitioners can contribute to improved effectiveness. The review focused on the implementation gap between theory and practice of EIA and what makes EIA effective. The key issues to emerge during workshop discussions at the Conference included the need to follow up post-EIA (eg, were predictions accurate) and sharing of information with other practitioners on how to improve effectiveness and engaging EIA early in the process.

These procedural improvements are important, however, are they the key to making EIA more effective? In reviewing international EIA activity, Jay et al (2007) conclude that while the quality of decisions has improved, as a consequence of the availability of EIA processes and their improved technical abilities, the influence of EIA on decisions about development proposals is less than desirable. As they say, "The achievement of (EIA's) substantive aim, contributing to more sustainable patterns of activity, although difficult to assess, appears to be even more elusive" (2007: 298). The way forward, they suggest, requires not just the improved procedural approaches of the past, but giving EIA "a more determinative position" in planning and decision-making processes. They have reflected that part of the problem has been the "rationalist beginnings" of EIA, where the focus was on collecting objective information that would, by itself almost, lead to better environmental protection. Instead, we need to redirect attention to the initial political purposes of EIA and assert its role as a major factor when development decisions are being made.

10.2 Technical Issues

Previous chapters have concentrated on the general philosophy and approach of EIA rather than its technical details. The literature abounds with discussion of issues of EIS accuracy and methods of assessment; detail which interested readers can pursue at their leisure. However, auditing and monitoring are related to both the technicalities and the philosophical base of EIA, and deserve some discussion.

The results of the monitoring processes provide information about the technical elements of EIA (such as methods for forecasting change). These results then form the basis of the audit of the overall EIA procedure. Munro, Bryant and Matte-Baker (1986) see that comprehensive environmental audits provide an opportunity for reviewing both the EIA process and EISs by examining:

- the accuracy of EIAs as forecasters of environmental consequences of proposals;
- the effectiveness of recommended mitigation procedures;
- the utility of monitoring techniques and programs; and
- the effectiveness of environmental management.

However, reviews of EIA procedures by these authors, and by Thomas, Instone and Durkin (1992) indicate that little auditing has been undertaken, in Australia at least. However, the establishment of a formal reporting mechanism as a requirement of the Commonwealth *Environment Protection and Biodiversity Conservation Act 1999* (EPBC Act) has led to formal monitoring programs and may facilitate auditing. Since commencement, auditing and monitoring under the EPBC Act has become an important feature of the Act, as reported in the Department of Sustainability, Environment, Water, Population and Communities (DSEWPC) annual report of 2011-2012. The Commonwealth DSEWPC reviews and investigates a number of reports of activities potentially in breach of provisions of the Act from the public, government and non-government organisations. The EPBC Act provides a range of approaches for compliance issues, including prevention, remediation, civil penalties and criminal prosecutions for more serious breaches, and reporting on compliance and enforcement issues.

DSEWPC also undertakes its own monitoring of development activities with a view to improving compliance with the Act. For example, it was reported in the 2011-2012 annual report that 412 actions were referred to the Australian government for a decision on whether approval was required under the EPBC Act. Thirty-two of these referrals were the result of compliance action taken by the department (DSEWPC (2012)).

The auditing of the EIA process has also been promoted by Bisset and Tomlinson (1988), as they see that EIA is intended to provide decision-makers with understanding of the environmental consequences of actions. Hence, the experiences obtained from EIAs should be used to improve the processes. In effect, this would create a feedback mechanism for transferring knowledge of actual environmental effects of proposals to future EIAs. Sadler (1988) goes somewhat further by arguing for the development of a mechanism for national and international dissemination of the results of audits and evaluation programs. His proposal is for particular attention to be given to impact prediction and mitigation, procedural effectiveness and the contribution of EIA to decision-making.

Contributions are being made to the quality of EISs by approaches such as that outlined by Lee and Colley (1990), where criteria for an EIS have been developed along with process and assessment levels. More specifically, in their review of ten assessments, Munro, Bryant and Matte-Baker (1986) were able to highlight several areas for improvement, including monitoring, ecological understanding and the integration of EIA with project planning.

The issue of monitoring was reviewed in detail by Martyn, Morris and Downing (1990). In particular, they proposed that monitoring should occur throughout the life of the proposal, and is especially important where the proposal is controversial or where there are uncertainties of data or impacts. Consideration of general principles for monitoring and what has occurred in practice leads them to develop proposals for the support of monitoring.

Monitoring has been recognised as part of the EIA process, but is seldom emphasised. Martyn, Morris and Downing suggest that monitoring is needed to:

- enforce conditions and standards associated with approvals;
- prevent environmental problems which result from inaccurate predictions, inadequate mitigation or factors unforeseen at the time of the assessment;
- minimise errors in future assessments and impact predictions;
- make future assessments more efficient, cost-effective and timely;
- provide opportunity for mitigation to be on an as-required basis where impact predictions are uncertain;

- provide ongoing management information about the project and its environmental effects; and
- provide for retrospective research to improve EIA and monitoring.

Similar objectives were identified in the International Study of the Effectiveness of Environmental Assessment, presented by Sadler (1997a), but with the additional clarification of verifying environmental compliance and performance.

Martyn, Morris and Downing (1990) also discuss the principles of monitoring and point out that the requirement for monitoring, as part of EIA procedures, varies. Sadler (1997a) has also reached the same conclusion that follow-up activities need to be specifically related to issues associated with the EIA. He goes on to propose that there are four main components to the follow-up phase:

- Surveillance to ensure terms and conditions are being followed in project construction.
- Monitoring to check for compliance with standards to test the effectiveness of mitigation and other protective measures, and to detect potentially damaging changes.
- Management to respond to unforeseen events or to offset larger-than-predicted effects.
- Auditing to review aspects of EIA practice and performance and to provide feedback for process improvement.

Internationally, Sadler (1997a) concluded that monitoring and EIA follow-up mechanisms remain poorly developed, especially by comparison to pre-decision activities. Australian experience relates closely to this conclusion and continues to be the case. Even where monitoring is specified as one of the matters to be dealt with, surveillance of the monitoring programs is rare (Martyn, Morris and Downing (1990)). The record for monitoring is considered to be poor, a conclusion which was supported in a survey of impact assessment procedures in Australia. From this survey, Thomas, Instone and Durkin (1992) found that there was little opportunity to see how the assessment processes were working, and the situation was little better for the monitoring of project effects. Even where effects were monitored, there was seldom any power to take remedial action if necessary. Overall, there appeared to be negligible external review of assessments. It is fair to say this issue remains current for EIA practice in Australia. Internationally, however, Wood (1995) notes that there are some countries where the monitoring of the EIA system is required, and some where informal monitoring and review takes place. With these examples to look towards, and as experience is gained, the extent of monitoring may increase, but incentive, or direction may be required.

Clearly, there are several elements and stages to monitoring. Shepard (1998) proposes that the key is the development of a comprehensive monitoring program that emphasises the integration of the "checking process" into the overall EIA process. She suggests there are several methods for "Post-Project Impact Assessment and Monitoring (PPIAM)", that is:

- Baseline monitoring – establishing a database during a representative pre-project period.
- Effects or impact monitoring – measuring changes during the project.
- Compliance monitoring – linked to baseline and impact monitoring, it is most closely associated with assessing compatibility with regulations.
- Scientific monitoring – checking accuracy in the EIA and explaining errors.
- Management monitoring – determining if mitigation approaches were used and if they were effective.
- Enforcement monitoring – ensuring that mitigation is being performed as specified.
- Effectiveness monitoring – measuring the extent of success of mitigation approaches.

Even with the application of this range of monitoring approaches, Shepard suggests that there are several requirements for monitoring to achieve its potential. Principally encouragement and commitment to monitoring is required, but so too is the sharing of information to make it more attractive, and the planning of monitoring to ensure it is integrated throughout the EIA.

As with others, Martyn, Morris and Downing (1990) have called for more effort to be given to monitoring, through the mandatory condition of an approved program for proposals considered under EIA. Likewise Malone (1997) supports the formalisation of monitoring. She identifies the barriers to effective monitoring programs and discusses what should be monitored, and how the monitoring should be carried out. In addition, she proposes a statutory model for "environmental impact monitoring" to formally build monitoring into EIA processes.

As an expansion of the notion of monitoring, Martyn, Morris and Downing have also proposed additional attention be given to the monitoring of the EIA process, or auditing the process, to improve its methodological basis and predictive capacity. These points are taken up in principles incorporated in the Australian and New Zealand Environment and Conservation Council national approach to EIA (ANZECC (1991a)).

An example of auditing of the EIA process and which leads to suggestions for future improvements is provided by Jones and Woods (1995). Their survey of EIAs that had involved public inquiries in Great Britain indicated the advantages of having public involvement early in the EIA process. Scoping and other ways of consulting with the community were seen to offer potential reductions in the time and costs associated with an EIA. These findings are in line with the directions of the Australian national approach (see 6.2).

When Sadler (1997a) conducted his initial study into EIA effectives, he was able to draw together the various proposals for monitoring EIA. The resulting "sound practice" principles outlined in Figure 10.1 *below* provide a clear rationale and framework for follow-up procedures. The effectiveness depends on the establishment of appropriate inspection and enforcement mechanisms. To date in Australia, such mechanisms usually have been limited to being associated with planning or pollution permits for a proposal, rather than being tied to the EIA process. The practice for EIA in Australia is that it is a State or Territory government department/minister who writes an assessment of the project, but it will often be a different department/agency or local government where actual development conditions are prepared and responsibility for implementation of these conditions lie. However, the Commonwealth's *Environment Protection and Biodiversity Conservation Act 1999* (EPBC Act) has recognised the need to follow-up on EIAs and may provide a lead for other procedures.

Figure 10.1 Sound Practice Principles for Follow-up in Environmental Assessment

Premise:
Follow-up activities on a scale consistent with the estimated significance of project impacts should be undertaken to ensure compliance with approvals, and to facilitate environment management and performance review.

Requirements:
These include appropriate legislation, regulations and/or administrate provisions and mechanisms to provide for inspection, enforcement of terms and conditions, monitoring and control of unanticipated impacts.

Follow up Actions and Principles:
Major functions and activities include:

- inspection and surveillance to check that terms and conditions are met;
- effects monitoring to measure environmental changes resulting from project implementation;
- compliance management to ensure that regulatory standards and requirements are being met;
- impact management to respond to adverse changes and environmental performance audit to verify effectiveness of EIA, mitigation and other components.

Inspection and surveillance should be undertaken as either a routine or periodic activity depending on project terms and conditions.

Monitoring, audit and other follow-up activities should be undertaken when:

- potential effects are uncertain or unknown but likely to be significant
- species or areas of concern act as a trigger
- changes can be realistically detected against natural variability.

Source: Sadler 1997a

In the future we can expect to see more attention given to monitoring specific proposals that have undergone EIA. This will provide data for auditing the overall EIA process with the intention of providing for improvements in practice. However, this level of auditing requires the commitment of resources by those administering EIA and recognising that EIA does not (or should not) finish with the release of an assessment report. It remains to be seen if these resources are made available.

We had emphasised the moves towards better monitoring and auditing, but virtually all other elements of EIA activity have been subject to discussion to seek more efficient and effective use of the process. For example, risk assessment (see chapter 3.5) has its own literature but in relation to the environment there has been increasing interest in its use in Ecological Risk Assessment. Suter (2008) provides an overview of the development of this approach, especially in the USA, and makes the point that its greater use can be expected; however, this will be dependent on the extent to which ecological (or natural environment) issues are included in the scope of an assessment. We are also seeing interest in the application of EIA, and SEA, to the investigation of climate change issues and consideration of what issues should be reported on. In this regard, Burdge (2008) argues that EIA and SEA must both assist our understanding the consequences of human communities of the effects of global climate change, and makes suggestions about how this would be facilitated, including making environmental justice a focus. Taking a similar view on the need to make decisions based on knowledge of climate change effects, Koornneef et al (2008) have specifically looked at the care that needs to be exercised in the screening and scoping stages, and suggest closer connections between SEA and EIA together with a strong role for community participation in the processes. These improvements may be able to improve the standard of the EISs produced, as there continues to be concern over the quality of many. Following a review of EISs in the USA, Tzoumis (2007) concluded that there continues to be a decline in both information adequacy and investigation of the impact of the preferred alternative on the environment.

In addition to what is covered in an EIA, another consideration is how they are covered. In this context, Pastakia and Jensen (1998) have proposed use of their "rapid impact assessment matrix", designed to provide a more numeric approach to the evaluation of options. The matrix allows a transparent and permanent record of the judgments, however, to convert the estimates of change into numeric results, the method continues to rely on judgments made by those conducting the EIA. This highlights a fundamental difficulty with EIA, where we have to combine information that is both quantitative and qualitative, essentially having to "compare apples with oranges" to come to an overall decision about the changes a project may cause. Development of "multi-criteria decision" aids[1] aims to help, but these processes are frequently complex, sometimes depending on computer software, and are not yet reliable. As Kain and Söderberg (2008: 7) report, there remain issues "regarding

1 Multi-criteria decision-making tools include the tabular and weighted matrix evaluation methods, discussed in Chapter 8.3.2.

transparency and understandability; qualitative/quantitative knowledge input; switching between different modes of weighting; software flexibility; as well as graphic and user interfaces".

Improvements in our ability to combine information from a range of data sets (ie, the apples and the oranges) and to remove human judgment from this and the subsequent evaluation of options is the "holy grail" of EIA. Efforts in this direction will continue, and rightly so, however, we must reiterate that EIA is a social and political process (certainly not just a technical process), and we cannot see how value judgments will be removed from the process. Yet consideration has been given to how to make these issues more transparent. Specifically, those with EIA experience have been concerned with how the test of significance is used to determine the issues to be included in the EIA. For instance, Lawrence (2007a) has investigated the positive and negative implications of three impact significance determination approaches; technical, collaborative and reasoned argumentation. Neither approach, nor their variants, provides "the answer", however:

> With an enhanced understanding of these approaches ... EIA practitioners, working in concert with interested and affected parties, can be in a better position to design and apply an impact significance determination approach that addresses the needs and preferences of stakeholders, is consistent with good practice, reinforces the positive tendencies of available approaches, offsets the negative tendencies of available approaches and facilitates the effective match of process and methods to context. (2007a: 753)

This advice leads him, in a subsequent article, to propose that an important matter in determining significance is the scope of the socio-environmental issues to be included in the EIA. He emphasises that this scope should not be narrow, and that the significance of issues needs to be guided by consideration of: cumulative as compared with individual impacts; socio-economic as compared with biophysical impacts; integrating the Precautionary Principle into the process; and when ESD/sustainability contributions drive the EIA process (Lawrence (2007b)). The emphasis on breadth is supported by Wood (2008) who, with experience relating to impacts on landscape and on noise, makes several suggestions for EIA practitioners. Of note is the need to give consideration to timing of impacts, duration, reversibility, and their cumulative potential; baseline considerations and mitigation opportunities; and uncertainty and limitations of the assessment process. His proposal for ensuring comparability of significance assessment between impacts echoes the concerns noted above (Pastakia and Jensen (1998); and Kain and Söderberg (2008)).

These insights are valuable, but need to be put in the political context that operates with any social process like EIA. While EIA practitioners may like to feel that the results of objective investigation will be accepted by decision-makers, a clear "reality-check" is provided by the review of Swedish practice by Engström et al (2008). They summarised their findings as follows:

> A main finding from the study is that issue importance is a necessary but not a sufficient condition for policy attention. Other explanations are needed to understand which environmental issues get attention in sectoral policy. Our assessment showed that while the level of knowledge does not provide an explanation, the presence of strong and well-organised stakeholders within the sector, with an interest in having a certain issue on the agenda, might be decisive for issue attention. Path dependency and limited attention capacity are other important factors. (2008: 241)

EIA practitioners cannot afford to ignore these political factors if they are to produce effective work that assists to identify impacts and mitigate the worst effects.

10.3 Administration

Administration of EIA procedures worldwide is a huge topic. The discussion here is confined to trends in Australia, although some trends would also be relevant for other countries.

Both White (1987) and the Bureau of Industry Economics (1990) have identified specific difficulties with the EIA processes in Australia. The concerns about duplication, direction for proponents and delays have been recognised in the national approach for EIA (ANZECC (1991a)) and the InterGovernmental Agreement on the Environment (IGAE) (National Strategy for Ecologically Sustainable Development (1996)). The principles contained in these initiatives are designed to develop certainty in the way EIA procedures are applied across Australia. While this is not yet evident, the relative frequency with which procedures are reviewed should provide opportunities for procedures to converge. Specifically, the national approach may also lead to formalisation of the negotiation process, which was sought by Woodhead (1987), as public participation becomes more formally built into EIA procedures.

The idea of a common approach to EIA within Australia was proposed by Martyn, Morris and Downing (1990), along with consideration of the establishment of an independent body to supervise EIA. They identified at least four tasks that such a body could perform: initiating EIA; supervision and assessment of the impact study; linking assessment recommendations and decision-making; and constituting public inquiries for an independent forum for assisting the EIA process.

The national approach to EIA is helping to ensure that the various procedures become essentially the same in practice, even if the administrative details are different. The Commonwealth EPBC Act has been an important factor in aligning EIA procedures around the country. Most States and Territories have undertaken comprehensive reviews of their EIA procedures in the last five years.

A key provision of the EPBC Act is the establishment of assessment bilateral agreements between the Commonwealth and each State or Territory (see 6.3.4). With the purpose of reducing the potential for duplication and delay in the assessment of proposals, these agreements would "accredit" the EIA procedures of the States and Territories. The effect would be that a proponent would need to complete only the accredited State assessment, although there would still be a requirement for the State/Territory and the Commonwealth to separately approve decisions, unless an approval bilateral is in place. In an approval bilateral agreement, the State assessment and decision process has to be carried out under an accredited management plan that is in force under the law of the State and which complies with Commonwealth regulations. The accredited management plan will be developed through negotiations between Commonwealth and State officials, with input from the public. Further, the plan may be disallowed by either House of the Commonwealth Parliament. Hence, there will be strong pressures to develop plans and agreements that are based on the general principles of the national approach, as these provide an already agreed set of criteria for EIA.

Development and adoption of the EPBC Act also indicates an evolution in the scope of environmental assessment. The expansion of EIA to embody SEA has been noted (see 3.10), and also associated with the Commonwealth's directions is the clearer recognition of the ways in which environment assessment relates to resources assessment. The Act identifies the links between assessment and resources like forests and marine areas, and specifically relates to activities on Commonwealth land. This is similar to the changes that have taken place in State and Territory procedures where, as noted in Chapter 6, many of these procedures have become closely aligned with resource management and land-use planning. This is a trend we can expect to see continue as the integration of environmental assessment into policy and plan making is needed at the earliest points if we are going to move towards ecologically sustainable development. In particular, Conacher (1994) argues that EIA on its own is an inadequate means of improving environmental quality, and that we need to integrate environmental assessment with land-use planning and management. While EIA is certainly still required at the project level, Marsden and Dovers (2002) argue that it fails critically to address likely significant environmental impacts at higher levels, or to deal with the dangers resulting from cumulative environmental effects.

10.4 Assessment Activity Outside the Formal Processes of Government

In discussing the current directions of assessment (see 2.3) we looked briefly at the extension of EIA into the internal planning processes of organisations (businesses, corporations, local governments, non-government groups, and the like). These organisations are not relying on the formal government procedures, such as those outlined in Chapter 6, to tell them when an environmental assessment is needed. Rather, they have developed "in-house" arrangements to identify when their activities may have environmental impacts, and the safeguards needed to reduce significant impacts. Many (and certainly the more experienced) proponents have already undertaken some type of feasibility assessment in determining a final development concept, before entering the formal EIA process. Encouragement for the development of informal arrangements is given by the increasing adoption of Environmental Management Systems (see 10.5) where the organisation has to identify the environmental impacts that result from its operations. Yet other organisations have developed their approaches to satisfy internal environmental requirements.

As an example of an informal EIA procedure, City West Water uses a Preliminary Environmental Assessment (PEA) that is integrated into its overall management procedures (see Figure 10.2 *over page*). The substance of the PEA is:

- a checklist that ensures the PEA considers issues such as compliance with legislation, consultation with affected communities, and typical environmental matters like amenity and air quality;
- a tabular checklist that identifies issues associated with specific environmental matters, such as sites of identified floral significance; and
- an Environmental Assessment/Summary sheet which itemises: a description of the project; options considered; description of environmental impacts and costs; evaluation process and expert advice obtained; actions to minimise impacts and further actions; formal approvals required; consultation process – this sheet ends with an assurance statement that all reasonable steps have been taken to recognise and reduce impacts, and is signed by the Project, Business Function and Environmental Programs Managers.

The emphasis of this assessment process is the projects that are initiated and developed by City West Water; a similar form of informal EIA has been considered by the City of Melbourne. In this case the concern is environmental impacts that can come from the decisions made by the council regarding planning applications (Leeson (2000)). These applications relate to projects that are initiated by individuals and organisations with no connection to the City, and where there may be implications for the overall environmental quality of the council's area (such as the implications for energy use and greenhouse gas emissions of allowing residential development that is not based on public transport). Here a process based on general checklists has been trialled alongside a process similar to that of City West Water to assess the impacts of projects initiated by the council (eg, road construction).

In addition to informal processes that provide for the direct assessment of environmental impacts, and which largely follow very similar stages to formal EIA procedures, over the last decade several procedures have evolved to identify environmental impacts. Often these are used alongside other

Figure 10.2 City West Water management procedures with integrated environment assessment

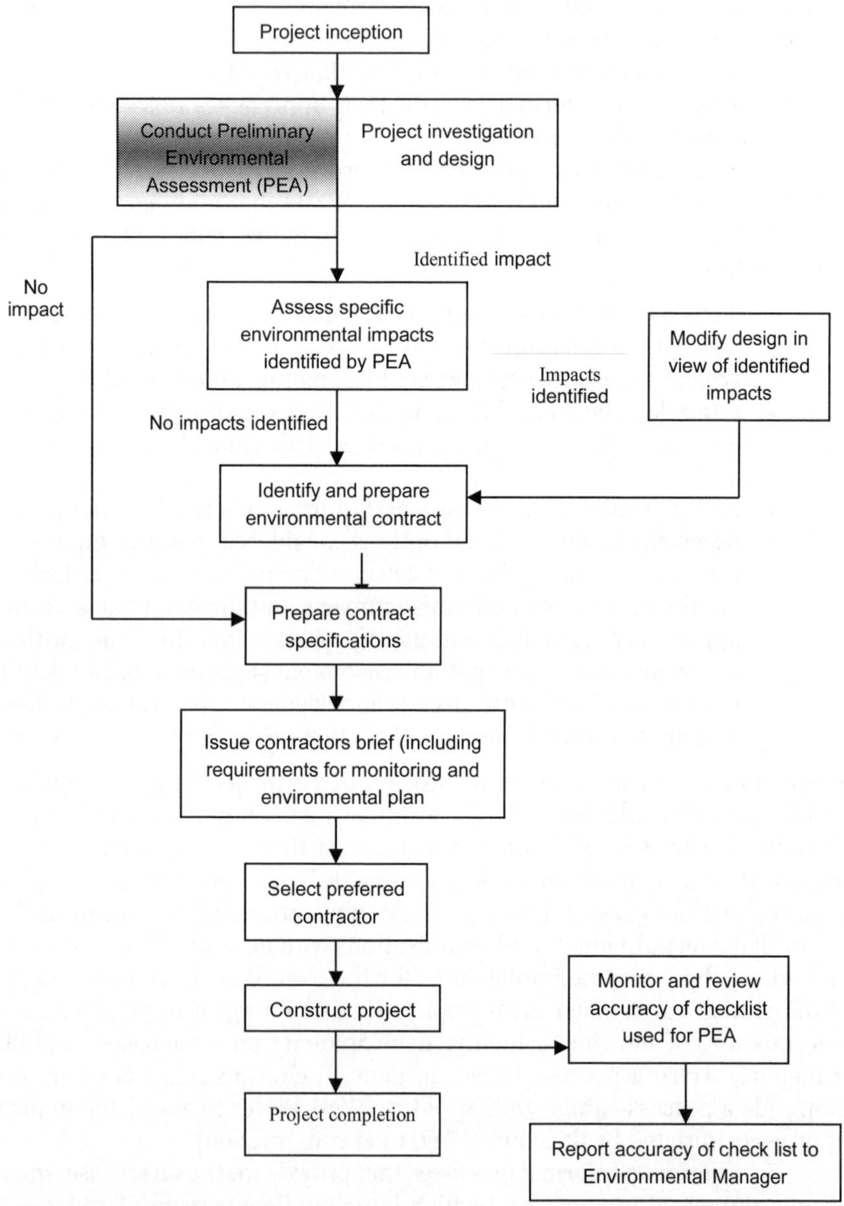

Source: City West Water (1999)

procedures, or are associated with broad environmental management and associated systems (particularly discussed in 10.5).

One important development, Munchenbery (2002) believes, is that of the triple-bottom line approach undertaken by business and governments. This seeks to ensure environmental, economic and social issues will be taken into account in all decision-making, regardless if any formal EIA process is undertaken. He emphasises this point with reference to the Business Council of Australia's *Statement on Strategies for Sustainable Development* that includes a commitment to integrate sustainable development considerations into all aspects of business planning and operations for its members. The Statement says this will be achieved by "incorporating environmental and social considerations into management systems, business planning and decision-making processes, along with conventional financial considerations and by applying risk management techniques that recognise financial, environmental and social risks" (2002: 192). Triple-bottom line reporting is something that we have seen emerging over the past two decades and is mostly used by business to demonstrate a concern for economics, environment and society although it can be used by governments to voluntarily communicate information on social, environmental and economic performance to their stakeholders.

Internationally, the Global Reporting Initiative *Sustainability Reporting Guidelines* (G3 Guidelines launched in 2006 and G3.1 Guidelines in 2011) is fast becoming the accepted framework for triple-bottom line reporting and organisations are rapidly adopting it to both improve corporate sustainability performance and communicate this to stakeholders. The growth in the uptake of corporate reporting is consistent with the current drive for good corporate governance, heightened transparency and sound risk management (Department of Environment, Water, Heritage and the Arts (2008)). The Global Reporting Initiative reports that statistics and analysis indicates that there continues to be a growth in sustainability reporting at local and global levels (GRI (2011)).

Like EIA, the concept of life cycle analysis (LCA) has developed from the need to take a holistic view of a project or activity, but with particular emphasis on industrial activity. LCA has been applied to industrial processes, such as the production of steel, to determine the environmental implications of all stages of the process. According to Flood (1994), the analysis is a means of assessing the effects of a business's products as well as the effects of obtaining the raw materials and components purchased from suppliers. Hence, LCA would take into consideration the impacts of procuring the basic materials (eg, mining), transport, manufacture, distribution and, ultimately, disposal of the used items.

Rubik (1993) notes that another term for LCA is the "eco-balance approach", which has been used in a number of contexts: the eco-labelling of products; eco-marketing of products; design of new products; investment and purchasing decisions; determining future development strategies; and designing environmental policy instruments. In all these situations the desire is to be able to provide an overview of the impacts to guide decisions-makers, whether they are shoppers deciding to buy a product, directors choosing between production processes or bureaucrats considering planning alternatives.

As with other forms of impact assessment, the difficulty with LCA is in defining what is to be included in the analysis and obtaining relevant data. In this vein, Rubik has identified several areas requiring further research: scope of the study; system description; comparison of alternatives; quality and comparability of data; and appropriate aggregation and validation methods.

Unlike EIA, the assessments undertaken within the framework of LCA are not currently required by legislation, nor are they always available for investigation by the public. Nonetheless, the LCA "tool" is essentially a procedure for assisting decision-making, and ensuring that environmental considerations are a main aspect. As with EIA, there is no way of ensuring that the environmentally best activity is undertaken, but LCA has the potential for alerting designers and planners to environmental considerations, and highlighting the impacts of particular decisions. Consequently, it has a strong educational element, which could ultimately lead decision-makers to aim for reductions in resource consumption and pollution loads.

In this respect, LCA, and the broad philosophy of EIA, has a close association with the notion of cleaner production. Cleaner production is "the continuous application of an integrated preventative environmental strategy to processes and products so as to reduce the risks to humans and the environment" (United Nations Environment Programme (nd: 2)). However, no matter what title is given to the processes, each involves the basics of EIA, being to identify the environmental issues (eg, use of resources, or pollution), forecast the impacts, consider alternatives and present the information in a way that aids the decision-makers.

An initial stage in cleaner production and similar assessment procedures is that of environmental auditing. The Institute of Environmental Assessment comments that environmental auditing has emerged as an important instrument for environmental protection. This is the term "given to the process for evaluating the environmental impacts during the operational phase of existing developments" (1993: 5). Now there are individuals accredited as environmental auditors, but the process of an audit is fairly clear and is discussed in many manuals and texts, such as those published by the International Chamber of Commerce. Auditing typically involves three steps: pre-audit activities (obtaining background data); activities at the site (observation of compliance and what is happening and discussions/interviews with operational personnel); and developing the report with its action plan.

In many respects, auditing complements EIA. While EIA typically focuses on the potential environmental effects during the planning stage, before operations begin, environmental auditing undertakes a monitoring role during the operational stages, to provide feedback for improvement. It is important to note that auditing is often about compliance with legislation and policies, although with the increased use of environmental management systems, continuous improvement through monitoring and review is becoming the main focus, rather than simply compliance. According to the Institute of Environmental Assessment, links between EIA and environmental auditing should become firmly established within the next decade, providing an "integrated approach to facility stewardship" (1993: 5). In other words, EIA coupled with environmental auditing support the principles of sustainable development.

10.5 EIA's Role in Environmental Management Systems

Since the early 1990s, interest in environmental management has opened up a new path for EIA. Whereas much of the previous running of EIA had been taken by government, with non-government organisations being forced to respond, this situation has changed somewhat since parts of the business community have begun to develop environmental policies to guide their operations.

The previous emphasis on efficiency would be superseded by a focus on excellence, where environmental considerations would be included in the thinking of the organisation. This approach requires the organisation to develop a strategy which includes the production of an environmental policy, and management systems to achieve it. Environmental assessment would play a part in the system through the organisation undertaking specific environmental audits and developing environmental indicators, and its involvement in environmental problem solving.

In 1996, the international standard for environmental management systems (EMS) was launched, and revised in 2004 (this version is currently under review). ISO 14001 is designed to enable an organisation to "formulate a policy and objectives taking into account legislative requirements and information about significant environmental impacts" (Standards Australia and Standards New Zealand (1995: 3)). This standard is based on the philosophy of:

- pollution prevention;
- compliance with environmental laws; and
- continual improvement (improvement in environmental performance up to the limit set by the best available technology.

There are over 250,000 organisations certified to ISO 14001 (BSI group).

Sutton (1997) suggests that the core features of the ISO 14001 EMS model is that the organisations working under the standard:

- develop an awareness about their environmental impact;
- develop a policy that sets –

  - the level of environmental performance they desire,
  - the parameters for how they will manage themselves to achieve this level;
- commit themselves to –
  - continual improvement in line with their policy and pollution prevention,
  - comply with relevant regulation and guiding principles (eg, codes of practice);
- establish mechanisms to follow through on the policy commitments;
- audit and evaluate their performance and take corrective actions; and
- review the appropriateness of their management system and make modifications if needed.

He suggests there are a number of benefits in using ISO 14001 as the basis of an EMS, particularly that the EMS is driven by the enthusiasm of the implementing organisation because it is a voluntary code; the system is focused on implementation (not just planning); it is flexible enough to accommodate all organisations; ISO 14001 is understood across industry sectors and across countries.

EMSs may give the impression that they provide security for the environment. But from the description of the stages it can be seen that the EMS is essentially a system for managing an organisation so that it minimises and deals with environmental impacts and ensures compliance with environmental and other relevant legislation. It is important to recognise that the primary focus is not on the details of how the environment is managed. Through evaluation, monitoring and review, EMSs should strive for continuous improvement – not only of the system, but to the operation of the organisation as a whole.

As a consequence, Sutton (1997) argues that organisations should aim to develop "leading-edge" EMSs that go beyond compliance.

A leading-edge EMS is one which, according to Sutton, actively seeks to contribute to environmental and social sustainability. These "sustainability-seeking" EMSs can be evolved from the ISO 14001 process, but have some key differences, essentially in the way in which the scope and possibilities of the EMS are conceived. He suggests that "a sustainability-seeking firm must much more rigorously set the framework for setting objectives and targets than is the case for a 'default'-style firm" (1997: 235).

The number of EMSs being implemented in government and industry has been increasing especially amongst large organisations. As of December 2006, Corporate Risk Management (formerly ISO World) noted that some 129,031 EMSs were certified worldwide, Japan having the most with 21,779. Australia was noted as having 1964 accredited EMSs developed (up from 835 in 2003). We can expect that the number of accredited EMSs worldwide will expand considerably through peer pressure (competition) within an industry, and the "ripple" effect from large organisations, and indeed an increase from over 18,000 accredited EMSs worldwide in mid-2000 to over 61,000 worldwide which was reported at the end of 2003! This "ripple" effect was seen when, in the late 1990s, the Ford Motor Company produced a guide for the development of its suppliers' EMSs. Since many suppliers were small enterprises the guide outlines what an EMS should contain and assists the process of getting the EMS accredited to ISO 14001 standard. However, in addition to the accredited EMSs, there are likely to be many organisations that have developed their EMS, but have not sought accreditation.

A more tangible example of the use of environmental management systems is provided by Victoria's corporatised Melbourne Water. Responsible for water supply and associated activities, Melbourne Water initially adopted the EMS standard, but included some local perspectives so that eight programs comprise its environmental and public health management system (Melbourne Water (2008)):

- An environmental policy committed to ISO 14001. Melbourne Water has an environment policy and this is supported by "Sustainable Water – A Strategic Framework", which shows real commitment to environmental sustainability.
- An Aspects and Impacts register which identifies all the positive and negative impacts on the environment that arise from our activities and risk ranks them.
- A system for ensuring that we know and meet our legislative and other requirements.
- A program for environmental improvement that is contained within our Annual Corporate Plan, the Water Plan and individual Performance Plans.
- Defined resources, roles and responsibility and authority;
- Systems for competence, training and awareness.
- A manual describing how the EMS works and system procedures explaining how the system is implemented to protect the environment across all activities, including those delivered by contractors.
- Regular monitoring and reporting of environmental impacts through internal quarterly reports and Melbourne Water's Annual Reports;
- External and internal audits.
- Regular management reviews to look at effectiveness of the Environment and Public Health Management System.

Embedded in the EMS concept is an implication that environmental assessment will be undertaken by a variety of people in an organisation. Also, those involved in the formal legislated EIA process will not necessarily be consulted. However, Melbourne Water's guidelines make it clear that EIAs for a wide range of the organisation's activities will be carried out by informed staff, but who will not necessarily be specifically trained in the specifics of EIA.

Such a change raises exciting possibilities. In particular, it provides the opportunity to decentralise EIA so that EIA is not focused on particular government departments and firms of consultants. Further, EIA does not have to be enshrined in legislation for its principles to be applied. It also opens up the chance to demystify EIA in that, rather than there being a selection of "experts", there is the possibility of many people in various positions to apply EIA. As a result, the community generally will have the opportunity to develop an "assessment literacy" and hence expand the practice of EIA past the current focus on projects, to review the whole gamut of activities in which organisations are involved (eg, plans, budgets, investments, policies).

Fundamentally, the expansion of environmental assessment through environmental management systems will ensure that many in the community become more aware of environmental issues, and gain the tools to assess impacts and develop responses. With the improved level of environmental education that this will provide, the goals of EIA will be better served. In particular, there will be a better chance of avoiding environmentally damaging activities, and the community may be more actively involved due to the environmental reporting of the organisation as part of the EMS.

10.6 Revisiting the Philosophy of EIA

Perceptions of the future of impact assessment, and particularly EIA, are mixed, but generally a continuing role is predicted and the increased use of SEA is progressing. The position of EIA appears more secure than other forms of assessment. Taylor (1984) sees that the EIA process benefited from the expansion of environmental interest groups, particularly those which were legally and scientifically well resourced. It remains to be seen whether other assessment approaches find similar forms of assistance.

As many countries and professions have institutionalised EIA, the chance of it losing importance in the near future is most unlikely. Institutionalisation, according to Taylor, allows EIA to bridge the gap between the "old" institutional constraints and the "new" regulatory problems. In particular, it provides a means for one agency to more easily regulate other public bodies when policies and proposals have inter-agency effects, rather than giving the agency regulatory power over others (an unacceptable situation for most organisations).

Specifically, Caldwell (1989) sees that the future of EIA will be tied to its integration in project planning. Then procedures for screening, scoping, external review and public participation must be part of EIA so that EIA takes place automatically to ensure that environmental values are part of decision-making. This point is taken up by Hollick (1986), who considers that the EIS is but one part of the EIA process. In particular, he sees that much of the success of the process depends on the modifications to the proposal undertaken before the EIS is submitted (refer to the Maud Creek gold mine case study in 6.6) on the consultations and information used during the EIS period, and on the monitoring of effects after the beginning of the proposal.

However, experience indicates that political pressures have been the driving force behind EIA. As a result, in future EIA will be most effective where environmental values are integrated into a nation's culture and public law and policy. The challenge for EIA then is obviously political rather than technical. Caldwell comments, "If EIA is to be more than a ritual, further change in the attitudes and behaviours of political leaders and public officials will be necessary" (1989: 4). He suggests that these changes could come from education or indoctrination, a stronger base in law for environmental policy (eg, environmental rights), and internationalisation of EIA to take account of effects which pass national frontiers. The recent tendency of decision-makers to espouse the idea of ESD (see 2.3) may be the beginning of a greater commitment to taking environmental issues into account as part of decisions. Certainly, the introduction and application of Sustainability Impact Assessments in trade negotiations is a good example of this expansion of EIA in decision-making (see 3.12).

Ross, Morrison-Saunders and Marshall (2006) identify a number of issues relating to improving the quality of EIA and suggest that inserting common sense back into EIA may improve practice. The main tools suggested by the authors for such improvements involve "more effective signals from government EIA regulators to project proponents, greater ownership of the EIA process by proponents and, with reservations, the provision and application of EIA guidance" (2006: 3). Essentially, setting the direction for EIA at an early stage will aid in ensuring a more effective process for government, community and the proponent.

However, the role of EIA remains a vexed question. For example, Wildavsky (1988) argues the case for a "trial and error" approach to environmental policy, and consequently to the assessment of proposals (see 3.5). In this situation EIA would have a greatly reduced role as a predictive tool. But it would remain an essential procedure for the identification of key environmental issues, the formulation of monitoring programs to check the rate of development of environmental impacts, and for the identification of related amelioration or remediation measures to take account of the impacts.

From the opposite point of view, Vlachos (1985) sees that assessment processes are a tool of policy analysis where a major interest is the reduction of uncertainties to probabilities. He also proposes that the main purpose of EIA is to provide a form of focused disagreement. In this vein, Rogers (1988) argues that while the purpose of EIA is to enable better informed decisions, decision-makers, proponents and the public profit differently from the process. She notes that criticism of EIA has come from both developers and protectionists: "[EIA] is one of the deceitful cooptions of the concept of ecology and environment. While sanctimoniously reciting the catechism of 'environmentalism' it anoints and blesses the 'process' of development" (quoting J Livingston (1988: 23)).

Expanding on this position, O'Riordan (1990) sees EIA located only towards the middle of the spectrum of philosophies which recognise the importance of the environment. He suggests that "dry greens" believe that science and market forces can bring about the environmental connection. For them EIA is a nuisance, but they accept that EIA can be institutionalised to assist development.

"Shallow greens" want to share the natural world with development, but intervention is needed to achieve this. They are unsure about EIA. Some see that it provides the opportunity to promote better management, which could be passed onto green consumers with the necessary publicity. Others see this as "ecological prostitution, the selling of ecological services to clients for money and favour" (O'Riordan (1990: 14)).

"Deep greens", however, put the earth first, and people second. They see policy in global terms, work with long-term perspectives and seek decentralised power structures. EIA then is an anathema because of its cosmetic significance, where development for human gain has priority.

While recognising these positions, O'Riordan notes that there is an even broader issue:

> Green capitalism, like green consumerism, is geared to maintaining commerce, profit and good public relations. EIA in this context acts as a signal and a warning for potentially unavoidable conflict. It also provides a basis for negotiation leading to what can be described as "green plus" marketing where the local environment and community services are "bettered" or improved by marketing and planning strategies ... [Then] EIA is moving away from being a defensive tool ... to a potentially exciting environment and social betterment technique that may well come to take over the 1990s. (1990: 12)

He suggests that the dry-shallow green form of EIA will continue for some time, but will move towards scientific variations of "green plus" management. Local communities will be more involved and environmental protection will be expanded. EIA can then be considered to be a process rather than a simple technique, a "process that is constantly changing in the face of environmental politics and managerial capabilities ... [acting] as a sensitive barometer to environmental values in a complex industrial society" (1990: 15). If O'Riordan's expectations for the community to become increasingly involved in EIA processes eventuate, the social relevance of EIA has a good chance of being maintained, if not increased.

Continual questioning of the role of EIA is needed, whether the community is heavily involved or not. As O'Riordan (1990) points out, there are still differing expectations for EIA; such as developers seeing it as a mechanism for assisting the approval of proposals, while many in the community expect that it will protect the environment, or at least determine the environmental impacts of a proposal to aid the decision makers.

However, the role of broad-ranging input to assessments is crucial, as EIA is a social tool and occupies a recognised place in the politics of decision-making. It must be remembered that all assessment procedures are value-laden, so without high levels of participation EIA would run the risk of slipping back to being a tool of bureaucrats, where technocrats assume the role of interpreting society's values. Also, since EIA has become institutionalised, it has spawned a range of professionals to look after it. In this situation, the community may feel that it should first accept the worth of EIA, and secondly, simply leave it to the professionals to undertake assessments. On both points the community would be selling itself short, as this situation would indicate an even greater need for community participation in EIA.

10.7 Sustaining EIA

Given that EIA is enshrined in legislation around the world, it is safe to assume that EIA will not disappear. However, its likely potential is of interest. We have already seen changes in the application of EIA, from projects to policies and plans, and its scope, from a concentration on the biophysical environment to an inclusion of social, economic, landscape values, climate change and other matters. In many EIA procedures these broader views have been possible almost since EIA became formalised in legislation. However, only now are we seeing that the practice of EIA is catching up with the intent of the process.

The developing relationship between EIA and ESD was touched upon in Chapter 2. Increasingly discussion of EIA is presented in the context of its role in sustainable development. As discussed in 3.12, sustainability impact assessments are occurring for policies, plans and programs but also in some trade negotiations in Europe. The relationship of EIA and sustainable development was a key aspect of the International Study of the Effectiveness of Environmental Assessment (currently in the process of being reviewed). Here it was recognised that sustainable development, or sustainability, is based on simple but key principles like those outlined by the World Bank "Guidelines on Environmental Sustainability" (noted in Sadler (1997a)):

> Output Guide: Waste emissions from a project should be within the assimilative capacity of the local environment to absorb without unacceptable degradation of its future waste absorptive capacity or other important services.
>
> Input Guide: Harvest rates of renewable resource inputs should be within regenerative capacity of the natural system that generates them; depletion rates of non-renewable resource inputs should be equal to the rate at which renewable substitutes are developed by human invention and investment."

These guides are derived from the more general principles associated with sustainability. In the context of environmental assessment Sadler (1997a) discusses the principles as:

- Precautionary principle – to emphasise conservation as a hedge against irreversible or highly damaging changes.

- "Anticipate and prevent", rather than adopting the costly and risky approach of "react and cure".
- Stay within source and sink constraints.
- Maintain natural capital at or near current levels, that is, there should be no aggregate/net loss or reduction of resource stocks or ecological diversity.
- Avoid conversion of land use from less intensive to more intensive uses.
- Polluter-pays principle – full costs for environmental damage must be borne by users.

As an illustration of the connection of environmental assessment to sustainability, Sadler (1997a) reports on the process for planning for climate change and for maintaining biodiversity. Figure 10.3 *opposite* presents the strategy proposed for each issue. Importantly, these issues are broad policies so the examples illustrate the connection of an SEA level analysis with sustainability.

In Australia there are examples of EIA explicitly being linked to sustainability particularly through ESD, the term adopted in Australia. The objectives of the Commonwealth's EPBC Act relate to ESD, specifically s 3(1)(b) states the Act is to "promote ecological sustainable development through the conservation and ecological sustainable use of natural resources". The EPBC Act also requires all Commonwealth bodies to report annually on their implementation of ESD, through s 516A. Similarly, the Tasmanian EIA processes are embedded in a broad resource management approach which is intended to support ESD principles (see 6.9). Even more directly linked to ESD are the NSW procedures which state that ecological sustainability requires a combination of good planning and an effective and environmentally sound approach to design, operation and management (Department of Urban Affairs and Planning (1997)). As a result, under regulations of the *Environmental Planning and Assessment Act 1979*, proponents of projects are required to have regard to the principles of ESD throughout the whole project life cycle. They are also advised that continual reference should be made to the question "is the proposal ecologically sustainable?" Nonetheless, these are only broad directions for proponents. Given the complexity of the concepts of sustainability, and the variety of ways in which sustainability has been interpreted, it remains to be seen how proponents will be able to live up to the relatively vague expectations of the regulations, and hence how EIA and ESD may be linked in practice.

In this context, Court et al (1996) provides a suggestion and even bigger challenge. They propose that to achieve ESD it is essential to incorporate Cumulative Impact Assessment (CIA) and Strategic Environmental Assessment (SEA) (see 3.9 & 3.10) into the EIA process. However, although the concepts have been around for at least two decades, formal CIA and SEA are still not widely used outside of the EU (although we have seen an increase in Australia under the EPBC Act in recent years). As it is unlikely that we will see rapid progress in their development, we will have to press on without them if a start to the achievement of ESD is to be made. With the introduction of the European Directive on SEA, we have seen an increased interest and applica-

Figure 10.3 Examples of Adapting Environmental Assessment to Sustainability Issues

In applying Environment Assessment to global issues the starting point is to build on the existing process to address global change, for example biodiversity and climate change, by:

- taking the UN conventions on climate change and biological diversity as policy references and legal commitments (for signatory countries);
- developing national guidance and interpretation as to the use of EA as an implementing mechanism;
- using existing methods and procedures to the fullest extent possible;
- building more integrative approaches as required; and
- recognising the specific and differentiated problems and policy characteristics of biodiversity and climate change in process design and application.

Framework approach to climate change:
A four step strategy is proposed comprising:

- focusing on long-term energy efficiencies which represent the biggest single area for greenhouse gas emissions reduction;
- undertaking resource management actions to "offset" emissions, such as reforestation initiatives to increase CO_2 sinks;
- looking for cost-effective, "no regrets" measures that are worth doing anyway; and
- preparing for the possibility that new scientific information will require nations to act more quickly.

Framework approach to biodiversity:
A quick start approach can be taken by applying appropriate time-space frameworks, giving attention to cumulative effects, and identifying key ecological processes and relationships. In this context, five aspects are important:

- establishing the impact zone and ecosystem context;
- identifying the issues and factors of concern and the ecological objectives for managing them;
- data gathering and understanding baseline conditions;
- identifying biodiversity effects and elements at risk; and
- establishing mitigation objectives and measures.

Source: Sadler (1997a)

tion of SEA internationally. As discussed in 5.5, New Zealand has embedded SEA into its EIA procedures and Western Australia has introduced SEA provisions into its *Environmental Protection Act*. Discussion of SEA appears to be moving from being academic to practical since the Commonwealth has entered into SEAs for a number of projects with the State governments in recent years (see chapter 6.3.4). Victoria, Tasmania, ACT and NSW also have some form of SEA in legislation, although not necessarily within EIA. For example, SEA occurs in the *State Environment Protection Policy* process in Victoria under the *Environment Protection Act 1970* where draft Policy Impact Assessments (PIA) are prepared and placed on exhibition. The PIA must include (amongst other matters), an assessment of the possible financial, social and environmental impacts of the alternative means by which the purposes of the policy or variation may be achieved (Marsden and Ashe (2006)).

In New South Wales, the *Environmental Planning and Assessment Act 1979* provides some mechanisms for SEA through its integrated land-use planning and environmental assessment system. However, as Marsden and Ashe (2006: 207) document, the use of SEA is mostly undertaken via other legislation, such as the *Fisheries Management Act* where an EIS is required for each of the state's significant fisheries. Draft fishery management strategies are prepared and the EIA is undertaken of the draft fishery management plan.

Frequently it seems we are linking the move towards ESD to the use of SEA. As the discussion in Chapter 3.10 indicated there has been an evolving debate on the role of SEA. In the words of Wallington et al (2007) this has covered the spectrum of interests ranging from those advocating a focus on "political" (role and place in decision-making) versus a focus on "technical" methods. They see the:

> recent and ongoing discourse on SEA's theoretical frameworks and assumptions is structured ... in terms of three vital elements, which together constitute a conceptual framework for SEA: the substantive purpose and values associated with SEA, the strategies chosen to achieve that purpose, and the mechanisms for operationalising SEA. (2007: 571)

This discourse has provided a considerable literature on the "political" methods, and more those that are "technical". For example, Tao et al (2006) propose that SEA provides benefits as a mechanism to promote environmental considerations in the land-use planning process, while Hanusch and Glasson (2008) investigate the need for more specific attention to be given to the monitoring of SEA processes and outcomes. Through Kørnøv (2009) we also see the broadening of the scope of SEA, past a focus on biophysical environmental issues (as summarised by Wallington et al (2007)) to include the public health aspects of noise, drinking water, air pollution, recreation/outdoor life and traffic safety. A broader perspective is of concern to Nilsson (2009) who argues that a renewed interest in refocusing SEA on the natural environment may reflect interest at the technical level (ie, the preferences of practitioners), but in doing so ignores the users in planning and policy making (ie, those officials who have the problem of having to make a decision about some proposal). Such discussion and debate is an essential aspect of the continuing development of SEA. If the current poor performance of SEA in South Africa (Retief (2007)) can be used as a guide to SEA in developing countries, then it shows the importance of monitoring processes (as Hanusch and Glasson (2008) note) and the need for work on both the political and technical aspects of SEA.

In other areas of environmental management the interconnections with EIA appear to be more evident. Consideration of LCA, and particularly consideration of SEA, focuses attention on the links that have been made with planning and decision-making processes at a variety of levels, from the technical to the political. Clarke provides a thoughtful summary of the interconnections:

> At the project level [EIA] should help project managers and the ultimate beneficiaries to design and implement programs which minimise environmental damage. But that does not mean they will be sustainable. At the program level, EIA can help decision-makers examine a range of policy options and weigh environmental concerns with other economic, political, social or technical concerns. Since the purpose of EIA is to facilitate the decision-making process, it will usually lead to a compromise. Here again, there is no guarantee that an EIA will result in a sustainable development program. People should be better informed. That is all ... EIA cannot make development sustainable. But it can help push decision-makers along the path towards sustainability. (1991: 3)

As we have pointed out frequently, EIA is essentially a social and therefore a political process. EIA is certainly a tool which provides a chance of environmental concerns being given consideration when decisions are made, but it does not constitute a universal law to ensure that the environment is protected. Rather, there has to be a desire to give our environment due respect. Nonetheless, EIA has served as an important vehicle for developing awareness of the environment among the general community. Because planners and designers have had to screen their proposals for environmental impacts, they have had to become more skilful with their designs to reduce impacts. These two educational effects have helped to provide a climate for the discussion and now the acceptance of sustainable development concepts.

Broadly, readers will notice that in most discussion of ESD the emphasis has become that of sustainable development. The loss of a focus on the ecological means that decision-makers do not yet make decisions on the basis of ecological or environmental considerations. However, compared with the situation before EIA, there have been vast improvements. As Liou and Yu (2004) suggest when assessing SEA in Taiwan and its contribution to ESD, full adoption and implementation of SEA in the policy making system requires greater support from a wide range of stakeholders.

If the momentum to give importance to environmental effects of policies, plans and projects can be maintained, we should get closer to the end point on Clarke's (1991) path to sustainability.

There is another aspect to the concept of sustainability that is not encompassed by linking EIA to SEA and ESD. Specifically, we also need to consider the sustainability of EIA itself. As we have indicated previously there are many indications that EIA has become an established part of decision-making processes for projects, and being enshrined in legislation and other procedures it will be continued. There are also many indications that serious thought continues to be given to how EIA can be improved, thereby adding to its sustainability. However, André et al (2004) identify a critical aspect of sustainability when they discuss the importance of developing capacity to undertake EIA by developing the capabilities of people; ie, developing training and education about EIA. In this discussion they distinguish between industrialised and developing countries. In the former, EIA capacity building is seen to be part of the broader field of capacity development in the environment field generally, and would be provided in a range of ways, from formal programs at universities and other higher education institutions, through conferences, workshops and other opportunities to exchange information amongst professionals; whereas, in developing countries the issue is more about developing the understanding of the need for EIA among the political decision-makers. André et al (2004) suggest that while many developing countries have EIA legislation and procedures, if the leaders "see EIA as a constraint in obtaining finance for a project, then EIA is seen as cosmetic" (2004: 415) and the resources to implement EIA effectively are unlikely to be provided. However, the extent of donor country training programs, scholarships to universities and regional meetings involving professionals from developing countries indicate a growing capacity. With this activity, there is no obvious reason to suspect that EIA will not be sustained into the foreseeable future.

10.8 Final Words

Overall, EIA remains a successful policy and environmental assessment tool that enables environmental outcomes to be properly assessed before decisions are made and actions are undertaken. It also facilitates public participation in decision-making in various forms and throughout the different stages of EIA. At its best, it is a "logical, pragmatic and fit-for-purpose" process (Ross, Morrison-Saunders, and Marshall (2006: 9)).

EIA is also an integrating process. We only have to look at the literature about EIA to appreciate the extent to which a growing range of theories and methods are being used to develop effective EIA processes. A typical volume of a journal such as the *Environmental Impact Assessment Review* may include articles about EIA procedures, life cycle analysis, quantitative methods, social impact assessment, economic effects, and SEA. Equally, the International Association for Impact Assessment has forums covering agriculture, forestry and fisheries, biodiversity and ecology, disasters and conflict, corporate stewardship and risk management, Health Impact Assessment, impact assessment law, policies and practice, indigenous peoples, integrated assessment of trade-related policies, public participation, Social Impact Assessment, and Strategic Environmental Assessment. While all this seems to add to the complexity of EIA, and lessen the chance of developing simple effective processes, the diversity directly supports the objectives of EIA (see Chapter 2.2).

In addition to the discussion above about adoption of SEA, incorporating CIA and ESD principles into EIA decision-making, the other issue to emerge as one that needs to be incorporated into project EIA is that of climate change and the potential for sea level rise. The Intergovernmental Panel on Climate Change (Bates et al (2008)) makes projections of global sea-level rise between the late 20th century and the end of this century to be between 0.18 and 0.59 metres, based on various scientific modelling results and observations. Developers, planners and bureaucrats are starting to model/predict such increases and the impacts to their proposals as well as in government land-use planning. It is certainly an issue that will need to be considered early in the environmental assessment process so that sound development decisions can be made. It is important for such issues to be included in the scoping stage of EIA.

The inclusion of climate change and sea level rise into project EIA also demonstrates the flexible nature of the impact assessment system to adapt to changes in current policy priorities. This example reinforces the social and political nature of EIA.

We have also seen the increase in the use of risk assessment tools in the development of EIA, specifically in identifying the potential effects or "risks" to be investigated for the EIS and assigning weightings to the various 'risks' and then reporting these in the EIS.

Although there are limitations to EIA, overall, it has been demonstrated that the use of EIA as an environmental assessment and management tool is entrenched in Australia's environmental and planning regimes to assist in identifying environmental impacts of development proposals and to generally provide for better decision making, improved proposal design and enhanced protection of environmental, social and economic values and assets.

Appendix A

Australian and New Zealand Environment and Conservation Council (ANZECC) Guidelines and Criteria for Determining the Need for and Level of Environmental Impact in Australia

Prepared by the ANZECC Working Group on National Environmental Impact Assessment, June 1996

Introduction

This document aims to achieve greater consistency in the application of the environmental impact assessment (EIA) process within Australia. It provides a common starting point for the process around the country. It sets out a national framework to guide those decisions on whether and at what level formal EIA procedures will apply to potentially significant proposals.

EIA legislation and procedures in Australia differ from one State or Territory to another. However all jurisdictions have agreed to apply common principles to the practice of EIA.

This agreement is recognized in the Intergovernmental Agreement on the Environment [IGAE], in particular in Schedule 3 referring to EIA. The IGAE also envisages criteria for guiding decisions about what classes and types of proposals would normally attract formal EIA procedures and on the level of assessment required.

The guidelines and criteria in this document are also predicated on "A National Approach to EIA in Australia", ANZECC October 1991, and the "National Strategy for Ecologically Sustainable Development", December 1992, as well as the provisions of the IGAE.

Use of the approach outlined here will enhance openness, as it will assist proponents and the wider community to understand the thought processes leading to recommendations and decisions on the application of the process. This will provide for greater accountability on the part of assessing authorities. It should also reduce uncertainties about the process.

Criteria for assessment

The criteria outlined below require the assessing authority to carefully evaluate each of the following factors. These criteria are expanded in Attachments 1 and 2 which follow. The criteria are:

1. The character of the receiving environment.
2. The potential of the proposal.
3. Resilience of the environment to cope with change.
4. Confidence of prediction of impacts.
5. Presence of planning or policy framework or other procedures which provide mechanism for managing potential environmental impacts.
6. Other statutory decision making processes which may provide a forum to address the relevant issues of concern.
7. Degree of public interest.

Application of the criteria

Most jurisdictions in Australia require decisions on the need for and level of EIA to be made on a case by case basis. Under these arrangements each proposal is evaluated on its merits following its referral to the relevant assessing authority.

Other jurisdictions choose to designate certain classes of development or receiving environments in the form of class lists.

The criteria are applicable in both situations.

In jurisdictions where decisions are made on a case by case basis the criteria may be used:

1. By proponents, local authorities or other government agencies, when deciding whether or not to bring the proposal to the attention of the assessing authority;
2. By the assessing authority when deciding whether or not to recommend or require that the proposal be subject to the formal EIA process; and
3. By the assessing authority when recommending or deciding on the level of EIA that is required.

In jurisdictions where class lists are used to designate requirements for EIA, the criteria may be used for the process of developing or reviewing the lists.

How to use the criteria

The criteria are directed principally to practitioners of EIA who are employed within the assessing authorities for the various jurisdictions in Australia.

They pose a series of questions which should be considered by the assessing authority. A check list of issues to be taken into account when answering each question is also included.

After all relevant questions have been answered, based on the best available knowledge at the time, assessing authorities will be expected to make judgements on the level of EIA (if any) required for a proposal or a range of designated proposals.

In exercising their judgement, assessing authorities will be attempting to estimate the potential for an individual proposal or a range of proposals to have a significant effect on the environment.

The criteria themselves do not identify thresholds which automatically determine the need for a formal EIA process. A proposal or a class or proposal may for instance have a number of potentially significant impacts identified, but not require a formal EIA process due to the presence of other suitable statutory approval processes. On the other hand a high level of uncertainty or a large number of unknowns may be sufficient to justify initiating the formal process.

When using the criteria to evaluate individual proposals, assessing authorities should undertake the evaluation jointly with the proponent and in conjunction with other relevant parties or agencies. Once the evaluation has been completed, the assessing authority will be responsible for making recommendations or decisions on the appropriate level of assessment.

If the criteria are being used to develop or review class lists assessing authorities should consult with proponent associations, community groups and other relevant agencies before making recommendations on the content of the lists.

Practitioners of EIA may also find the criteria useful for purposes other than their primary application. They may assist in identifying issues that require further investigation during the formal EIA process, or they may assist in training staff or in the preparation of EIA manuals.

Proponents can use the criteria to anticipate the need for a formal EIA process before referring details of a proposal to the assessing authority. If proponents cannot provide satisfactory answers to some of the questions raised in the criteria, it is very likely the assessing authority would have the same concerns. In these circumstances the proponent may consider refining the proposal, obtaining further information or introducing additional ameliorative measures to overcome problems identified during the initial evaluation.

Proponents are encouraged to consider these issues and to canvass them informally with the assessing authority early in the planning phase of a proposal.

Where a discretionary approach is applied, it will be possible to review and revise conclusions about the need for EIA and an iterative process may be followed to complete the documentation.

Proponents and the community can both use the criteria to hold the assessing authorities more accountable. All parties will know the questions to be considered by the assessing authority when determining the need for and level of formal EIA. Further, the questions being asked will be the same in each jurisdiction.

This does not guarantee that the answers will always be the same. It does mean, however, that the process is more open and inconsistency and uncertainty should be reduced.

The decision on the level of assessment should be made publicly available. Answers to the questions set out in the Attachments could provide the basis for this public explanation.

In addition, the evaluation of the information on the proposal by the assessing authority should ideally be associated with a process which incorporates public consultation in the preparation of the advice on the level of assessment, or includes the opportunity for the initial decision on the level of assessment to be reviewed.

Proponents should be made aware that a decision about the EIA process for a proposal does not obviate the need to pursue approvals required under other processes or legislation.

[Note: Attachment 1 has been reproduced in Table 7.1 on *page 179*.]

Attachment 2
Guidelines and criteria for determining of the need for and level of environmental impact assessment in Australia: checklists

1. *Character of receiving environments (natural and human)*

When evaluating questions under this heading consider the following:

(1) Question: Is it or is it likely to be part of the conservation estate or subject to treaty? Cover –

- national park;
- conservation park;
- wilderness area;
- marine park;
- aquatic reserve;
- heritage/historic area or item;
- national estate;
- world heritage listing;
- area of cultural significance; and
- scientific reserve.

(2) Question: Is it an existing or potential environmentally significant area? Cover –

(a) *Geomorphological Characteristics*

- wetlands;
- lakes;
- coastlines and dunes;
- islands;
- rivers or estuaries;
- plateaus;

- alpine areas;
- desert areas; and
- karst.

(b) *Ecological Systems*
- flora and fauna communities which are uncommon, threatened or endangered;
- mangroves;
- environmentally sensitive marine localities;
- saltmarshes
- coral and seagrass meadows;
- rainforests (and old growth indigenous forests);
- desert communities;
- urban bushland;
- alpine meadows
- remnant vegetation;
- wilderness; and
- wildlife corridor.

(3) Question: Is it vulnerable to major natural or induced hazards? Cover –
- areas subject to erosion;
- steep slopes;
- catchments for ground and surface water resources;
- earthquake, bushfire, flood or cyclone prone areas; and
- salinisation.

(4) Question: Is it a special purpose area? Cover –
- aesthetic or high scenic value;
- land use designation;
- recognised tourist area;
- area targeted for growth; and
- other special purpose eg for defence, telecommunication.

(5) Question: Is it an area where human communities are vulnerable? Cover –
- previous pollution/contamination;
- social, cultural and economic factors; and
- ability to absorb change.

(6) Question: Does it involve a renewable or a non renewable resource? Cover –
- oil, gas, coal, gravel, sand, mineral;
- forest or woodland area;
- fish or crustacean breeding area and fishing ground;
- quality farmland; and
- ground and surface water use.

(7) Question: Is it a degraded area, subject to significant risk levels, or a potentially contaminated site? Cover –

- air quality/pollution;
- surface water quality/pollution;
- groundwater quality/contamination;
- risk levels;
- soil quality/contamination;
- geological stability; and
- rehabilitation potential.

2. *Potential project impacts*

(1) Question: Will construction, operation and/or decommissioning of the proposal have the potential to cause significant changes to the receiving environment? Cover –

(a) *Physical Factors*

- significant land disturbance;
- erosion, subsidence and instability;
- alteration of water courses and drainage patterns;
- effects on quantity, quality or availability of surface or groundwater;
- alteration of wetlands, mangroves, lakes or estuaries;
- excavation, dredging, filling or reclamation of any area including those areas subject to regular inundation by water or for the purpose of creating islands or bodies of water;
- likelihood of salinisation;
- effect on coastal processes, including wave action, sediment movement or accretion, or water circulation patterns; and
- microclimatic effects.

(b) *Biological Factors*

- threatens biodiversity;
- threatens maintenance of ecological processes;
- involves extensive clearing, burning or modification of vegetation;
- threatens ecological processes or life support systems;
- displacement of fauna or creation of significant barriers to fauna movement;
- introduction of noxious weeds, vermin, feral species, disease or the release of genetically modified organism;
- the risk of fire;
- use of, and generation of, pesticides, herbicides, fertilizers or other chemicals which may build up residues in the environment; and
- change of hydrological regime.

(c) *Land Use*

- major changes of land use;
- land use which is significantly different from surrounding uses and/or inconsistent with the planning or development objectives of the locality;
- involves the reservation of, alteration to or alienation of Crown Land;
- involves substantial change to the economic value of land or water;
- may curtail alternative beneficial uses of an area eg for conservation recreation or transport; and
- may limit use by other users or place increased demands on natural resources in short supply eg water.

(d) *Resource Use*

- involves foregoing the development of alternatives uses of a natural resource;
- may seriously affect the livelihood of existing users; and
- may cause disruption to industries involving use of renewable resources (fishing) or loss of species (damage to nursery area).

(e) *Community*

- involves large population movement including influx or departure of construction or operational workforces to or from existing towns, or the creation of new towns;
- may cause substantial changes to the demographic structure of a community;
- may cause the economic stability of a community or its public sector revenue or expenditure base to change markedly;
- may affect health, safety, welfare or quality of life of individuals and communities through factors such as odour, dust, noise, physical dislocation, relocation of people for environmental reasons;
- may cause significant disruption to existing communities through community concern, change to personal vulnerability, loss of security, privacy, safety and amenity;
- may bring about a significant change in the level or nature of community resources (such as cultural character, occupations/labour force, distribution of jobs, distribution of incomes, facilities, local leadership and community identity).

(f) *Infrastructure*

- significant increase in the demand on services and infrastructure including roads, waste management, power supplies, water drainage and demand for housing, medical, hospital, education and social services.

(g) *Heritage*

- may cause adverse effects on Aboriginal and Torres Strait Island communities by restricting access to land, disturbing sacred sites or causing changes to life styles or affecting other cultural values;
- may cause damage to Aboriginal and Torres Strait Island anthropological or archaeological site or relics or result in increased visitation to areas likely to contain archaeological sites, relics or artefacts; and
- may adversely affect sites or items of historic significance.
- obstruction of views and sunlight;
- exceeding local planning authority or other height or gross floor area restrictions;
- degradation of scenic amenity by extensive land disturbance, vegetation clearance or incompatible building form;
- major illumination or reflection impacts on adjacent properties;
- may cause the creation of adverse wind effects; and
- building form.

(h) *Aesthetics*

- obstruction of views and sunlight;
- exceeding local planning authority or other height or gross floor area restrictions;
- degradation of scenic amenity by extensive land disturbance or vegetation
- clearance or incompatible building form;
- major illumination or reflection impacts on adjacent properties;
- may cause the creation of adverse wind effects; and
- building form.

(2) Question: Could implementation of the proposal give rise to unhealthy or unsafe conditions? (on site or off site, short term or long term)? Cover –

(a) *Air*

- may cause large scale generation of dust, smoke, grit, odorous fumes or other toxic or radioactive gaseous emissions; and
- may increase substantially atmospheric concentrations of gases which contribute to the greenhouse effect. or ozone damage.

(b) *Water*

- may cause deterioration of water quality and quantity by significant changes in levels of salinity, colour, odour, turbidity, temperature, dissolved oxygen, nutrients, pH factors, or

pollutants such as oil, toxins, antifouling compounds (TBT etc) or heavy metals; and
- may cause damage to marine environment through accidental spills of oil, fuel, cargo, waste or sewage.

(c) *Wastes*
- involves disposal of significant volumes of sewage, industrial or domestic wastes; and
- involves disposal of significant or deleterious volumes of spoil, overburden or process wastes.

(d) *Hazards*
- may create hazards due to the use, storage, disposal or transportation of flammable, explosive, toxic, radioactive, carcinogenic or mutagenic substances, or substances suspected of being carcinogenic or mutagenic;
- involves the emission of electromagnetic or other radiation which may adversely affect nearby electronic equipment or human health; and
- may lead to traffic hazard.

(e) *Noise Factors*
- may cause substantial increase in traffic noise and vibrations (road, rail, aircraft, marine vessels); and
- involves increase in noise and vibration from stationary plant and other on-site activities and equipment including construction and operational activities affecting people working and residing nearby.

(f) Other
- may result from accidental releases; and
- may involve cumulative impacts.

(3) Question: Will the project significantly divert resources to the detriment of other natural and human communities? Cover –
- access to services and infrastructure; and
- opportunity costs/options foregone.

3. *Resilience of environment to cope with change*

(1) Question: Can the receiving environment absorb the level of impact predicted without suffering irreversible change?

(2) Question: Can land uses at and around the site be sustained? Cover –
- viability and competitiveness of existing neighbouring development.

(3) Question: Can sustainable uses of the site be achieved beyond the proposal life? Cover –

- future uses; and
- issues of intergenerational equity.

(4) Question: Will contingency/emergency plans be in place to deal with accidental events during the construction, operation and decommissioning of the facility? Cover –

- response mechanisms; and
- consequences of higher than normal emissions, for short-term/ long-term releases.

4. *Confidence of prediction of impacts?*

(1) Question: What level of knowledge do we have on the resilience of a given significant ecosystem? Cover –

- adequacy of base line data;
- level of certainty attached to any management or rehabilitation program; and
- relevance of comparable situations.

(2) Question: Is the proposal design and technology sufficiently detailed and understood to enable impacts to be established? Cover –

- previous experience with design;
- relevant models;
- degree of accuracy desired; and
- degree of accuracy achievable.

(3) Question: Is the level and nature of change on the natural and human environment sufficiently understood to allow the impacts of the proposal to be predicted and managed. Cover –

- adequacy of baseline data.

(4) Question: Is it practicable to monitor predicted effects? Cover –

- frequency and duration of monitoring;
- feedback loops; environmental management plans; and
- community involvement.

(5) Question: Are present community values of land use and resource use likely to change? Cover –

- source of values; and
- degree of stress and change likely in community.

5/6. *Presence of a planning policy framework or other statutory approval processes*

(1) Question: Is the proposal consistent with existing zoning or the long-term policy framework for the area?

(2) Question: Do other statutory approval processes exist to adequately assess and manage proposal impacts? Cover –

- level of investigation and documentation;
- assessment processes in other agencies or jurisdictions; and
- whether process involves public participation.

(3) Question: What legislation, standard codes or guidelines are available to properly monitor and control operations on site and the type or quantity of impacts? Cover –

- legislative powers to require unsafe situations to be immediately rectified; and
- proponent liability.

7. *Degree of public interest*

(1) Question: Is the proposal controversial or could it lead to controversy or concern in the community? Cover –

- evidence of existing public interest;
- perceived environmental values and risk;
- public interest groups;
- proposals that might become more controversial as more information is made available over time;
- proposals involving raw materials or end products that are controversial in their own right (eg radioactive materials, hazardous waste); and
- proposals requiring movement of materials which may prove to be controversial en route (transporting radioactive materials through a country town).

(2) Question: Will the amenity, values or lifestyle of the community be adversely affected?

(3) Question: Will the proposal reduce, maintain or generate new inequities in the community?

Appendix B

Intergovernmental Agreement on the Environment, Schedule 3

Source: Department of the Environment, Sports and Territories, 1996

Environmental Impact Assessment

1. The parties agree that it is desirable to establish certainty about the application, procedures and function of the environmental impact assessment process, to improve the consistency of the approach applied by all levels of Government, to avoid duplication of process where more than one Government or level of Government is involved and interested in the subject matter of an assessment and to avoid delays in the process.

2. The parties agree that impact assessment in relation to a project, program or policy should include, where appropriate, assessment of environmental, cultural, economic, social and health factors.

3. The parties agree that all levels of Government will ensure that their environmental impact assessment processes are based on the following:

(i) the environmental impact assessment process will be applied to proposals from both the public and private sectors;

(ii) assessing authorities will provide information to give clear guidance on the types of proposals likely to attract environmental impact assessment and on the level of assessment required;

(iii) assessing authorities will provide all participants in the process with guidance on the criteria for environmental acceptability of potential impacts including the concept of ecologically sustainable development, maintenance of human health, relevant local and national standards and guidelines, protocols, codes of practice and regulations;

(iv) assessing authorities will provide proposal specific guidelines or a procedure for their generation focussed on key issues and incorporating public concern together with a clear outline of the process;

(v) following the establishment of specific assessment guidelines, any amendments to those guidelines will be based only on significant issues that have arisen following the adoption of those guidelines;

(vi) time schedules for all stages of the assessment process will be set early on a proposal specific basis, in consultations between the assessing authorities and the proponent;

(vii) levels of assessment will be appropriate to the degree of environmental significance and potential public interest;

(viii) proponents will take responsibility for preparing the case required for assessment of a proposal and for elaborating environmental issues which must be taken into account in decisions, and for protection of the environment;

(ix) there will be full public disclosure of all information related to a proposal and its environmental impacts, except where there are legitimate reasons for confidentiality including national security interests;

(x) opportunities will be provided for appropriate and adequate public consultation on environmental aspects of proposals before the assessment process is complete;

(xi) mechanisms will be developed to seek to resolve conflicts and disputes over issues which arise for consideration during the course of the assessment process;

(xii) the environmental impact assessment process will provide a basis for setting environmental conditions, and establishing environmental monitoring and management programs (including arrangements for review) and developing industry guidelines for application in specific cases.

4. A general framework agreement between the Commonwealth and the States on the administration of the environmental impact assessment process will be negotiated to avoid duplication and to ensure that proposals affecting more than one of them are assessed in accordance with agreed arrangements.

5. The Commonwealth and the States may approve or accredit their respective environmental impact assessment processes either generally or for specific purposes. Where such approval or accreditation has been given, the Commonwealth and the States agree that they will give full faith and credit to the results of such processes when exercising their responsibilities.

References

Note: The following list is intended to provide the reader with sources of specific information. Because EIA ranges over a number of traditional subject areas, there are many publications which relate to aspects of EIA. While those noted below generally represent a comprehensive grouping, they are not a total listing. The reader is sure to find other suitable references using the subject headings from the text.

Abaza, H, Bisset, R and Sadler, B, 2004, *Environmental Impact Assessment and Strategic Environmental Assessment: Towards an Integrated Approach*, UNEP, pp 39-44.

Able, M and Stoking, M, 1981, "The Experience of Underdeveloped Countries", in O'Riordan, T and Sewell, WRD (eds), *Project Appraisal and Policy Review*, Wiley, Chichester.

ACT Planning and Land Authority, 2005, *Planning System Reform Project: Technical Paper 4 – Review of Environmental Impact Assessment in the ACT*, <www.actpla.act.gov.au> (accessed November 2007).

ACT Planning and Land Authority website, *Environmental Impact Statements*, <www.actpla.act.gov.au> and follow links to "Design and Build" under "Browse by Topic", then "Environmental Assessment" (accessed August 2013).

African Development Bank, 2000, *Coastal and Marine Resources Management Guidelines*, <www.afdb.org/about/oesu/coastal.html>.

Ahmad, YJ (ed), 1983, *Environmental Decision-making*, vol 1, Hodder & Stoughton, London.

Ahmad, YJ (ed), 1984, *Environmental Decision-making*, vol 2, Hodder & Stoughton, London.

Alberti, M and Susskind, L, 1996, "Managing Urban Sustainability: An Introduction to the Special Issue", *Environmental Impact Assessment Review*, vol 16, no 4-6, pp 213-221.

Alcoa Aust, 1980, *Alcoa Portland Aluminium Smelter Environment Effects Statement*, Melbourne.

Anderson, E and Sadler, B, 1996, *International Study of the Effectiveness of Environmental Assessment*, Final report of the EIA Tripartite Workshop, Canberra, 21-24 March 1996.

André, P, Delisle, CE and Revéret, J-P, 2004, *Environmental Assessment for Sustainable Development*, Presses Internationales Polytechnique, Montréal.

Anon, 2000, "Industry Concerns Grow Over EPBC Approvals", *Newsletter – An Environment Institute of Australia Publication*, issue 28, October, p 1.

Anton, DK, 1999, "New Spheres of Public Environmental Influence: Community Access Points Under the EPBC Act 1999", Conference paper to *Navigating the New Environment Act*, Environment Defenders' Office (Vic), 13 November, Melbourne, pp 1-7.

Appleyard, D, 1979, "The Environment as a Social Symbol", *APA Journal*, April, pp 143-153.

Arbter, K, 2008, "Strategic Environmental Assessment (SEA) and Sustainability Impact Assessment (SIA) Common Features, Differences and Practical Experience", Presentation to the OECD Workshop on Sustainability Assessment Methodologies, 14-15 January 2008, Royal Tropical Institute, Amsterdam, the Netherlands, <www.oecd.org/dataoecd/16/12/39924683.pdf> (accessed August 2008).

Arce, R and Gullón, N, 2000, "The Application of Strategic Environmental Assessment to Sustainability Assessment of Infrastructure Development", *Environmental Impact Assessment Review*, vol 20, no 3, pp 393-402.

Armour, A, 1983, "Current Methods in Environmental Impact Assessment", in Armour, A (ed), *Issues in Environmental Impact Assessment*, Working Paper 4, Faculty of Environmental Studies, York University, Toronto.

Armour, A, 1990, "Integrating Impact Assessment in the Planning Process: From Rhetoric to Reality", *Impact Assessment Bulletin*, vol 8, nos 1 and 2, pp 3-13.

Armstrong, AF, 1982, *First Directory of Australian Social Impact Assessment*, Program in Public Policy Studies, University of Melbourne, Melbourne.

Armstrong, JE and Harman, WW, 1980, *Strategies for Conducting Technology Assessments*, Westview Press, Boulder.

Arnstein, SR, 1971, "Eight Rungs on the Ladder of Citizen Participation", in Cahn, ES and Passett, BA (eds), *Citizen Participation: Effecting Community Change*, Praeger, New York.

Ashby, E, 1976, "Background to Environmental Impact Assessment", in O'Riordan, T and Hey, RD, *Environmental Impact Assessment*, Saxon House, Farnborough.

Ashby, S, 2000, "Environment Protection and Biodiversity Conservation Act, 1999: Implementation Issues at the Victorian State Level, in Environment Defenders' Office (Victoria) Ltd", *Environment Protection and Biodiversity Conservation Act 1999 (Cth) Seminar Papers*, 3 May, pp 34-43.

Asian Development Bank, 2002, *Environment Policy*, <www.adb.org/> and follow links to "Publications" and "Policies, Strategies and Plans" (accessed January 2009).

Asian Development Bank, 2003, *Environment Related Documents*, <www.adb.org/> and follow links to "Projects" (accessed April 2004).

Asian Development Bank, 2004a, *Environment*, <www.adb.org/> and follow links to "Environmental Sustainability" under "Projects".

Asian Development Bank, 2004b, *Office of Environment and Social Development*, <www.adb.org/> and follow links to "Environmental Sustainability" under "Projects" (accessed April 2004).

Asian Development Bank, 2008, *Strategy 2020: The Long-Term Strategic Framework of the Asian Development Bank 2008-2020*, <www.adb.org/> and follow links to "Policies and Strategies" under "About".

Asian Development Bank, 2009, Environment Policy homepage, <www.adb.org/> and follow links to "Publications" and "Policies, Strategies and Plans" (accessed 10 January 2009).

Atkins, R, 1984, "A Comparative Analysis of the Utility of EIA Methods", in Clark, BD et al (eds), *Perspectives on Environmental Impact Assessment*, D Reidel, Dordrecht.

Attorney-General (SA), 1987, *Regulation Review Procedures*, Adelaide.

Australian and New Zealand Environment and Conservation Council, 1991a, *A National Approach to Environmental Impact Assessment in Australia*, Canberra.

Australian and New Zealand Environment and Conservation Council, 1991b, *A National Approach to Environmental Impact Assessment in Australia: Background Paper of the Working Group*, Canberra.

Australian and New Zealand Environment and Conservation Council, 1996, *Guidelines and Criteria for Determining the Need for and Level of Environmental Impact Assessment in Australia*, Prepared by the ANZECC Working Group on National Environmental Impact Assessment, Canberra.

Australian Environment Council, 1984, *Guide to Environmental Legislation and Administrative Arrangements in Australia*, Report no 16, AGPS, Canberra.

Australian Manufacturing Council, 1992, *The Environmental Challenge: Best Practice Environmental Management*, AMC, Melbourne.

Bagri, A and Vorhies, F, 1999, *The RAMSAR Convention and Impact Assessment*, IUCN, 4 March, <http://biodiversityeconomics.org>.

Bailey, PD, 1997, "IEA: A New Methodology for Environmental Policy?", *Environmental Impact Assessment Review*, vol 17, no 4, pp 221-226.

Barbour, IG, 1980, *Technology, Environment and Human Values*, Praeger, New York.

Barlett, RV, 1989, "Impact Assessment as a Policy Strategy", in Barlett, RV (ed), *Policy through Impact Assessment*, Greenwood Press, Westport Ct, pp 1-4.

Bates, BC, Kundzewicz, ZW, Wu, S and Palutikof, JP (eds), 2008, *Climate Change and Water: Technical Paper of the Intergovernmental Panel on Climate Change*, IPCC Secretariat, Geneva.

Bates, G, 1983, "Environment Protection: A New Role for the Courts?", *Environment Victoria*, December, p 3.

Beanlands, GE and Duinker, PN, 1983, *An Ecological Framework for Environmental Impact Assessment in Canada*, Institute for Resource and Environmental Studies, Dalhousie University, Halifax.

Becker, HA, 1991, "The Future of Impact Assessment in the European Communities and Their Member Countries", *Impact Assessment Bulletin*, vol 9, no 4, pp 69-80.

Beder, S, 1990, "Environmental Impact Statements – The Ethical Dilemma", Paper presented to National Engineering Conference, Canberra.

Beder, S, February 1993, "Bias and Credibility in Environmental Impact Assessment", *Chain Reaction*, No 68, pp 28-30.

Beer, T and Ziolkowski, F, 1996, "Environmental Risk Assessment – An Australian Perspective", in Norton, TW, Beer, T and Dovers, SR (eds), *Risk and Uncertainty in Environmental Management*, Proceedings of the 1995 Australian Academy of Science Fenner Conference on the Environment, Centre for Resource and Environmental Studies, Australian National University, Canberra, pp 3-13.

Bektashi, L and Cherp, A, 2002, "Evolution and Current State of Environmental Assessment in Azerbaijan", in *Impact Assessment and Project Appraisal*, vol 20, no 4, pp 1-11.

Bendix, S, 1984, "How to Write a Socially Useful EIS", in Hart, SL, Enk, GA and Hornick, WF (eds), *Improving Impact Assessment*, Westview Press, Boulder.

Berger, TR, 1981, "Public Inquiries and Environmental Assessment", in Clark, SD (ed), *Environmental Assessment in Australia and Canada*, Ministry for Conservation, Melbourne.

Bisset, R, 1988, "Developments in EIA Methods", in Wathern, P (ed), *Environmental Impact Assessment*, Unwin Hyman, London, pp 47-61.

Bisset, R and Tomlinson, P, 1988, "Monitoring and Auditing of Impacts", in Wathern, P (ed), *Environmental Impact Assessment*, Unwin Hyman, London, pp 117-128.

Black, P, 1981, *Environmental Impact Analysis*, Praeger, New York.

Blumm, M.C, 1988, The Origin, Evolution and Direction of the United States National Environmental Policy Act, *Environmental and Planning Law Journal*, vol 5, no 3, pp 179-193.

Borouse, MA, Chen, K and Christakis, AN, 1980, *Technology Assessment*, Creative Futures North Holland, New York.

Bosward, JH and Staveley, P, 1981, "Integration of Environmental Planning and Impact Assessment Procedures in New South Wales", in Clark SD (ed), *Environmental Assessment in Australia and Canada*, Ministry for Conservation, Melbourne.
Bowles, RT, 1981, *Social Impact Assessment in Small Communities: An Integrative Review of Selected Literature*, Butterworths, Toronto.
Briffett, C, 1999, "Environmental Impact Assessment in Southeast Asia: Fact or Fiction", *GeoJournal*, vol 49, no 3, pp 333-338. Available ProQuest Information and Learning Company [May 2004].
Briffett, C, Obbard, JP and Mackee, J, 2003, "Towards SEA for the Developing Nations of Asia", *Environmental Impact Assessment Review*, vol 23, issue 2, pp 171-196.
British Standards Institute, 1994, *Environmental Management Systems*, BS7750, London.
Brooke, C, 1998, *Biodiversity and Impact Assessment*, Paper prepared for the conference on Impact Assessment in a Developing World, Manchester, October, IUCN website, <http://biodiversityeconomics.org/>.
Brown, AL and McDonald, GT, 1989, *To Make Environmental Assessment Work More Effectively*, Institute of Applied Environmental Research, Griffith University, Brisbane.
Brown, AL and Nitz, T, 2000, "Where Have All the EIAs Gone?", *Environmental and Planning Law Journal*, vol 17, no 2, pp 89-98.
Brown, AL and Thérivel, R, 2000, "Principles to Guide the Development of Strategic Environmental Assessment Methodology", *Impact Assessment and Project Appraisal*, vol 18, no 3, pp 183-189.
Brown, JM and Campbell, EA, 1990, "Risk Perception and Risk Communication", *Planner*, 13 April, pp 13-14.
Brown, L, 1997, "Experience in SEA under the UNDP Environmental Management Guidelines", in Harvey, N and McCarthy, M, *Environmental Impact Assessment in the 21st Century*, Conference Proceedings, University of Adelaide, April, pp 110-115.
Brownlea, AAB, 1988, "Risk Assessment", in Hindmarsh, RA, Hundloe, TJ, McDonald, GT and Rickson, RE (eds), *Papers on Assessing Social Impacts of Development*, Institute of Applied Environmental Research, Griffith University, Brisbane, pp 78-87.
BSI group, *ISO 14001 Environmental Management*, <http://www.bsigroup.com/> and follow links to "Standards", (accessed 12 July 2013).
Buckley, R, 1987, *Environmental Planning Techniques*, Department of Mines and Energy (SA), Adelaide.
Buckley, R, 1997, Strategic Environment Assessment, *Environmental and Planning Law Journal*, vol 14, no 3, pp 174-180.
Buckley, R, 1998, "Cumulative Environmental Impacts", *Environmental Methods Review: Retooling Impact Assessment for the New Century*, AEPI and IAIA, The Press Club, Fargo (Nth Dakota), pp 95-99.
Buckley, R, 2000, "Strategic Environmental Assessment of Policies and Plans: Legislation and Implementation", *Impact Assessment and Project Appraisal*, vol 18, no 3, pp 209-215.
Bujagali Hydropower project webpage "Environmental Responsibility", <www.bujagali-energy.com/> and follow the relevant links (accessed June 2008).
Burch, WR, 1976, "Who Participates? A Sociological Interpretation of Natural Resource Decisions", *Natural Resources Journal*, vol 16, no 1, pp 41-54.
Burdge, RJ, 1988, "An International History of Social Impact Assessment", in Hindmarsh, RA, Hundloe, TJ, McDonald, GT and Rickson, RE (eds), *Papers on Assessing Social Impacts of Development*, Institute of Applied Environmental Research, Griffith University, Brisbane, pp 14-21.

Burdge, RJ, 1989, "Utilizing Social Impact Assessment Variables", *Impact Bulletin*, vol 8, nos 1 and 2, pp 85-99.

Burdge, RJ, 2008, "The Focus of Impact Assessment (and IAIA) Must Now Shift to Global Climate Change!", *Environmental Impact Assessment Review*, vol 28, pp 618-622.

Bureau of Industry Economics, 1990, *Environmental Assessment Impact on Major Projects*, Research Report no 35, AGPS, Canberra.

Burnside International Limited, 2006, *Bujagali Hydropower Project: Social and Environmental Assessment Report, Executive Summary*, Canada.

Cahn, ES and Cahn JC, 1971, "Maximum Feasible Participation: A General Overview", in Cahn, ES and Passett, BA (eds), *Citizen Participation: Effecting Community Change*, Praeger, New York.

Caldwell, LK, 1989, "Understanding Impact Analysis: Technical Process, Administration, Reform, Policy Principle", in Barlett, RV (ed), *Policy through Impact Assessment*, Greenwood Press, Westport Ct, pp 7-16.

Canadian Environmental Assessment Agency, 1996, *Cumulative Environmental Effects – Referenced Annotated Bibliography*, <www.ceaa-acee.gc.ca/017/0005/0001/biblio_e.htm> (accessed December 2004).

Canadian Environmental Assessment Agency, 1997, *Canadian Environmental Assessment Act and Regulations*, <www.ceaa.gc.ca> and follow links to "Acts and Regulations" (accessed August 2008).

Canadian Environmental Assessment Agency, 1999, *International Summit on Environmental Assessment – Final Report*, Quebec City, June 12-14, 1999, <www.ceaa-acee.gc.ca/017/012/summit_e.htm> (accessed January 2009).

Canadian Environmental Assessment Agency, 2004a, *Introduction and Features: Canadian Environmental Assessment Act*, <www.ceaa.gc.ca/013/intro_e.htm#4> (accessed April 2004).

Canadian Environmental Assessment Agency, 2004b, *Legislation and Regulations*, <www.ceaa.gc.ca/013/index_e.htm> (accessed April, 2004).

Canadian Environmental Assessment Agency, 2004c, Review of the *Canadian Environmental Assessment Act*, <www.ceaa.gc.ca/013/001/index_e.htm> (accessed April 2004).

Canadian Environmental Assessment Agency, 2007, "Participant Funding for the Joint Review Panel Process of the Proposed Mackenzie Gas Project", <www.ceaa.gc.ca/010/0001/0002/index_e.htm> (accessed November 2007).

Canadian Environmental Assessment Agency, 2010, *SEA: Cabinet Directive on the Environmental Assessment of Policy, Plan and Program Proposals*, Ottawa.

Canadian Environmental Assessment Agency, May 2013, *Operational Policy Statement: Assessing Cumulative Environmental Effects under the Canadian Environment Assessment Act 2012*, <https://www.ceaa-acee.gc.ca> and follow links to "Statements" and then "Policy and Guidance" (accessed 26 June 2013).

Canadian Environmental Assessment Agency, 2013, *Overview: Canadian Environmental Assessment Act, 2012* <http://www.ceaa-acee.gc.ca> and follow links to the "Overview of the Canadian Environmental Assessment Act, 2012".

Canter, LW, 1977, *Environmental Impact Assessment*, McGraw-Hill, New York.

Canter, LW, 2000, "Methods for Effective Environmental Information Assessment", in Porter, AL and Fittipaldi, JJ (eds), *Environmental Methods Review: Retooling Environmental Impact Assessment for the New Century*, AEPI and IAIA, The Press Club, Fargo (Nth Dakota), pp 58-68.

Canter, LW and Kamath, J, 1995, "Questionnaire Checklist for Cumulative Impacts", *Environmental Impact Assessment Review*, vol 15, no 4, pp 311-339.

Carley, MJ and Bustfeld, ES, 1984, *Social Impact Assessment: A Guide to the Literature*, Westview Press, Boulder.

Cass, M, 1975, "The Federal Government's EIS Proposals", in *The EIS Technique*, Australian Conservation Foundation, Melbourne.
Catlow, J and Thirwall, CG, 1976, *Environmental Impact Analysis*, Research Report II, Department of the Environment, HMSO, London.
Centre for Environmental Studies, 1976, *Land Transport Alternatives for Webb Dock*, University of Melbourne, Melbourne.
Cherp, A, 2001a, EA Legislation and Practice in Central and Eastern Europe and the Former USSR: A Comparative Analysis, *Environmental Impact Assessment Review*, vol 21, pp 335-361.
Cherp, A, 2001b, *EA Legislation and Practice in Countries in Transition*, <www.personal.ceu.hu/departs/envsci/eianetwork/publications/ cherp01.html> (accessed March 2004).
Cherp, A, 2004, "Editorial", *Impact Assessment and Project Appraisal*, vol 22, no 2.
Chuen, NY, 1989, "Legal and Institutional Arrangements for EIA in Malaysia", *Impact Bulletin*, vol 8 nos 1 and 2, pp 309-318.
CITET, 2004, *Regional Program For Environmental Impact Assessment in the METAP Countries*, <www.citet.nat.tn/english/citet/metap.html> (accessed April 2004).
City West Water, 1999, *Preliminary Environment Assessment*, part of unpublished Environmental Management System, City West Water, St Albans.
Clark, BD, Bissett, R and Wathern, P, 1981, "The British Experience", in O'Riordan, T and Sewell, WRD (eds), *Project Appraisal and Policy Review*, Wiley, Chichester.
Clark, M, 1976, *The Environmental Impact Statement as an Aid to Tasmanian Developers*, Environmental Studies Occasional Paper no 2, Centre for Environmental Studies, University of Tasmania, Hobart.
Clark, SD, 1981, "Public Participation: A Discussion Paper", in Clark, SD (ed), *Environmental Assessment in Australia and Canada*, Ministry for Conservation, Melbourne.
Clarke, T, 1991, "EIA and Sustainable Development", *EIA Newsletter*, no 6, pp 17-18.
Clean Energy Regulator <http://www.cleanenergyregulator.gov.au/> and follow links to "National Greenhouse and Energy Reporting" and then "Published Information", (accessed 1 July 2013).
Clement, K, 2000, *Economic Development and Environmental Gain*, Earthscan, London.
Coates, VT and Coates, JF, 1989, "Making Technology Assessment an Effective Tool to Influence Policy", in Barlett, RV (ed), *Policy through Impact Assessment*, Greenwood Press, Westport, Ct, pp 17-25.
Colley, R and Lee, N, 1990, "Reviewing the Quality of Environmental Statements", *Planner*, 13 April pp 12-13.
Commission of the European Communities, 2009, *Report from the Commission to the Council the European Parliament, The European Economic and Social Committee and the Committee of the Regions on the application and effectiveness of the Directive on Strategic Environmental Assessment (Directive 2001/42/EC)*, <http://eur-lex.europa.eu/en/index.htm> and follow the relevant links, accessed 12 March 2013
Commonwealth Of Australia, 1975, *Administrative Procedures under the Environment Protection (Impact of Proposals) Act 1974-1975*, AGPS, Canberra.
Commonwealth of Australia, 1992, *National Strategy for Sustainable Development*, AGPS, Canberra.
Conacher, A, 1994, "The Integration of Land-use Planning and Management with Environmental Impact Assessment: Some Australian and Canadian Perspectives", *Impact Assessment*, vol 12, no 4, pp 347-372.

Conacher, A and Conacher, J, 2000, *Environmental Planning and Management in Australia*, Oxford University Press, South Melbourne.

Conrad, J (ed), 1982, *Society, Technology and Risk Assessment*, Academic Press, London.

Conservation Commission, NT, 1984, *A Guide to the Environmental Assessment Process in the Northern Territory*, Darwin.

Convention on Biological Diversity, 2007, <www.cbd.int/impact/background.shtml> (accessed June 2008).

Cook PL, 1981, "Costs of Environmental Impact Statements and the Benefits They Yield in Improvements to Projects and the Opportunities for Public Involvement", in United Nations Economic Commission for Europe, *Environmental Impact Assessment,* Pergamon, Oxford.

Cooper, C, 1981, *Economic Evaluation and the Environment*, Hodder & Stoughton, London.

Cooper, TA and Canter, LW, 1997, "Substantive Issues in Cumulative Impact Assessment: A State-of-Practice Survey", *Impact Assessment,* vol 15, no 1, pp 15-31.

Coordinator-General's Department, Qld, 1979, *Impact Assessment of Development Projects in Queensland*, Coord General's Dept Brisbane.

Couch, WJ, 1991, "Recent EIA Developments in Canada", *EIA Newsletter* no 6, EIA Centre, University of Manchester, p 3.

Council on Environment Quality website, <www.whitehouse.gov/administration/eop/ceq> (accessed 8 April 2008).

Council on Environmental Quality, 2013. <http://www.whitehouse.gov/administration/eop/ceq> and follow links to "Press Releases" and see 18 September 2013 release.

Council on Environmental Quality, 2007, *A Citizen's Guide to the NEPA: Having Your Voice Heard*, Washington, USA, <www.nepa.gov/nepa/nepanet.htm> (accessed January 2009).

Council on Environmental Quality, 1997, *The National Environmental Policy Act: A Study of Its Effectiveness After Twenty-five Years*, Washington, US.

Country Roads Board, 1980, *Omeo-Mitta Mitta Road Link Environment Effects Statement*, Melbourne.

Court, J, Wright, C and Guthrie, A, 1996, "Environment Assessment and Sustainability: Are We Ready for the Challenge?", *Australian Journal of Environmental Management,* vol 3, no 1, pp 42-55.

Craig, D, 1990, "Social Impact Assessment: Politically Orientated Approaches and Applications", *Environmental Impact Assessment Review*, vol 10, nos 1 and 2, pp 37-54.

Cullen, P, 1975, "Techniques for Evaluating Environmental Impacts", in *The EIS Technique,* Australian Conservation Foundation, Melbourne.

Cullingworth, JB, 1974, *Town and Country Planning in Britain*, 5th edn, Allen & Unwin, London.

Curi, K, 1983, "Environmental Impact Assessment from the Point of View of a Developing Country", in PADC (ed), *Environmental Impact Assessment*, Martinus Nijhoff, Hingham.

De Jough, P, 1988, "Uncertainty in EIA", in Wathern, P (ed), *Environmental Impact Assessment*, Unwin Hyman, London, pp 62-84.

Denholm, M, 2007, "Gunns Go-ahead Despite Failures", *The Australian*, 6 July 2007, <www.theaustralian.news.com.au> (accessed July 2007).

Department of Arts, Heritage and Environment, 1985a, *Annotated Bibliography of Publications on Cost-Benefit Analysis and Environmental Impact Assessment*, AGPS, Canberra.

Department of Arts, Heritage and Environment, 1985b, *Compendium of Case Studies on Using Cost-Benefit Analysis in the Environmental Impact Assessment Process*, AGPS, Canberra.

Department of Climate Change, <www.greenhouse.gov.au/emissionstrading/index.html>.

Department of Conservation and Environment, WA, 1978, *Procedures for Environmental Assessment of Proposals in Western Australia*, Perth.

Department of the Environment, 2003, *EIA*, <www.jas.sains.my/> (accessed April 2004).

Department of the Environment (Cth) *Environment Protection and Biodiversity Conservation Act 1999 – Environment Assessment Manual, Implementing Chapter 4 EPBC Act*, <http://www.environment.gov.au/> and follow links to "Legislation", (accessed October 2013).

Department of the Environment, Melbourne <http://www.environment.gov.au/> and follow links to the "Environment Protection and Biodiversity Conservation Act 1999" under "Legislation", and then "Public Notices and Invitation to Comment", then "Strategic Assessment Notices".

Department for the Environment, SA, 1980, *Generic Guidelines for an Environmental Impact Statement*, Adelaide.

Department of the Environment, Tas, 1974, *Guidelines and Procedures for Environmental Impact Studies*, Hobart.

Department of Environment and Heritage, 2003, *Annual Report 2002-03*, <www.deh.gov.au/index.html> (accessed May 2004).

Department of Environment and Heritage, 2004, "Home Page" <www.deh.gov.au/index.html>.

Department of Environment and Heritage, *About the EPBC Act*, <www.deh.gov.au/epbc/about/index.html> (accessed May 2004).

Department of Environment and Heritage, 2006, *Changes to the EPBC Act – What You Need to Know*, <www.environment.gov.au/epbc/publications/2006-amendments-brochure.html> (accessed June 2007).

Department of Environment and Primary Industries, Vic, 2012, *Time-stamping Native Vegetation Data Public Consultation 2010-12 Final Report*.

Department of Environment and Primary Industries, Vic, May 2013, *Draft Habitat Compensation Under the Biodiversity Conservation Strategy: Melbourne's Strategic Assessment*, Melbourne.

Department of the Environment and Water Resources, 2007, *EPBC Act – Environment Assessment Process*, <www.environment.gov.au> and follow links to "Environment Protection" under "Topics" and then "Environment Assessments" (accessed September 2007).

Department of the Environment and Water Resources, 2007, *EPBC Act – Compliance Auditing*, <http://www.environment.gov.au/> and follow links to the "Environment Protection and Biodiversity Conservation Act 1999" under "Legislation", and then "Compliance and Enforcement", and then "Compliance Auditing" (accessed June 2007).

Department of the Environment and Water Resources, *EPBC Act – Protecting Heritage Places*, February 2007, <http://www.environment.gov.au/> and follow links to the "Environment Protection and Biodiversity Conservation Act 1999" under "Legislation" and then "EPBC Act Publications" (accessed January 2009).

Department of the Environment and Water Resources, 2007, *Operation of the EPBC Act Annual Report 2006-07*, <www.environment.gov.au/epbc/index.html>.

Department of the Environment, Land and Planning, 1992, *The Guide to the Australian Capital Territory's Land (Planning and Environment) Act 1991*, ACT Government Printer, Canberra.

Department of the Environment, Sports and Territories, 1996, *InterGovernmental Agreement on the Environment*, <www.environment.gov.au/ and follow links to "Ecologically Sustainable Development" under "Topics"> (accessed 21 May 2008).

Department of the Environment, Sports and Territories, 1997, *Government Environment Departments on the Net*, June, <www.environment.gov.au/portfolio/epg/other_govt.html> Department of Environment, Water, Heritage and the Arts, *Operation of the Environment Protection and Biodiversity Conservation Act 1999*, Annual Report 2006-2007, Canberra.

Department of the Environment, Water, Heritage and the Arts, *Corporate Sustainability Reporting*, <www.environment.gov.au/settlements/industry/corporate/reporting/index.html> (accessed August 2008).

Department of the Environment, Water, Heritage and the Arts, 2008, *Independent Review of the EPBC Act 1999: Discussion Paper*, Commonwealth of Australia.

Department of Environmental Affairs and Tourism, 2004, *Guideline Document EIA Regulations*, <www.environment.gov.za/PolLeg/GenPolicy/eia.htm#top> (accessed April 2004).

Department of Housing and Urban Development, 1997, *Guide to the Assessment of Major Development Projects*, 2nd edn, Publishing and Promotions Unit, Planning Division, Adelaide.

Department of Human Services, 2008, *Victorian Environmental Health Indicator Project* (VEHIP), <www.health.vic.gov.au/> and follow the relevant links under "Publications" (accessed 12 July 2008).

Department of Infrastructure, 2002, *Environment Assessment Review: Issues and Options Paper*, Victorian Government, Melbourne.

Department of Infrastructure and Planning, 2007, *Guidelines for EIS under the Integrated Planning Act 1997 Version 2*, Queensland Government, <www.dip.qld.gov.au> and follow the relevant links (accessed 16 May 2008).

Department of Infrastructure and Planning (Qld), *Planning and Development laws*, <www.dip.qld.gov.au> and follow the links under "Infrastructure and Planning" (accessed December 2008).

Department of Infrastructure, Planning and Environment, 1996, *Guide to Environmental Impact Assessment in the Northern Territory*, <www.nt.gov.au/nreta/environment/assessment/eiaguide> (accessed July 2008).

Department of Infrastructure and VicRoads, 1998, *Scoresby Transport Corridor Environment Effects Statement: Vol 1*, Prepared by Sinclair Knight Mertz, Melbourne.

Department of Justice, "Pulp Mill Assessment", <www.justice.tas.gov.au/justice/pulpmillassessment> (accessed August 2007).

Department of Local Government, Sport and Recreation, "Home Page", <www.localgovernment.qld.gov.au> (accessed 20 July 2008).

Department of Natural Resources and Environment, 1997, *Regulatory Impact Statement: Proposed Fisheries Regulations 1998*, Melbourne.

Department of Natural Resources, Environment and The Arts, 1996, *Guide to the EIA Process in the Northern Territory*, <www.nt.gov.au/nreta/environment/assessment/eiaguide/index.html>.

Department of Natural Resources, Environment and the Arts, Home Page, <www.nt.gov.au/nreta/environment/assessment/index.html> (accessed October 2007).

Department of Planning, "Home Page", <www.planning.nsw.gov.au>.

Department of Planning, 1994, *Environmental Planning and Assessment (Part 5) Amendment Act 1993*, Circular no A25, April.

Department of Planning, December 2006, *Environmental Planning Legislation Amendment Act 2006 Fact Sheet*, <www.planning.nsw.gov.au/planning_reforms/> (accessed January 2008).

Department of Planning, March 2006, *A Community Guide: NSW Major Projects Assessment System*, <www.planning.nsw.gov.au/assessingdev/> (accessed February 2008).

Department of Planning, May 2007, *Commonwealth Environmental Protection and Biodiversity Conservation Act 1999: Guide to implementation in NSW*, <www.planning.nsw.gov.au/assessingdev/> (accessed February 2008).

Department of Planning (a), "Planning Reform", <www.planning.nsw.gov.au/planning_reforms/index.asp> (accessed January 2008).

Department of Planning (b), "How to Find Out Which Development Assessment Process Applies", <www.planning.nsw.gov.au/> (accessed on January 2008).

Department of Planning 2013(1) (NSW) <http://www.planning.nsw.gov.au> and follow the links to "Development Proposals" and then "Development Assessment Systems", and then "Local and Regional Development".

Department of Planning 2013(2) (NSW) <http://www.planning.nsw.gov.au> and follow the links to "Development Proposals" and then "Development Assessment Systems", and then "Part 5 Assessment System".

Department of Planning and Development, Vic, 1995, *Guidelines for Environment Assessment and Environment Effects Act 1978*, 5th edn, Government Printer, Melbourne.

Department of Primary Industries, Water and Environment, 1999, *Environmental Management and Pollution Control Act 1994: Regulatory Impact Statement*, <www.dpiwe.tas.gov.au/env/risempca.html>.

Department of Primary Industries, Water and Environment, Tas, "Home Page", <www.dpiwe.tas.gov.au/inter.nsf/Home/l?Open>.

Department of State Development, Infrastructure and Planning (1) Qld, <http://www.dsdip.qld.gov.au/> and follow links to "Infrastructure and Planning" and then "Development Applications", then "State and Assessment Referral Agency Applications", then "State and Assessment Referral Agency" (accessed October 2013).

Department of State Development, Infrastructure and Planning (2) Qld <http://www.dsdip.qld.gov.au/> and follow links to "Coordinator General Projects" and then "Assessments and Approvals", then "Coordinated Projects" (accessed October 2013).

Department of Sustainability, Environment, Water, Population and Communities (Cth) *Great Barrier Reef Strategic Assessment Fact Sheet*, (June 2012) <http://www.environment.gov.au/> and follow links to "Environment Protection" and then "Environment Assessments", then "Strategic Assessments" then "Strategic Assessment – Great Barrier Reef" (accessed 18 July 2012).

Department of Sustainability, Environment, Water, Population and Communities, 2011-12, "Annual Report" <http://www.environment.gov.au> and follow links to "Science and Research" under "Topics" and then "State of the Environment (SoE) Reporting" (accessed 15 September 2013).

Department of Sustainability, Environment, Water, Population and Communities, 2011, Australian Government Response to the Report of the Independent Review of the *Environment Protection and Biodiversity Conservation Act 1999*.

Department of Sustainability and Environment, 2004, *The Environment Impact Assessment Process in Victoria* <www.dse.vic.gov.au> and see DSE Home Page.

Department of Sustainability and Environment, 2005, *Assessment Guidelines: Ryan Corner Wind Farm Environment Effects Statement* <www.dse.vic.gov.au> (accessed May 2008).

Department of Sustainability and Environment, 2006, *Ministerial Guidelines for Assessment of Environmental Effects under the Environment Effects Act 1978*, 7th edn, DSE, Melbourne.

Department of Urban Affairs and Planning, 1996, *EIS Guidelines: Extractive Industries, Dredging and Other Extraction in Riparian and Coastal Areas,* Department of Urban Affairs and Planning, Sydney.

Department of Urban Affairs and Planning, 1997, *Integrated Development Assessment: White Paper and Exposure Draft Bill,* Department of Urban Affairs and Planning, Sydney.

Department of Urban Affairs and Planning, 2000, *Coal Mines and Associated Infrastructure: EIS Guideline,* Planning NSW.

Dingle, AE, 1984, *The Victorians: Settling*, Fairfax, Syme and Weldon, Sydney.

Division of Environmental Management, 1996, *Environmental Assessment Manual – Version 1,* Department of Environment and Land Management, Hobart.

Dixon, J 2002, "All at SEA? Strategic Environmental Assessment in New Zealand", in Marsden, S and Dovers, S (eds), *Strategic Environmental Assessment in Australasia*, Federation Press, Sydney, pp 195-210.

Dorney, LC (ed), 1989, *The Professional Practice of Environmental Management*, Springer-Verlag, New York.

Dover, S, 2002, "Too Deep a SEA? Strategic Environmental Assessment in the Era of Sustainability", in Marsden, S and Dovers, S (eds), *Strategic Environmental Assessment in Australasia*, Federation Press, Sydney, pp 24-46.

Doyle, T, 2000, *Green Power: The Environment Movement in Australia*, UNSW Press, Sydney.

Doyle, T and Kellow, A, 1995, *Environmental Politics and Policy Making in Australia*, MacMillan, South Melbourne.

DSEWPC, May 2013, *Major Project Development Assessment Processes Productivity Commission Inquiry*, Submission <http://www.pc.gov.au/projects/> (accessed November 2013).

DSEWPC, July 2012, *Draft Framework of Standards for Accreditation of Environmental Approvals under the EPBC Act 1999.*

Duncan, R, 2008, "Problematic Practice in Integrated Impact Assessment: The Role of Consultants and Predictive Computer Models in Burying Uncertainty", *Impact Assessment and Project Appraisal*, 26(1), March, pp 53-66

Ebisemiju, FS, 1993, "Environmental Impact Assessment: Making it Work in Developing Countries", *Journal of Environmental Management*, vol 38, issue 4, pp 247-273.

Elling, B, 2000, "Integration of Strategic Environmental Assessment into Regional Spatial Planning", *Impact Assessment and Project Appraisal*, vol 18, no 3, pp 233-243.

Engström, R, Måns Nilsson, M. and Finnveden, G, 2008, "Which Environmental Problems Get Policy Attention? Examining Energy and Agricultural Sector Policies in Sweden", *Environmental Impact Assessment Review*, vol 28, pp 241-255.

enHealth Council, 2000, *Health Impact Assessment Implementation Guidelines: Draft*, Department of Health and Aged Care, Canberra.

Environment and Planning Division, 1997, *Guidance Notes for Developers: Activities Which May Cause Environmental Harm*, Department of Environment and Land Management, Hobart.

Environment Australia, 1999, *Possible Application of a Greenhouse Trigger under the Environment Protection and Biodiversity Conservation Act 1999: Consultation Paper*, Environment Australia, Canberra.

Environment Australia, 2000, *Environment Protection and Biodiversity Conservation Act -Administrative Guidelines* <http://www.environment.gov.au/> and follow links to the "Environment Protection and Biodiversity Conservation Act 1999" under "Legislation" and then "EPBC Act Publications" (accessed January 2009).

Environment Australia, 2003a *ESD Reporting Guidelines*, <http://www.environment.gov.au/> and follow links to the "Environment Protection and Biodiversity Conservation Act 1999" under "Legislation" and then "EPBC Act Publications" (accessed Jan 2008).

Environment Australia, 2003b, *Generic ESD and Environmental Performance Indicators for Commonwealth Organisations*, <http://www.environment.gov.au/> and follow links to the "Environment Protection and Biodiversity Conservation Act 1999" under "Legislation" and then "EPBC Act Publications" (accessed May 2004).

Environmental Assessment in Countries in Transition, Home Page, <web.ceu.hu/envsci/eianetwork> (accessed Jan 2008).

Environmental Defenders Office Qld, *The Wilderness Society v Minister for Environment (Gunns pulp mill case 1)* <http://www.edo.org.au/edoqld/> and follow links to "Court Cases", (accessed Sept 2013).

Environmental Defenders Office, National Environment Laws, <http://www.edohandbook.org/> and follow the relevant links (accessed September 2013).

Environmental Defenders Office WA, May 2002, *Environmental Impact Assessment in Western Australia* <www.edo.org.au/edowa> (accessed April 2004).

Environmental Defenders Office (Qld) Inc, date unknown, *Full Federal Court Win for Conservationists in Nathan Dam Appeal!* <www.edo.org.au/edoqld/> (accessed July 2008).

Environment Protection Agency (Qld), *A Guide to Environmental Impact Assessment*, <www.epa.qld.gov.au/publications?id=703> (accessed December 2007).

Environment Protection Agency, 1994a, *Executive Summaries of Consultants' Reports*, Canberra.

Environment Protection Agency, 1994b, *Main Discussion Paper: Public Review of the Commonwealth Environmental Impact Assessment Process*, Canberra.

Environment Protection Agency, 2000, *Draft Summary of Impact Assessment Processes in Queensland*, EPA, Brisbane.

Environment Protection Authority (NT), 2007, *Guidelines for the Preparation of an Environmental Impact Assessment – Terra Gold Mining Ltd Maud Creek Gold Project* <www.nt.gov.au/nreta/environment/assessment/register/maudcreek/index.html> (accessed March 2008).

Environment Protection Authority (Tas), 2008, *Environmental Impact Assessment: A Guide* <www.epa.tas.gov.au/> (accessed January 2009).

Environment Protection Authority (Vic), 1993, *A Question of Trust: Accredited Licensee Concept*, Publication 385, Melbourne.

Environment Protection Authority (Vic), 1995, *Protecting Waters in Port Phillip Bay: State Environment Protection Policy (Waters of Victoria) draft Schedule F6 (Waters of Port Phillip Bay) and Draft Policy Impact Assessment*, EPA, Melbourne.

Environment Protection Authority (WA), 2002, *Environmental Impact Assessment (Part IV Division 1) Administrative Procedures*, Perth, <www.epa.wa.gov.au/docs/1139_EIA_Admin.pdf>.

Environment Protection Authority (WA), 2004, *Guide to EIA Environmental Principles, Factors and Objectives*, Version 2, Perth.

Environment Protection Authority (WA), 2007 *General Guide for State Government Departments and Local Authorities (as Decision-Making Authorities)*, Perth <www.epa.wa.gov.au/eia.asp> (accessed December 2007).

Environment Protection Authority (WA), 2012, *Environmental Impact Assessment Administrative Procedures* <http://www.epa.wa.gov.au/EIA/assessdev> (accessed September 2013).

Environmental Resources Ltd, 1983, *Environmental Health Impact Assessment of Irrigated Agricultural Development Projects: Guidelines and Final Report for World Health Organisation*, Regional Office for Europe, London.

Environmental Resources Management Australia, 2006, *Ryan Corner Wind Farm Environment Effects Statement*, Melbourne, pp 21:1-21:35.

Environmental Resources Management Australia, 2006, *Ryan Corner Wind Farm Environment Effects Statement – Summary Report*, Melbourne.

EPA WA Guidelines <http://www.epa.wa.gov.au/EIA/assessdev/>.

Erikson, PA, 1994, *A Practical Guide to Environmental Impact Assessment*, Academic Press, San Diego.

Europa, 2001, *31985L0337 Council Directive 85/337/EEC of 27 June 1985*, <http://europa.eu.int/> (accessed April 2004).

Europa, 2003, *Assessment of the Environmental Impact of Projects* <http://europa.eu.int/> (accessed April 2004).

European Commission, 1999, *Guidelines for the Assessment of Indirect and Cumulative Impacts as Well as Impact Interactions*, European Commission.

European Commission, 2004, <http://europa.eu.int/comm/environment/eia/home.htm> (accessed May 2004).

Evans, DG, 1982, "Energy Analysis as an Aid to Public Decision-making", in *Energy, Money and Engineering*, Institute of Chemical Engineering, Series, no 78, Pergamon, Oxford.

Eversley, DEC, 1976, "Some Social and Economic Implications of Environmental Impact Assessment", in O'Riordan, T and Hey, RD, *Environmental Impact Assessment,* Saxon House, Farnborough.

Ewan, C, Young, A, Bryant, E and Calvert, D, 1992, *National Framework for Health Impact Assessment in Environmental Impact Assessment*, vol 1, *Executive Summary and Recommendations*, University of Wollongong, Wollongong.

Federal Environmental Assessment Review Office, 1993, *Addressing Cumulative Environmental Effects: A Reference Guide for the Canadian Environmental Assessment Act*, March 2008, <www.ceaa.gc.ca>.

Feldmann, L, 1998, The European Commission's Proposal for a Strategic Environmental Assessment Directive: Expanding the Scope of Environmental Impact Assessment in Europe, *Environmental Impact Assessment Review*, vol 18, no 1, pp 3-14.

Finsterbusch, K, 1980, *Understanding Social Impacts*, Sage, Beverly Hills.

Finsterbusch, K, Llewellyn, LG and Wolfe, LP, 1983, *Social Impact Assessment Methods*, Sage, Beverly Hills.

Fisher, DE, 1980, *Environmental Law in Australia*, University of Queensland Press, Brisbane.

Flood, M, 1994, "Environmental Impact Study", *Warmer Bulletin*, no 40, February, pp 16-17.

Fookes, T and Schijf, B, 1997, Reflections on EIA within the New Zealand Resource Management Act, *Environmental Impact Assessment for the 21st Century, Conference Proceedings*, Adelaide, April, pp 57-73.

Ford Motor Company, no date, ISO 14001 Environmental Management System Workbook, Ford Motor Company.

Formby, J, 1981, "The Australian Experience", in O'Riordan, T and Sewell, WRD (eds), *Project Appraisal and Policy Review*, Wiley, Chichester.

Formby, J, 1986, *Approaches to Social Impact Assessment,* Working Paper 1986/8, Centre for Resources and Environmental Studies, Australian National University, Canberra.

Formby, J, 1987, *The Australian Government's Experience with Environmental Impact Assessment*, Working Paper 1987/9, Centre for Resources and Environmental Studies, Australian National University, Canberra.

Formby, J, 1989, "The Politics of Environmental Impact Assessment", *Impact Bulletin*, vol 8, nos 1 and 2, pp 191-196.

Fowler, RJ, 1981, "The Australian Experience", in Clark, SD (ed), *Environmental Assessment in Australia and Canada*, Ministry for Conservation, Melbourne.

Fowler, RJ, 1982, *Environmental Impact Assessment, Planning and Pollution Measures in Australia*, AGPS, Canberra.

Friends of the Earth, 1980, *The Environment Effects Act 1978: A Critical Review*, Melbourne.

Gabocy, T and Ross. T, 1998, Ecological Risk Assessment: A Guideline Comparison and Review, in Porter, AL and Fittipaldi, JJ (eds), *Environmental Methods Review: Retooling Environmental Impact Assessment for the New Century*, AEPI and IAIA, The Press Club, Fargo (Nth Dakota), pp 193-200.

Gale, RP, 1983, "The Consciousness Raising Potential of Social Impact Assessment", in Daneke, GA et al (eds), *Public Involvement and Social Impact Assessment*, Westview Press, Boulder.

Garcia, MW and Daneke, GA, 1983, "The Role of Public Involvement in Social Impact Assessment: Problems and Prospects", in Daneke, GA et al (eds), *Public Involvement and Social Impact Assessment*, Westview Press, Boulder.

Gariepy, M, 1991, "Toward a Dual Influence System: Assessing the Effects of Public Participation in Environmental Impact Assessment for Hydro-Quebec Projects", *Environmental Impact Assessment Review*, vol 1, no 4, pp 353-374.

Garrett, Hon. P, 4 July 2008, "Significant Step in Kimberley Hub Site Search", media release, <www.environment.gov.au/> and follow links to "Environment Home" and then "Ministers" then "Media Release Archive" (accessed July 2000).

GHD, November 2012, *VicRoads Western Highway Project – Section 3: Ararat to Stawell: Environmental Risk Assessment Report.*

GHK, September 2010, *Collection of information and data to support the Impact Assessment study of the review of the EIA Directive*, London.

Gilpin, A, 1995, *Environmental Impact Assessment: Cutting Edge for the Twenty-first Century*, Cambridge University Press, Cambridge.

Giroult, E, 1988, "WHO Interest in Environmental Health Impact Assessment", in Wathern, P (ed), *Environmental Impact Assessment*, Unwin Hyman, London, pp 257-271.

Gittle, M, 1980, *Limits to Public Participation*, Sage, Beverly Hills.

Glacken, CJ, 1967, *Traces on the Rhodian Shore*, University of California Press, Los Angeles.

Glasson, J, Thérivel, R and Chadwick, A, 1994, *Introduction to Environmental Impact Assessment*, UCL Press, London.

Glasson, J, Thérivel, R and Chadwick, A, 2005, *Introduction to Environmental Impact Assessment*, University College London Press, London.

Global Reporting Initiative, 2002, *Sustainability Reporting Guidelines*, Global Reporting Initiative, Boston.

Goodman, R, 1971, *After the Planners*, Simon and Schuster, New York.

Goodrich, C, Taylor, N and Bryon, H, 1987, "Theoretical Considerations for Social Impact Research", Paper presented to Annual Conference of Sociological Association of Australia and New Zealand, University of New South Wales, Sydney.

Gour-Tanguay, R (ed), 1977, *Environmental Politics in Developing Countries*, Erich Schmidt Verlag, Berlin.

Greene, D, 1984, "Review of Environmental Assessment Procedures and Legislation", Report of notes of points raised at EIA seminar held in Melbourne, 18 October 1984, Ministry for Planning and Environment, Melbourne.

Grima, AP, 1983, "Analysing Public Inputs to Environmental Planning", in Daneke, GA et al (eds), *Public Involvement and Social Impact Assessment*, Westview Press, Boulder.

GRI Sustainability, 2011, "Reporting Statistics Publication Year 2011", <https://www.globalreporting.org/> and follow links to "Resource Library" (accessed June 2013).

Guide to the EIA Process in the Northern Territory (2013) <http://www.ntepa.nt.gov.au/> and follow links to "Environmental Assessments".

Gunns Ltd, no date, "Bell Bay Pulp Mill- frequently asked questions", <www.gunnspulpmill.com.au/faqs.html> (accessed July 2007).

Gunns Ltd, 2006, *Bell Bay Pulp Mill – Executive Summary*, <www.gunnspulpmill.com.au/iis/default.php> (accessed 10 July 2007).

Hanusch, M and Glasson, J, 2008, "Much Ado About SEA/SA monitoring: The Performance of English Regional Spatial Strategies, and Some German Comparisons", *Environmental Impact Assessment Review*, vol 28, pp 601-617.

Hart, SL, 1984, "The Costs of Environmental Review: Assessment Methods and Trends", in Hart, SL, Enk, GA and Hornick, WF (eds), *Improving Impact Assessment*, Westview Press, Boulder.

Health Canada, 2000, *Canadian Handbook on Health Impact Assessment*, <http://www.hc-sc.gc.ca/index-eng.php> (accessed January 2009).

Health Department of Victoria, Western Metropolitan Region, 1989, *Environmental Health Strategy: Discussion Paper*, Melbourne.

Hey, RD, 1976, "Impact Prediction in the Physical Environment", in O'Riordan, T and Hey, RD, *Environmental Impact Assessment*, Saxon House, Farnborough.

Hickie, D and Wade, M, 1998, Development of Guidelines for Improving the Effectiveness of Environmental Assessment, *Environmental Impact Assessment Review*, vol 18, no 3, pp 267-288.

Hobbs, BF, Rowe, MD, Pierce, BL and Meier, B, 1984, "Comparisons of Methods for Evaluating Multi-attributed Alternatives in Environmental Assessments", in Hart, SL, Enk, GA and Hornick, WF (eds), *Improving Impact Assessment*, Westview Press, Boulder.

Hobbs, M, 2003, *Third Reading – Resource Management Amendment Bill (no 2),* <www.beehive.govt.nz/> and follow the links under "Speeches" (accessed May 2004).

Hogg, DMcC, 1975, "Tying the EIS into the Planning Process", in *The EIS Technique*, Australian Conservation Foundation, Melbourne.

Hollick, M, 1986, "Environmental Impact Assessment: An International Evaluation", *Environmental Management*, vol 10, no 2, pp 157-178.

Hong Kong Environmental Protection Department, 2004, *Environmental Impact Assessment Ordinance*, <www.epd.gov.hk/eia/> (accessed April 2004).

Horton, S and Memon, A, 1997, "SEA: The Uneven Development of the Environment?", *Environmental Impact Assessment Review,* vol 17, no 3, pp 163-175.

House of Representatives Standing Committee on Environment and Conservation, 1979, *Environmental Protection: Adequacy of Legislative and Administrative Arrangements*, AGPS, Canberra.

Hrezo, MS and Hrezo, WE, 1984, "The Role of Human Values, Attitudes, and Beliefs in Environmental Assessment", in Hart, SL, Enk, GA and Hornick, WF (eds), *Improving Impact Assessment*, Westview Press, Boulder.

Htun, N, 1988, "The EIA Process in Asia and the Pacific Region", in Wathern, P (ed), *Environmental Impact Assessment*, Unwin Hyman, London, pp 225-238.

Htun, N, 1989, "EIA and Sustainable Development", *Impact Bulletin*, vol 8, nos 1 and 2, pp 15-23.

Hufschmidt, MM, James, DE, Meister, AD, Bower, BT and Dixon, JA, 1983, *Environment, Natural Systems and Development: An Economic Valuation Guide,* John Hopkins University Press, Baltimore.

Hughes, JD, 1975, *Ecology in Ancient Civilisations*, University of New Mexico Press, Albuquerque.

Hundloe, T, McDonald, GT, Ware, J and Wilks, L, 1990, "Cost-Benefit Analysis and Environmental Impact Assessment", *Environmental Impact Assessment Review*, vol 10, nos 1 and 2, pp 55-68.

Hyman, EL, Moreau, DH and Stiftel, B, 1984, "Towards a Participant Value Method for the Presentation of Environmental Impact Data", in Hart, SL, Enk, GA and Hornick, WF (eds), *Improving Impact Assessment*, Westview Press, Boulder.

IAIA Health Impact Assessment page <http://www.iaia.org/iaiawiki/hia.ashx> (accessed 25 May 2013).

IAIA and IEA (UK), 1999, *Principles of Environmental Impact Assessment Best Practice* <www.iaia.org/> and follow links to "Publications and Resources" and then "Downloadable Publications" (accessed January 2008).

IAIA, April 2004, *Case Studies on Biodiversity in Impact Assessment Submitted to CBD*, IAIA Newsletter, vol 15, no 4, ND, USA, p 5.

IAIA, July 2005, *Biodiversity in Impact Assessment*, Special Publication Series no 3.

IAP2, date unknown, *Foundations of Public Participation* <http://www.iap2.org.au/documents/item/83>, (accessed 2 July 2013).

ICI Australia, 1979, *Point Wilson Environment Effects Statement*, Melbourne.

Industries Assistance Commission, 1989, *The Environmental Impacts of Travel and Tourism*, Inquiry into Travel and Tourism, Discussion Paper no 1, AGPS, Canberra.

Institute for Environmental Studies, 2000, *European Forum on Integrated Environmental Assessment*, 4 October 2008, <www.springerlink.com/> and follow links to "Articles".

Institute of Environmental Assessment, 1993, "EIA and Environmental Auditing", *EIA Newsletter* 8, EIA Centre, University of Manchester, Manchester, p 4.

International Association for Impact Assessment, 2000, *Principles of Environmental Impact Assessment Best Practice*, <www.iaia.org> (accessed October 2000).

International Association for Impact Assessment, 2007, *About IAIA Sections*, <www.iaia.org/> (accessed December 2008).

International Chamber of Commerce, 1989, *Environmental Auditing*, ICC, Paris.

International Chamber of Commerce, 1991, *An ICC Guide to Effective Environmental Auditing*, ICC, Paris.

International Finance Corporation, 2007, "IFC Provides Financing and Construction Begins on World Bank Group-Supported Bujagali Hydropower Project", <http://www.webwire.com> and follow links to "News by Industry" (accessed June 2008).

International Finance Corporation, 2008, *Bujagali Energy Ltd: Environmental and Social Review Summary*, <www.ifc.org/> (accessed 2 March 2008).

ISO World, 2000, *The Number of ISO 14001/EMAS Registration of the World*, <www.ecology.or.jp/isoworld/english/analy14k.htm>.

Israel Ministry of Foreign Affairs, 1994, *Environmental Impact Statements -EIS-Regulations*, <www.sviva.gov.il/> (accessed January 2009).

Jain, RK, Urban, LV and Stacey, GS, 1981, *Environmental Impact Analysis*, 2nd edn, Van Nostrand Reinhold, New York.

James, D and Boer, B, 1988, *Application of Economic Techniques in Environmental Impact Assessment*, Report for Australian Environment Council, Canberra.

Jay, S, Jones, C, Slinn, C and Wood, C, 2007, "Environmental impact assessment: Retrospect and prospect", *Environmental Impact Assessment Review*, vol 27, pp 287-300.

Jeffery MI, 1989, "Environmental Assessment Processes", *Impact Bulletin*, vol 8, nos 1 and 2, pp 289-307.

Jessee, L, 1998, "The National Environmental Policy Act Net (NEPAnet) and DOE NEPA Web: What They Bring to Environmental Impact Assessment", *Environmental Impact Assessment Review*, vol 18, no 1, pp 73-82.

Jones, CE and Woods, C, 1995, "The Impacts of Environment Assessment on Public Inquiry Decisions", *Journal of Planning and Environmental Law,* October, pp 890-904.
Jones, MG, 1984, "The Evolving EIA Procedure in the Netherlands", in Clark, BD et al (eds), *Perspectives on Environmental Impact Assessment,* D Reidel, Dordrecht.
Kalensnik, SV and Pavlenko, VF, 1976, *Soviet Union: A Geographical Survey,* Progress Publications, Moscow.
Kannegieter, T, 1991, "Environmental Risk: Trying to Quantify It Is Hard Enough, Deciding on Its Acceptability Is Even Harder", *Engineers Australia,* 23 August, pp 20-22.
Kain, J-H and Söderberg, H, 2008, "Management of Complex Knowledge in Planning for Sustainable Development: The Use of Multi-criteria Decision Aids", *Environmental Impact Assessment Review,* vol 28, pp 7-21.
Kates, RW, 1978, *Risk Assessment of Environmental Hazard,* SCOPE Report no 8, John Wiley and Sons, Chichesteen.
Kay, WD, 1989, "Impact Assessment and Regulating Technological Change: Why the Philosophy of Technology Is a Political Problem", in Barlett, RV (ed), *Policy through Impact Assessment,* Greenwood Press, Westport Ct, pp 121-127.
King, SC and Pushchak, R, 2008, "Incorporating Cumulative Effects into Environmental Assessments of Mariculture: Limitations and Failures of Current Siting Methods", *Environmental Impact Assessment Review,* vol 28, pp 572-586.
Kingsland, S, 2006, *The Victorian Environmental Health Indicator Project A Discussion Paper on the Development of Environmental Health Indicators for Victoria,* <www.health.vic.gov.au> (accessed February 2008).
Kinnaird, M, 1990, "The Environmental Impact Statement: Its Uses and Abuses", Paper presented to National Engineering Conference, Canberra.
Kjellerup, U, 1999, "Significance Determination: A Rational Reconstruction of Decisions", *Environmental Impact Assessment Review,* vol 19, issue 1, pp 3-19.
Koornneef, J, Faaij, A and Turkenburg, W, 2008, "The Screening and Scoping of Environmental Impact Assessment and Strategic Environmental Assessment of Carbon Capture and Storage in the Netherlands", *Environmental Impact Assessment Review,* vol 28, pp 392-414.
Kørnøv, L, 2009, "Strategic Environmental Assessment as Catalyst of Healthier Spatial Planning: The Danish Guidance and Practice", *Environmental Impact Assessment Review,* vol 29, pp 60-65.
Kriwoken, LK and Rootes, D, 1966, *Environmental Impact Assessment and Sustainable Antarctic Tourism,* Initial symposium draft for Polar Tourism: Environmental; Implications and Management, Scott Polar Research Institute University of Cambridge, August.
Lang, R, 1979, "Environmental Impact Assessment: Reform or Rhetoric?", in Leiss, W (ed), *Ecology vs Politics in Canada,* University of Toronto Press, Toronto.
Lang, R and Armour, A, 1981, *The Assessment and Review of Social Impacts,* Federal Environmental Assessment Review Office, Ottawa.
Lawrence, DP, 1997, "The Need for EIA Theory-Building", *Environmental Impact Assessment Review,* vol 17, no 2, pp 79-107.
Lawrence, DP, 2007a, "Impact Significance Determination – Designing an Approach", *Environmental Impact Assessment Review,* vol 27, pp 730-754.
Lawrence, DP, 2007b, "Impact Significance Determination – Pushing the Boundaries", *Environmental Impact Assessment Review,* vol 27, pp 770-788.
Lee, N and Colley, R, 1990, *Reviewing the Quality of Environmental Statements,* Occasional Paper no 24, EIA Centre, University of Manchester, Manchester.
Lee, N and Wood, CM, 1991, *EIA Legislation and Regulations in the EEC,* Leaflet 5, EIA Centre, University of Manchester, Manchester.

Lee, N, 1991, “Quality Control in the EIA Process”, *EIA Newsletter* 6, pp 22-23.

Leeson, R 1994, “EIA and the Politics of Avoidance”, *Environmental and Planning Law Journal*, vol 11, no 1, pp 71-92.

Leeson, R, 2000, personal communication, November.

Leistritz, FL, 1998, “Economic and Fiscal Impact Assessment”, in Porter, AL and Fittipaldi, JJ (eds), *Environmental Methods Review: Retooling Environmental Impact Assessment for the New Century*, AEPI and IAIA, The Press Club, Fargo (Nth Dakota), pp 219-227.

Leopold, LB, Clark, FE, Handshaw, BB and Balsley, JR, 1971, *A Procedure for Evaluating Environmental Impact*, Geological Survey Circular 645, US Geological Survey, Washington.

Leshinsky, R (2010) *Environmental Impact Assessment*, at 1. Accessed 3 August 2012 at <www.lawhandbook.org.au/handbook/ch11s01s03.php#>.

Leu, W-S, Williams, WP and Bark, AW, 1996, “Development of an Environmental Impact Assessment Evaluation Model and its Application: Taiwan Case Study”, *Environmental Impact Assessment Review*, vol 16, issue 2, pp 115-133.

Lind, T, 1991, “Nordic Co-operation on Environmental Impact Assessment”, *EIA Newsletter* 6, p 15.

Liou, M and Yu, Y, 2004, “Development and Implementation of Strategic Environmental Assessment in Taiwan”, *Environmental Impact Assessment Review*, vol 24, issue 3, pp 337-350.

Lockie, S, Franetovich, M, Sharma, S and Rolfe, J, 2008, “Democratisation Versus Engagement? Social and Economic Impact Assessment and Community Participation in the Coal Mining Industry of the Bowen Basin, Australia”, *Impact Assessment and Project Appraisal*, vol 26, issue 3, pp 177-187.

Lucas, AR, 1976, “Legal Foundations for Public Participation in Environmental Decision-making”, *Natural Resources Journal,* vol 16, no 1, pp 73-102.

Lyon, KA, 1989, “Factors Influencing Public Policy Makers’ Interpretation”, *Impact Bulletin*, vol 8, nos 1 and 2, pp 249-260.

Macintosh, A, 1 July 2009, *Australia’s National Environmental Legislation: A Response to Early Journal of International Wildlife Law & Policy*, vol 12, No 3, pp 166-179. Available at SSRN: <http://ssrn.com/abstract=1546134> (accessed October 2013).

Malcolm, J, 2002, “Strategic Environmental Assessment: Legislative Developments in Western Australia”, in Marsden, S and Dovers, S (eds), *Strategic Environmental Assessment in Australasia*, Federation Press, Sydney.

Malone, N, 1997, Environmental Impact Monitoring, *Environmental and Planning Law Journal*, vol 14, no 3, pp 222-238.

Marsden, S, 1997, “Applying EIA to Legislative Proposals: Practical Solutions to Advance ESD in Commonwealth and State Policy-Making”, *Environmental and Planning Law Journal*, vol 14, no 3, pp 159-173.

Marsden, S and Ashe, J, 2006, “Strategic Environmental Assessment Legislation in Australian States and Territories”, *Australasian Journal of Environmental Management*, vol 13, pp 205-215.

Marsden, S and Dovers, S (eds), 2002, *Strategic Environmental Assessment in Australasia*, Federation Press, Sydney.

Marsh, GP, 1967, *Man and Nature*, Oxford University Press, London.

Marshall, P, 1992, *Nature’s Web: An Exploration of Ecological Thinking*, Simon and Schuster, London.

Martinick Bosch Sell Pty Ltd, 2004, *Bootu Creek Manganese Project Northern Territory Public Environmental Report* (Bootu Creek Resources Pty Ltd).

Martyn, A, Morris, M and Downing, F, 1990, “Environmental Impact Assessment Process in Australia”, Discussion paper compiled by Environmental Institute of Australia, Canberra.

Matthews, WH, 1975, "Objective and Subjective Judgments in Environmental Impact Analysis", *Environmental Conservation*, vol 2, no 2, pp 121-131.

Maystre, L, 1991, "EIA in Switzerland", *EIA Newsletter* 6, p 16.

McEwan, P, 1996, *Guidelines for Responding to Requests to the minister for Planning for a Decision on the Need for an Environments Effects Statement (EES)*, May, Unpublished briefing memorandum, Planning division, Department of Infrastructure, Melbourne.

McGrath, C, "Avoid the Legal Pitfalls in the EPBC Act by Understanding its Key Concepts", 2005, *National Environmental Law Review*, Spring edition.

McGrath, C, date unknown, "Federal Environmental Laws Consider Direct and Indirect Impacts of an Action" *Environmental Law Publishing*, <http://envlaw.com.au/> (accessed 2 July 2008).

McHarg, I, 1971, *Design with Nature*, Doubleday, New York.

Meeker, JW, 1974, *The Comedy of Survival*, Charles Scribner's Sons, New York.

Melbourne Water, 1994a, *Managing Environmentally: A Guide to Melbourne Water's Environmental Management System*, Melbourne.

Melbourne Water, 1994b, *Environmental Evaluation Handbook*, Melbourne.

Melbourne Water, 2008, *Environment and Public Health Management System* <www.melbournewater.com.au/> (accessed 10 July 2008).

Mernitz, S, 1980, *Mediation of Environmental Disputes*, Praeger, New York.

Miller, RE, 1984, "The EIS and the Decision Maker: Closing the Gap", in Hart, SL, Enk, GA and Hornick, WF (eds), *Improving Impact Assessment*, Westview Press, Boulder.

Minister for the Environment, 1997, *Amendments to the Environment Protection Act 1986: A Public Discussion Paper*, Office of the Minister for the Environment, Perth.

Minister for Planning, 2008, *Ryan Corner Wind Farm Assessment* <www.dse.vic.gov.au/> (accessed 18 June 2008).

Ministry for Conservation (Vic), 1977 *Guidelines for Environment Assessment*, Melbourne.

Ministry for the Environment, 2001, "The Resource Management Act and You: Getting in on the Act" <www.mfe.govt.nz/publications/rma> (accessed April 2004).

Ministry for the Environment, 2003, "Fixing the Resource Management Act" <www.mfe.govt.nz/publications/rma/#pubgen> (accessed January 2009).

Ministry for the Environment, 2005, *Regulations on Environmental Impact Assessment* <www.regjeringen.no/> (accessed January 2009).

Molesworth, S, 1985, "The Case For and Against Community Participation", Seminar notes for Rivers and Community Resource: A Community Responsibility sponsored by MMBW, 11 April 1985, Melbourne.

Monbaillin, X, 1984, "EIA Procedures in France", in Clark, BD et al (eds), *Perspectives on Environmental Impact Assessment*, D Reidel, Dordrecht.

Moreira, IV, 1988, "EIA in Latin America", in Wathern, P (ed), *Environmental Impact Assessment*, Unwin Hyman, London, pp 239-253.

Morgan, R, 1998, *EIA: A Methodological Perspective*, Kluwer Academic Publishers, Boston.

Morrison-Saunders, A and Bailey, J, 2000, "Transparency in EIA Decision-making: Recent Developments in Western Australia", *Impact Assessment and Project Appraisal*, vol 18, no 4, pp 260-270.

Morrison-Saunders, A and Arts, J, 2005, "Learning from Experience: Emerging Trends in Environmental Impact Assessment Follow-up", *Impact Assessment and Project Appraisal*, vol 23, no 3, pp 170-174.

Moss, B, 1976, "Ecological Considerations in the Preparation of Environmental Impact Statements", in O'Riordan, T and Hey RD, *Environmental Impact Assessment*, Saxon House, Farnborough.

Munchenbery, S, 2002, "Strategic Environmental Assessment: A Business Perspective" in Marsden, S and Dovers, S (eds), *Strategic Environmental Assessment in Australasia*, Federation Press, Sydney, pp 192-193.

Munn, RE (ed), 1975, *Environmental Impact Assessment: Principles and Practice*, SCOPE Report no 5, John Wiley and Sons, Chichester.

Munro, DA, Bryant, TJ and Matte-Baker, A, 1986, *Learning from Experience: A State-of-the-art Review and Evaluation of Environmental Impact Assessment Audits,* Canadian Environmental Assessment Research Council, Quebec.

Nadeem, O and Hameed, R (2008) "Evaluation of Environmental Impact Assessment System in Pakistan", *Environmental Impact Assessment Review*, vol 28, pp 562-571.

Nilsson, M (2009) "On (Re)Focusing SEA on the Natural Environment", *Environmental Impact Assessment Review*, vol 29, pp 76-77.

National Advisory Body on Scheduled Wastes, 2000, *On Schedule Eventually: A Case Study of Problem Solving Through Effective Community Consultation*, Environment Australia, Canberra.

National Health and Medical Research Council, 1994, *National Framework for Environmental and Health Impact Assessment*, AGPS, Canberra.

National Park Service US Department of the Interior, *Yellowstone National Park*, <www.nps.gov/yell> (accessed July 2008).

National Strategy for Ecologically Sustainable Development, 1996, *Summary of the Main Points of The Intergovernmental Agreement on the Environment,* <www.environment.gov.au/esd/national/nsesd/strategy/index.html> (accessed 31 July 2008).

Neider, C (ed), 1954, *Man Against Nature*, Harper, New York.

Nelkin, D, 1975, "The Political Impact of Technical Expertise", *Social Studies of Science*, vol 5, pp 35-54.

New South Wales Parliament, 1980, *Environmental Planning and Assessment Act (1979) – Regulation*, Supplement to *Government Gazette*, no 120, 29 August, Sydney.

New South Wales Planning and Infrastructure, *Development Assessment Systems*, <http://www.planning.nsw.gov.au> and follow links to "Development Proposals" (accessed 2 September 2013).

New Zealand Ministry for the Environment, 1997, *The Resource Management Act*, <www.mfe.govt.nz> (accessed August 2008).

O'Connor, D, 1983, "New Approaches to Environmental Assessment", in *Proceedings from the Environmental Impact Assessment and Procedures Public Seminar*, Bulletin 142, Department of Conservation and Environment, WA, Perth.

O'Riordan, T, 1976, *Environmentalism*, Pion, London.

O'Riordan, T, 1985, "Political Decision-making and Scientific Indeterminacy", in Covello, VT, Mumpower, JL, Stallen, PJM and Uppuluri, VRR (eds), *Environmental Impact Assessment, Technology Assessment, and Risk Analysis*, Springer-Verlag, Berlin.

O'Riordan, T, 1990, "EIA from the Environmentalist's Perspective", *VIA*, 4 March, pp 10-15.

Office of Regulation Reform, nd, *Regulatory Impact Statement Handbook*, Office of Regulation Reform, Melbourne.

Olympic Games Social Impact Assessment Steering Committee, 1989, *Social Impact Assessment Olympic Games Bid Melbourne 1996*, Melbourne.

Organisation for Economic Co-operation and Development, 1988, *New Technologies in the 1990s: A Socio-economic Strategy*, Paris.

PADC (ed), 1983, *Environmental Impact Assessment*, Martinus Nijhoff, Hingham.

Parliament of Victoria, 2000, *Environment and Natural Resources Committee*, <www.parliament.vic.gov.au/enrc/default.htm>.

Parliament of Victoria, September 2011, *Environment and Natural Resources Committee: Inquiry into the Environment Effects Statement Process in Victoria* <http://www.parliament.vic.gov.au> and follow links to "Committees" and "Inquiry Status Table" (accessed 10 June 2012).

Parliament of Victoria, March 2012, *Victorian Government Response to the Report of the Environment and Natural Resources Committee on its Inquiry into the Environment Effects Statement Process in Victoria*, at 1. Accessed 10 June 2012 at <http://www.parliament.vic.gov.au/> and follow links to "Committees".

Pastakia, CMR and Jensen, A, 1998, The Rapid Impact Assessment Matrix (RIAM) for EIA, *Environmental Impact Assessment Review*, vol 18, pp 461-482.

Pearson, C and Pryor, A, 1978, *Environment: North and South*, John Wiley and Sons, New York.

Pennicuik, S, 2008, "Channel Deepening Economics Come under Spotlight", <www.vic.greens.org.au/> (accessed April 2008).

Peterson, GL and Gemmell, RS, 1981, "Social Impact Assessment: Comments on the State of the Art", in Finsterbusch, K and Wolfe, CP, *Methodology of Social Impact Assessment,* Hutchinson Ross Stroudsburg.

Physalia Ocean Sciences, 2004, *Environmental Impact Assessment (EIA),* <www.physaliaos.com/sect2-7.htm> (accessed May 2004).

Piddington, KW, 1981, "Environmental Impact Assessment under the New Zealand National Development Act", in Clark, SD (ed), *Environmental Assessment in Australia and Canada*, Ministry for Conservation, Melbourne.

Planning and Development Regulation (ACT), 2008, <www.legislation.act.gov.au/sl/2008-2/default.asp> (accessed June 2008).

Planning Assessment Commission (1) NSW *About the Planning Assessment Commission* <http://www.pac.nsw.gov.au/> and follow links to "About Us", (accessed October 2013).

Planning Assessment Commission (2) NSW, <www.pac.nsw.gov.au> (accessed September 2013).

Planning and Land Management (ACT), 2002, "Territory Plan", Canberra.

Planning SA, "Home Page", <www.planning.sa.gov.au/go/home>.

Planning SA, *How Major Development Applications are Assessed* <http://www.sa.gov.au/> and follow links to "Housing, Property and Land" and then "Building and Development", then "Building Development Applications", "Major Development Applications and Assessments".

Porter, AL and Rossini FA, 1983, "Why Integrated Impact Assessment", in Rossini, FA and Porter, AL, *Integrated Impact Assessment*, Westview Press, Boulder.

Porter, CF, 1985, *Environmental Impact Assessment: A Practical Guide*, University of Queensland Press, Brisbane.

Powell, JM, 1975, "Conservation and Resource Management in Australia 1788-1914", in Powell, JM and Williams, M (eds), *Australian Space: Australian Time*, Oxford University Press, Melbourne.

Premier's Department Queensland, 1992, *Impact Assessment in Queensland: Policies and Administrative Arrangements*, Brisbane.

Purnama, D, 2003, "Reform of the EIA Process in Indonesia: Improving the Role of Public Involvement", *Environmental Impact Assessment Review*, vol 23, issue 4, pp 415-439

Raff, M, 1997, "Ten Principles of Quality in Environmental Impact Assessment", *Environmental and Planning Law Journal*, vol 14, no 3, pp 207-221.

Raff, MJ, 2000, "Environmental Impact Assessment Under the Environment Protection and Biodiversity Conservation Act, 1999 (Cth)", in Environment Defenders' Office

(Victoria) Ltd, *Environment Protection and Biodiversity Conservation Act 1999 (Cth) Seminar Papers*, 3 May 2000, pp 44-52.

Ramsay, CG, 1984, "Assessment of Hazard and Risk", in Clark, BD et al (eds), *Perspectives on Environmental Impact Assessment*, D Reidel, Dordrecht.

Ramsay, R and Rowe, G, 1995, *Environmental Law and Policy in Australia: Text and Materials*, Butterworths, Sydney.

Rau, JG and Wooten, DC, 1980, *Environmental Impact Analysis Handbook*, McGraw-Hill, New York.

Resource Planning and Development Commission, "Gunns Pulp Mill Proposal", <www.rpdc.tas.gov.au/poss/pulp/whatsnew> (accessed August 2007).

Renton, S and Bailey, J, 2000, Policy Development and the Environment, *Impact Assessment and Project Appraisal*, vol 18, no 3, pp 245-251.

Rees, C, 1999, "Improving the Effectiveness of Environmental Assessment in the World Bank", *Environmental Impact Assessment Review*, vol 19, no 3, pp 333-339.

Retief, F, 2007, "A Performance Evaluation of Strategic Environmental Assessment (SEA) Processes within the South African Context", *Environmental Impact Assessment Review*, vol 27, pp 84-100.

RICS *Draft Guidance Note – Environmental Impact Assessment*, <https://consultations.rics.org/consult.ti/environ_impact/> and follow the relevant links (accessed 10 June 2013).

Richardson, G, 1994, *Whatever It Takes*, Bantam Books, Sydney.

Rickson, RE, Burdge, RJ, Hundloe, T and McDonald, GT, 1990, "Institutional Constraints to Adoption of Social Impact Assessment as a Decision-making and Planning Tool", *Environmental Impact Assessment Review*, vol 10, nos 1 and 2, pp 233-243.

Rickson, RE, Western, J and Burdge, RJ, 1988, "Theoretical Bases of Social Impact Assessment", in Hindmarsh, RA, Hundloe, TJ, McDonald, GT and Rickson, RE (eds), *Papers on Assessing Social Impacts of Development*, Institute of Applied Environmental Research, Griffith University, Brisbane, pp 1-13.

Roche, C, 1999, *Impact Assessment for Development Agencies: Learning to Value Change*, Oxfam GB, Oxford.

Roe, D, Dalal-Clayton, B and Hughes, R, 1995, *A Directory of Impact Assessment Guidelines*, International Institute for Environment and Development, London.

Rogers, J, 1988, "Environmental Impact Assessment: Does It Really Work?", *Habitat*, October, pp 22-23.

Ross, WA, 1987, "Evaluating Environmental Impact Statements", *Journal of Environmental Management*, vol 25, pp 137-147.

Ross, W, Morrison-Saunders, A and Marshall, R, 2006, "Common Sense in Environmental Impact Assessment: It is Not as Common as it Should Be", *Impact Assessment and Project Appraisal*, vol 24, no 1, pp 3-22.

Rubik, F, 1993, "Eco-balances and Life Cycle Analysis", *EIA Newsletter* 8, EIA Centre, University of Manchester, Manchester, p 4.

Rzeszot, U, 1991, "EIA in Poland", *EIA Newsletter*, no 6, p 17.

Sadar, MH, 1994, *Environmental Impact Assessment*, Carleton University Press, Ottawa.

Sadler, B, 1988, "The Evaluation of Assessment: Post EIS Research and Process Development", in Wathern, P (ed), *Environmental Impact Assessment*, Unwin Hyman, London, pp 129-142.

Sadler, B, 1997a, *International Study of the Effectiveness of Environmental Assessment Final Report: Evaluating Practice to Improve Performance*, Main Report, 15 July 1997 <www.iaia.org/> (accessed January 2009).

Sadler, B, 1997b, *International Study of the Effectiveness of Environmental Assessment Final Report: Evaluating Practice to Improve Performance*, Executive Summary, 15 July 1997 <www.ceaa-acee.gc.ca>.

Sadler, B and Verheem, 1996, *"Strategic Environmental Assessment: Status, Challenges and Future Directions"*, Ministry of Housing, Spatial Planning and the Environment of the Netherlands, Zoetermeer, p 27.
Sammarco, PW, 1990, The Resource Assessment Commission and Environmental Impact Assessment in Australia: Major Issues, Paper presented to National Conference of the Environmental Institute of Australia, Adelaide.
Sánchez, LE and Silva-Sánchez, SS, 2008, "Tiering strategic environmental assessment and project environmental impact assessment in highway planning in São Paulo, Brazil", *Environmental Impact Assessment Review*, vol 28, pp 515-522.
Sander, T, date unknown, "Environmental Impact Statements and Their Lessons for Social Capital Analysis" <www.ksg.harvard.edu/programs/saguaro> and follow links to "Articles and Books by Putnam and Sander" under "Resources" (accessed June 2004).
Savan, B, 1986, "Sleazy Science", *Alternatives*, vol 13, no 2, April, pp 11-17.
Save Bastion Point, "Friends v Madden" <http://savebastionpoint.org/> and follow links to "Resources" (accessed 2 July 2013)
Senate Environment, Communications, Information Technology and the Arts References Committee, 1999, *Commonwealth Environment Powers*, May, Parliament of the Commonwealth of Australia, Canberra, Senate Standing Committee on Environment, Recreation and the Arts, 1989, *Environmental Impact of Development Assistance*, AGPS, Canberra.
Senate Environment, Communications, Information Technology and the Arts Legislation Committee, 1999, *Environment Protection and Biodiversity Conservation Bill 1998 and Environmental Reform (Consequential Provisions) Bill 1998*, Parliament of the Commonwealth of Australia, Canberra.
Senate Standing Committee on Science, Technology and the Environment, 1987, *Technology Assessment in Australia*, AGPS, Canberra.
Sewell, GH and Korrick, S, 1984, "The Fate of EIS Projects: A Retrospective Study", in Hart, SL, Enk, GA and Hornick, WF (eds), *Improving Impact Assessment*, Westview Press, Boulder.
Sewell, WRD, 1981, "How Canada Responded to the Berger Inquiry", in O'Riordan, T and Sewell, WRD, *Project Appraisal and Policy Review*, Wiley, Chichester.
Sewell, WRD and O'Riordan, T, 1976, "The Culture of Participation in Environmental Decision-making", *Natural Resources Journal*, pp 1-21.
Sewell, WRD and Phillips, SD, 1979, "Models for the Evaluation of Public Participation Programmes", *Natural Resources Journal*, vol 19, no 2, pp 338-357.
Shepard, A, 1998, Post-Project Impact Assessment and Monitoring, *Environmental Methods Review: Retooling Impact Assessment for the New Century*, Army Environmental Policy Institute and International Association for Impact Assessment, The Press Club, Fargo (Nth Dakota), pp 164-170.
Shopley, JB and Fuggle, RF, 1984, "A Comprehensive Review of Current Environmental Impact Assessment Methods and Techniques", *Journal of Environmental Management*, vol 18 pp 25-47.
Short, M, 1991, "The Labyrinth of Approvals", *Age*, 31 May, p 11.
Shrader-Frechette, KS, 1985, *Risk Analysis and Scientific Method*, D Reidel, Dordrecht.
Shrybman, S, 1990, "International Trade and the Environment", *Alternatives*, vol 17, no 2, pp 20-28.
Sinclair, AJ, Diduck, A and Fitzpatrick, P, 2008, "Conceptualizing Learning for Sustainability Through Environmental Assessment: Critical Reflections on 15 years of Research", *Environmental Impact Assessment Review*, vol 28, pp 415-428.
Sinclair Knight Merz, 1999, *Darwin Joint Terminal Public Environment Report* (for Shell Australia), Darwin, Australia.

SMEC, 1999, *Auditor's Report on the Supplement to the Draft Environmental Impact Statement for the Second Sydney Airport Proposal*, Environment Australia, Canberra.

Smith, LG, 1991, "Canada's Changing Impact Assessment Provisions", *Environmental Impact Assessment Review*, vol 11, no 1, pp 5-9.

Smyth, RB, 1990, "EIA on EIA: Experience – A Need For Improvement", in *Independent Review of Environmental Impact Assessment in NSW – How Can It Be Improved?*, Proceedings of workshop/seminar, Environment Institute of Australia and Australian Association of Consultant Planners, pp 3-9.

Social Impact Unit, WA, 1990, *Annual Report*, Perth.

Soderstrom, EJ, 1981, *Social Impact Assessment*, Praeger, New York.

Sors, AI, 1984, "Monitoring and Environmental Impact Assessment", in Clark, BD et al (eds), *Perspectives on Environmental Impact Assessment*, D Reidel, Dordrecht.

Spry, A, 1976, "A Consultant's View on Environmental Impact Statements in Australia", *Search*, vol 7, no 6, pp 252-255.

Srinivasan, M (ed), 1982, *Technology Assessment and Development*, Praeger, New York.

Standards Australia and Standards New Zealand, 1995, *Environmental Management Systems – Specification With Guidance for Use: AS/NZS ISO 14001*, SAA, Sydney.

State of Queensland, 1999a, *Integrated Planning Act 1997: Explanatory Notes*, <www.dip.qld.gov.au/ipa> (accessed January 2009).

State of Queensland, 1999b, *State Development and Public Works Organisation Amendment Bill 1999: Explanatory Notes*, <www.austlii.org/au/legis/qld/bill_en/sdapwoab 1999575> (accessed February 2009).

State Pollution Control Commission, NSW, 1974, *Principles and Procedures for EIA*, Sydney.

Steinemann, A, 2000, Rethinking Human Health Impact Assessment, *Environmental Impact Assessment Review*, vol 20, no 6, pp 627-645.

Stern, A, 1991, "Using Environmental Impact Assessments for Dispute Management", *Environmental Impact Assessment Review*, vol 11, no 1, pp 81-87.

Stern, PC and Fineberg, HV (eds), 1996, *Understanding Risk: Informing Decisions in a Democratic Society*, National Academy Press, Washington.

Street, P, 1997, "Scenario Workshops: A Participatory Approach to Sustainable Urban Living", *Futures*, vol 29, no 2, pp 139-158.

Stinchcombe, K and Gibson, RB, 2001, "Strategic Environmental Assessment as a Means of Pursuing Sustainability: Ten Advantages and Ten Challenges", *Journal of Environmental Assessment Policy and Management*, vol 3, no 3 pp 343-372.

Stockland and Maunsell Australia Pty Ltd, 2007, *Point Lonsdale Residential and Waterways Proposal – Environment Effects Statement*, Melbourne.

Strohmann, CH, 1998, "Training Approaches in Environmental Technology Assessment (EnTA)", in Porter, AL and Fittipaldi, JJ (eds), *Environmental Methods Review: Retooling Environmental Impact Assessment for the New Century*, AEPI and IAIA, The Press Club, Fargo (Nth Dakota), pp 186-192.

Sudara, S, 1984, "EIA Procedures in Developing Countries", in Clark, BD et al (eds), *Perspectives on Environmental Impact Assessment*, D Reidel, Dordrecht.

Sustainability Energy Authority Victoria, 2002, *Policy and Planning Guidelines for Development of Wind Energy Facilities in Victoria*, Melbourne.

Sustainability Energy Authority Victoria, 2003, *Victorian Wind Atlas*, Melbourne.

Suter, GW, 2008, "Ecological Risk Assessment in the United States Environmental Protection Agency: A Historical Overview", *Integrated Environmental Assessment and Management*, vol 4, issue 3, pp 285-289.

Suter, GW and Cormier, SM, 2008, "A Theory of Practice for Environmental Assessment", *Integrated Environmental Assessment and Management*, vol 4, issue 4, pp 478-485.

Sutton, P, 1997, "Targeting Sustainability: The Positive Application of ISO 14001", in Sheldon, C (ed), *ISO 14001 and Beyond: Environmental Management Systems in the Real World*, Greenleaf Publishing, Broom Hall (Sheff), pp 211-242.

Talbot, AR, 1983, *Settling Things*, Conservation Foundation, Washington.

Tao, T, Tan, Z and He, X, 2006, "Integrating Environment into Land-use Planning Through Strategic Environmental Assessment in China: Towards Legal Frameworks and Operational Procedures", *Environmental Impact Assessment Review*, vol 27, pp 243-265.

Taplin, R, 1998, "Climate Impacts and Adaptation Assessment", in Porter, AL and Fittipaldi, JJ (eds), *Environmental Methods Review: Retooling Environmental Impact Assessment for the New Century*, AEPI and IAIA, The Press Club, Fargo (Nth Dakota), pp 234-246.

Tasmania EPA Guide to EIA, April 2013, <http://epa.tas.gov.au/regulation/document?docid=1222> (accessed September 2013).

Taylor, S, 1984, *Making Bureaucracies Think*, Stanford University Press, Stanford.

Thérivel, R, 1993, "Systems of Strategic Environmental Assessment", *Environmental Impact Assessment Review*, vol 13, no 3, pp 145-168.

Thérivel, R. and Ross, B, 2007, "Cumulative Effects Assessment: Does Scale Matter?", *Environmental Impact Assessment Review*, Vol 27, 365-385.

Thomas, I, 2005, *Environmental Management: Processes and Practices for Australia*, The Federation Press, Sydney.

Thomas, IG, Instone, L and Durkin, P, 1992, *Impact Assessment In Australia: A Survey*, Internal report to Faculty of Environmental Design and Construction, RMIT, Melbourne.

Tomlinson, P, 1984, "The Use of Methods in Screening and Scoping", in Clark, BD et al (eds), *Perspectives on Environmental Impact Assessment*, D Reidel, Dordrecht.

Town and Country Planning Directorate, nd, *Social Impact Assessment in New Zealand: A Practical Approach*, Ministry of Works and Development, Wellington.

Toyne, P, 1994, *The Reluctant Nation*, ABC Books, Sydney.

Treweek, J, 1995, "Ecological Impact Assessment", *Impact Assessment*, vol 13, no 3, pp 289-316.

Treweek, J and Hankard, P, 1998, "Ecological Impact Assessment", in Porter, AL and Fittipaldi, JJ (eds), *Environmental Methods Review: Retooling Environmental Impact Assessment for the New Century*, AEPI and IAIA, The Press Club, Fargo (Nth Dakota), pp 263-272.

Tzoumis, K, 2007, "Comparing the Quality of Draft Environmental Impact Statements by Agencies in the United States Since 1998 to 2004", *Environmental Impact Assessment Review*, vol 27, pp 26-40.

Tzoumis, K and Finegold, L, 2000, "Looking at the Quality of Draft Environmental Impact Statements Over Time in the United States: Have Ratings Improved?", *Environmental Impact Assessment Review*, vol 20, no 5, pp 557-578.

Underwood, RT, 1978, "Community Participation and the Consideration of Alternatives in Major Works Studies", Paper presented to Ministry for Conservation Seminar Environment Assessment in Victoria, Melbourne.

Union Fenosa Wind Australia website, <www.unionfenosa.com.au/f/news/032012.pdf> (accessed 14 June 2013).

United Nations Environment Programme, date unknown, *The UNEP Cleaner Production Programme*, Industry and Environment Programme Activity Centre, UNEP, Paris.

United Nations Environment Programme "Sustainability Impact Assessment in the European Union", <www.unep.org/> (accessed July 2008).

United Nations Environment Programme, 1997, *Environmental Impact Assessment Training Resource Manual*, 1st edn, United Nations Environment Programme, Geneva.

United Nations Environment Programme, 2002, *Environmental Impact Assessment Training Resource Manual*, 2nd edn, <www.unep.ch/etb/publications/EIAMan2editionToc.php> (accessed 20 January 2009).

United States Environment Protection Agency, 1992, "The Guardian: Origins of the EPA", <www.epa.gov/history/publications/origins.htm> (accessed September 2007).

United States Environment Protection Agency, 1994, *Executive Order 12898 on Environmental Justice*, <www.epa.gov/oswer/ej/ejndx.htm#12898> (accessed March 2008).

United States Environment Protection Agency, 1997, *The National Environmental Policy Act: A Study of Its Effectiveness After Twenty-five Years*, Council on Environmental Quality, Executive Office of the President, Washington, USA.

United States Environment Protection Agency, 2007, "A Citizen's Guide to the NEPA – Having Your Voice Heard", Council on Environmental Quality, Executive Office of the President, Washington, USA.

United States Environment Protection Agency, 2008, "Environmental Justice: How does the EPA define Environmental Justice?" <www.epa.gov/compliance/resources/faqs/ej/index.html#faq2> (accessed June 2008).

University of Manchester EIA Centre, 1997, "Home Page", <www.art.man.ac.uk/EIA/EIAC.htm>.

URS, 2008, *Maud Creek Project – Executive Summary*, Darwin, <www.nt.gov.au/nreta/ environment/assessment/register/index.html> (accessed July 2008).

Vanclay, F, 2004, "The Triple Bottom Line and Impact Assessment: How do TBL, EIA, SIA, SEA and EMS relate to each other?, *Journal of Environmental Assessment Policy and Management*, vol 6, no 3, pp 265-288.

VCEC, 2009, "A Sustainable Future for Victoria: Getting Environmental Regulation Right, Overview and Recommendations", <http://www.vcec.vic.gov.au/>, (accessed August 2013).

Vellinga, P, 1998, "European Forum on Integrated Environmental Assessment (EFIEA)", *Global Environmental Change*, vol 9, no 1, pp 1-3.

VicRoads, 1989, *Western Ring Road – Environment Effects Statement: Supplementary Report no 10 Social Impact Assessment*, Roads Corporation, Melbourne.

VicRoads, 2000, *Calder Highway Keynton to Faraday: Environment Effects Statement, Vol 1*, VicRoads, Melbourne.

Victorian Government Response to the Report of the Environment and Natural Resources Committee on its Inquiry into the Environment Effects Statement Process in Victoria, March 2012, media release.

Vlachos, E, 1985, "Assessing Long-range Cumulative Impacts", in Covello, VT, Mumpower, JL, Stallen, PJM and Uppuluri, VRR (eds), *Environmental Impact Assessment, Technology Assessment, and Risk Analysis*, Springer-Verlag, Berlin, pp 49-79.

Von Moltke, K, 1984, "Impact Assessment in the United States and Europe", in Clark, BD et al (eds), *Perspectives on Environmental Impact Assessment*, D Reidel, Dordrecht.

WA EPA, November 2012, "Guidelines for Preparing a Public Environmental Review", Government of Western Australia.

Wallace, PJ, Strong, D and Luckeneder, T, 1990, "Community Involvement in Strategic Planning for the Transmission System", in *Proceedings of the Institution of Engineers Australia Electrical Energy Conference*, Canberra, pp 179-182.

Wallington, T, Bina, O and Thissen, W, 2007, "Theorising Strategic Environmental Assessment: Fresh Perspectives and Future Challenges", *Environmental Impact Assessment Review*, vol 27, pp 569-584.

Wandesforde-Smith, G, 1980, "Environmental Impact Assessment and the Politics of Development in Europe", in O'Riordan, T and Turner, RK (eds), *Progress in Resource Management and Environmental Planning*, vol 2, Wiley, Chichester.

Wang, Y, Morgan, RK and Cashmore, M, 2003, "Environmental Impact Assessment of Projects in the People's Republic of China: New Law, Old Problems", *Environmental Impact Assessment Review*, vol 23, issue 5, pp 543-579.

Ward C, 1976, "Education Requirements for Impact Review", in O'Riordan, T and Hey, RD, *Environmental Impact Assessment*, Saxon House, Farnborough.

Wathern, P, 1988a, "An Introductory Guide to EIA", in Wathern, P (ed), *Environmental Impact Assessment*, Unwin Hyman, London, pp 1-30.

Wathern, P, 1988b, "The EIA Directive of the European Community", in Wathern, P (ed), *Environmental Impact Assessment*, Unwin Hyman, London, pp 192-209.

Wells, C, 1991, "Impact Assessment in New Zealand: The Resource Management Act", *EIA Newsletter*, no 6, pp 19-20.

Wengert, N, 1976, "Citizen Participation: Practice in Search of a Theory", *Natural Resources Journal*, vol 16, no 1, pp 23-40.

Weston, H and Crawshaw, D, 1984, *A Review of the Coverage of Social Issues in Environment Effects Statements Completed in Victoria 1974-1983*, School of Environmental Planning, University of Melbourne, Melbourne.

White, B, 1987, "The Private Developers' Viewpoint", in *Seminar on Environmental Legislation and Its Impact on Management*, Institution of Engineers (Aust) and Department of Environment and Planning, Sydney, pp 33-40.

Wildavsky, A, 1988, *Searching for Safety*, Transaction, New Brunswick (USA).

Wildman, P, 1988, "Social Impact Assessment Methodological and Social Policy Issues", in Hindmarsh, RA, Hundloe, TJ, McDonald, GT and Rickson, RE (eds), *Papers on Assessing Social Impacts of Development*, Institute of Applied Environmental Research, Griffith University, Brisbane, pp 31-51.

Wildman, PH and Baker, GB, 1985, *The Social Impact Assessment Handbook*, Social Impact Publications, Armidale.

Wilkinson, CH, 1998, "Environmental Justice Impact Assessment: Key Components and Emerging Issues", in Porter, AL and Fittipaldi, JJ (eds), *Environmental Methods Review: Retooling Environmental Impact Assessment for the New Century*, AEPI and IAIA, The Press Club, Fargo (Nth Dakota), pp 273-282.

Wolfe, CP, 1983, "Social Impact Assessment: Methodological Overview", in PADC (ed), *Environmental Impact Assessment*, Martinus Nijhoff, Hingham.

Wood, C, 1993, "Strategic Environmental Assessment", *EIA Newsletter* 8, EIA Centre, University of Manchester, Manchester, pp 23-24.

Wood, C, 1995, *Environmental Impact Assessment: A Comparative Review*, Longman, Harlow (Essex).

Wood, C, 2003, "Environmental Impact Assessment in Developing Countries: An Overview", Presented at *Conference on New Directions in Impact Assessment for Development: Methods and Practice* 24-25 Nov 2003, EIA Centre, University of Manchester, Manchester.

Wood, G, 2008, "Thresholds and Criteria for Evaluating and Communicating Impact Significance in Environmental Statements: 'See No Evil, Hear No Evil, Speak No Evil'"?, *Environmental Impact Assessment Review*, vol 28 , pp 22-38.

Woodhead, WR, 1987, "The Consultant's Perspective", in *Seminar on Environmental Legislation and Its Impact on Management*, Institution of Engineers (Aust) and Department of Environment and Planning, Sydney, pp 52-63.

Woodside (May 2013) *Factsheet: Browse LNG* <http://www.woodside.com.au/> and follow links to "Browse LNG" under "Our Businesses" (accessed 14 June 2013).

Woodward, AE, 1981, "The Public Inquiry: Its Purpose and Conduct", in Clark, SD (ed), *Environmental Assessment in Australia and Canada*, Ministry for Conservation, Melbourne.

World Bank, 2003, *Accountability at the World Bank: The Inspection Panel 10 Years On* Washington, DC.

World Bank, 2009, *Environment Assessment*, <http://web.worldbank.org/> (accessed January 2009).

World Bank Group, 1999, *The World Bank Operational Policies: Environment Assessment, OP 4.01*, January, <http://web.worldbank.org/> (accessed January 2009).

World Bank Group, 2003, *Accountability at the World Bank: The Inspection Panel 10 Years On*, The World Bank, Washington, USA, pp vii-xvi and pp 80-88.

World Commission on Dams, 2000, *Dams and Development*, <www.dams.org/report> (accessed November 2000).

World Commission on Environment and Development, 1987, *Our Common Future*, Oxford University Press, Oxford.

World Resources Institute, 2004, *Environmental Impact Assessments in Latin America,* <http://earthtrends.wri.org/> (accessed February 2009).

Yap, NT, 1989, "Round the Peg or Square the Hole", *Impact Bulletin*, vol 8, nos 1 and 2, pp 69-84.

Zerbe, N and Dedeurwaerdere, T, 2003, "Sustainability Impact Assessment: Policy Brief Paper" seminar on *Sustainability Impact Assessment* 26-27 March 2003, Center for Philosophy of Law, Louvain, Belgium, <www.agro-montpellier.fr/sustra> (accessed June 2008).

Index

ALSO AVAILABLE FROM THE FEDERATION PRESS

Environment & Sustainability
A Policy Handbook
2nd edition

Stephen Dovers and Karen Hussey

Dovers and Hussey argue that better public policy is the key to creating a more sustainable environment and show what this might involve. This is an intensely practical book, intellectually rigorous, and written in an accessible style.

They meld a deep knowledge of traditional public policy with a close understanding of sustainability problems and an appreciation of the complex institutional systems which make up modern civil society. As well as establishing an operational framework for policy analysis and prescription, they provide ways of fulfilling key elements of an effective policy process with chapters on:

- Problem-framing
- Policy-framing
- Policy implementation
- Policy monitoring and evaluation
- Public participation
- Institutional settings and reform for effective policy.

This edition includes: new perspectives from policy theory and practice; expanded coverage of international dimensions of trade and regulation; and, coverage of emerging issues and policy challenges such as climate mitigation and adaptation, urban sustainability, and the nexus between climate change, energy and water.

Reviews of the 1st edition

"Stephen Dovers' new book on creating, implementing and evaluating environmental policy is an important book that deserves a wide readership. … It should be mandatory reading for every minister responsible for an environment or natural resource portfolio, every government policy officer responsible for environmental issues, and every environmental management student."

Chris McGrath, Hearsay

"This is a very valuable book. Almost every point Dovers made prompted reflection … it is one of the most engaging books I have read for some time."

Arthur Lucas, Reviews in Australian Studies

2013 • 9781862879348 • paperback • 251 pages • $59.95

ALSO AVAILABLE FROM THE FEDERATION PRESS

Climate Change and Australia
Warming to the Global Challenge

Ben Saul, Jane McAdam, Steven Sherwood, James Slezak and Tim Stephens

Climate change is complex and confusing. Media reports are inconsistent, scientific data and terminology are complex, predictions are alarming and often paralysing, and, sometimes, facts are deliberately obfuscated.

The authors draw on expertise from science, economics, geography and international law to demystify the confusion. They examine the science of global warming and the responses required; Australia's vulnerabilities and reactions to date; global climate law and politics from Rio to Cancún and beyond; food insecurity, water scarcity and conflicts over resources; climate-related displacement and migration; and global governance of climate change.

This book provides a clear, balanced account of the science on climate change, how it is affecting Australia and the world, and the policy options available.

"Succinct and avoiding detailed analysis ... the likely reader will be someone wanting a concise summary of the state of science, law and policy in order to arm themselves for debate with facts, context and an understanding of some of the complex implications of climate change."

Law Institute Victoria Journal

"This is an engaging and well-written book. ... It is refreshingly honest ... For anyone interested in climate change and the debate surrounding it, this eclectic and wide ranging work is definitely worth reading."

NSW Bar News

2012 • 9781862878723 • paperback • 255 pages • $39.95

ALSO AVAILABLE FROM THE FEDERATION PRESS

Legal Responses to Climate Change

Nicola Durrant

The challenges of climate change pose problems requiring new and innovative legal responses by legal practitioners, government officials and corporate officers.

This book addresses a broad range of areas where climate change has an impact and analyses the key legal responses to climate change, both at the international level and within Australia at federal, State and local levels.

In particular, it critically examines:

- the rights, duties and market mechanisms established under the international climate change regime;
- the effect of climate change policies on the implementation of environmental and planning laws;
- new regimes for the implementation of renewable energy and energy efficiency initiatives;
- legal frameworks for the implementation of biological and geological sequestration projects (including forest projects and carbon rights); and
- legal principles for the design of an effective carbon trading scheme for Australia.

It also considers the role of the common law including:

- the likely response of the law of torts to emerging forms of climate change harm; and
- potential liabilities for professionals who must take climate change into account in their decision-making and advice.

"This book by legal academic Durrant cannot be ignored. While she comprehensively argues that there is a place for regulation in reducing greenhouse gas emissions, Durrant does more. She shows how the law is central to whatever 'solutions' we devise to the problem of climate change."

Law Institute Victoria Journal

2010 • 9781862878082 • paperback • 365 pages • $59.95

Environmental Decision Making
Exploring Complexity and Context
2nd edition

Ronnie Harding, Carolyn Hendriks and Mehreen Faruqi

Contemporary environmental decisions have to meet integrated ecological, economic and social goals. They are made within the context of sustainability and most involve a complex mix of actors and institutions – differing values and differing interests.

This book explores environmental decision-making processes and contexts, and their effect on sustainability outcomes. For example:

- How do different value systems influence what environmental issues come onto the public agenda, and their management?
- What institutions and actors are involved in the processes?
- What tools are available and what are their limitations?
- How should we deal with uncertainty and risk?
- How do we incorporate relevant but different forms of knowledge?

The authors take a multidisciplinary approach and engage in themes from political science, law, economics, philosophy, natural sciences, geography, engineering and sociology. Their book is rich with practical examples, including three contemporary case studies.

The book is aimed at the ever-widening range of people who are, or are hoping to become, environmental professionals, whether from the scientific, technical or social science fields.

"A great strength of the book is that it comprehensively demonstrates the wickedness of environmental challenges and the need for an interdisciplinary approach to environmental decision-making. …

This is a 'must-have' book for students in environmental and natural-resource management, and related fields of geography and planning. It would also be of benefit for those studying in the social and political domains where their likely career path will intersect with environmental issues."

Australasian Journal of Environmental Management

2009 • 9781862877481 • paperback • 415 pages • $57.95

ALSO AVAILABLE FROM THE FEDERATION PRESS

Renegotiating The Environment

Jenny Stewart and Grant Jones

Renegotiating the Environment shows how managers can learn to acknowledge, understand and use politics to achieve more consensual and sustainable solutions.

'This book is a 'must read' for anyone involved in natural resources … It is a refreshingly honest analysis of the roles that politics, power, leadership, governance and conflict play in decision-making.'

Leith Boully, Chair, Community Advisory Council, Murray-Darling Basin Commission

"… an original book, with important messages about how to implement change."

Professor Peter Cullen, Chair, Wentworth Group of Concerned Scientists

"The nine compelling case studies, encompassing aspects of water management, forestry and urban development, give the analysis a thoroughly practical and contemporary flavour. … clearly written and reassuringly uncluttered by ideological presumptions. I recommend it strongly."

Professor Andrew Parkin Editor, Australian Journal of Political Science

2004 • ISBN 978 1 86287 473 2 • paperback • 192 pages • $39.95

The Precautionary Principle in Practice

Jacqueline Peel

"... brings together in one publication many real world examples of how the precautionary principle has been applied. ... an amazing achievement [which] is likely to be regarded as one of the foundation studies on the implementation of the precautionary principle."

Alternative Law Journal

"... useful [wherever] expert evidence is complex, conflicting or inadequate and values are diverse."

Law Institute Victoria Journal

"... will assuredly become an essential reference work."

Resource Management Journal (NZ)

2005 • ISBN 978 1 86287 519 7 • paperback • 256 pages • $64.95